INFRARED RADIATION

Edited by **Vasyl Morozhenko**

Infrared Radiation
Edited by Vasyl Morozhenko

Published by InTech
Janeza Trdine 9, 51000 Rijeka, Croatia

Publishing Process Manager Sasa Leporic

Technical Editor Teodora Smiljanic

Cover Designer InTech Design Team

Additional hard copies can be obtained from orders@intechopen.com

Infrared Radiation, Edited by Vasyl Morozhenko
p. cm.
ISBN 978-953-51-0060-7

For EU product safety concerns:
IN TECH d.o.o., Prolaz Marije Krucifikse Kozulić 3, 51000 Rijeka, Croatia,
info@intechopen.com or visit our website at intechopen.com.

Meet the editor

Vasyl Morozhenko graduated from Taras Shevchenko National University of Kyiv, Ukraine and received diploma in Optics of solid in 1983. After that he joined V. Lashkaryov Institute of Semiconductor Physics in Kyiv as a researcher. In 1994 he received his PhD degree in physics on semiconductors and dielectrics. Dr. Morozhenko's dissertation covered research on the effect of external influences on infrared emission of semiconductors. From August 1998 to November 1999 he worked as NAS/NRC resident researcher in the NASA Goddard Space Flight Center, Greenbelt, MD, USA. During this period of time he worked on infrared heterodyne spectroscopy of gases. After completing this research program he returned to the V. Lashkaryov Institute of Semiconductor Physics, where he works until now. Currently he holds an appointment as a senior staff scientist. His research interests are infrared spectroscopy, optical and emitting properties of semiconductors and semiconductor resonant structures and effect of external influences on them, optical, optoelectronic and magnetooptical devices for infrared optics.

Contents

Preface

Infrared (IR) radiation is a type of electromagnetic radiation that lies between visible light and radio waves on the electromagnetic spectrum. It has wavelengths from 0.74 μm (red edge of the visible light) to about 100 μm (origin of the shortwave radio band).

The IR radiation goes a long way in people's life both in a global and in a domestic sense. Approximately half of the energy in the solar spectrum falls on the IR radiation and it is exaggeration to say that it is the source of life on the Earth as well as a source of global catastrophes. It is not without reason that IR radiation is also called "thermal". Without it our Earth would be like lifeless snowball. On the other hand, the excess absorption of solar radiation in the IR region in different geological periods of the Earth led to overheating and desiccation of the planet and caused the mass extinction of every living thing. Destructive typhoons, hurricanes, drought and hail in the middle of summer – these are the results of imbalance of earth-radiation budget in the IR region.

IR radiation was discovered by British astronomer William Herschel in his study of the Sun in 1800. But he was not able to interpret it correctly and this discovery was forgotten. More than a half-century passed before J. C. Maxwell deduced the equations which comprise the basic laws of electromagnetism (1864). They show that infrared radiation has the same nature and properties as the visible light. The appearance of the theory of electromagnetism and development of the measuring instruments gave rise to rapid progress of research in this area. During the second half of the XIX century, it was investigated the absorption spectra of some compounds and it was found the connection between the substance and its unique pattern of the IR absorption (W. de W. Abney, E. R. Festing, W.H. Julius et al.). These researches laid down the foundations of the IR spectroscopy. And the investigations of the energy distribution in the spectrum of the blackbody by R. Kirchhoff, J. Stefan, L. Boltzmann, W, Wien et al. were crowned with the theory of Max Planck (1990), which initiated the beginning of the quantum mechanics. At the beginning of the XX century, A.A Glagoleva-Arkadieva got the radio waves with wavelength equal to 80 μm (1923), i.e. waves corresponding with the infrared wavelength band. Thus, it was experimentally proved that there exists a continuous transition from the IR radiation to radio waves.

In the present time research of the IR radiation and implementation of systems related to it in various areas of life are very important for scientific and technical progress and for the further progress of our civilization. Since IR radiation is invisible to the human eye, many do not realize how deeply it has come into our living. Remote-control boxes, fiber-optic communications, wireless local area networks, security systems, fire sensors etc. - all of them are based on transmitting, receiving and analyzing the IR signals. The IR radiation helps to warm our dwellings and to prepare food.

Infrared radiation is widely used in industry, scientific research, medicine and military technology. Recording of emission and absorption spectra in the IR range (the IR spectroscopy) is used in study of structure of the electron shell of atoms, for the purpose to determine the molecular structure, as well as for the qualitative and quantitative analysis of complex mixtures of molecular composition. This method is widely used as in manufacture as well as in the laboratory in physics and astronomy, biology, medicine, chemistry and forensic science.

It is impossible not to mention an important area as the infrared thermography. Image analysis in the IR rays can be called as the universal way to get various information about the world around us. Recording of thermal image allows to see and analyze the distribution pattern of temperature fields. Firefighters use this method to establish the hotbed of fire and people retrieval. On the thermal images it is possible to reveal local overheating in the various systems and structures. In medicine thermal imaging allows visualization of inflammatory processes, combining absolute harmlessness, speed and ease way of obtaining information. At the present time any geological prospecting can not do without an analysis of satellite images in IR rays.

All of the facts given above are only a small part of a wide spectrum of applications of the IR radiation. At present it is impossible to cover in one book the whole range of questions related to the IR. The Editorial Board and the Publisher of the book did not confine to some narrow area but tried to represent as wide as possible information about different research and development policy of the contemporary studies in this field. The book contains as researches covered the development of new infrared optical components and devices, research procedures and obtaining information about a test object, as well as the latest basic researches of the peculiarities of the IR radiation and possibilities of their application. All articles have passed the strict review and have a high scientific level.

Structurally, the book is divided into the chapters each of them is an original scientific research. The chapters contain a short historical introduction substantiating the urgency of the research, the description of theory and procedure of the study, discussion of the obtained results and a brief conclusion with reference of their the possible application. The chapters conclude with a list of references. Thematically the chapters are arranged in any order, since all of researches are equally interesting and important.

The book provides with a wide range of issues related to IR radiation. It is of interest to researchers and design engineers working in this area. In addition, it contains an extensive list of references that can certainly be useful for researchers beginning to master this important and interesting area and also for students.

On behalf of the Editorial Board and the Publisher I would like to thank the authors who have sent the results of their investigation. Unfortunately, by various reasons, not all the research papers have been published. Wish the all authors all of the best in their research and engineering work.

I want to thank the Editorial Board of the book for the opportunity to take part in the work on this book. The collaboration with the team of the InTech was pleasant and useful for me.

Vasyl Morozhenko
Institute of Semiconductor Physics
National Academy of Sciences of Ukraine
Ukraine

1

Nanoscale Radiative Heat Transfer and Its Applications

Svend-Age Biehs[1], Philippe Ben-Abdallah[2] and Felipe S.S. Rosa[2]
[1]*Institut für Physik, Carl von Ossietzky Universität Oldenburg, D-26111 Oldenburg*
[2]*Laboratoire Charles Fabry, Institut d'Optique, CNRS, Université Paris-Sud, Campus Polytechnique, RD128, 91127 Palaiseau Cedex*
[1]*Germany*
[2]*France*

1. Introduction

Heat radiation at the nanoscale is a relatively young but flourishing research field, that has attracted much attention in the last decade. This is on the one hand due to the fact that this effect is nowadays experimentally accessible (Hu et al. (2008); Kittel et al (2005); Narayanaswamy et al. (2008); Ottens et al. (2011); Rousseau et al. (2009); Shen et al. (2009)), and on the other hand due to the unusual properties of thermal radiation at nanometric distances, which makes it highly promising for future applications in nanotechnology. Among these near-field properties (i.e., properties at distances smaller than the thermal wavelength), we can mention: (i) the energy exchange is not limited by the well-known Stefan-Boltzmann law for black bodies and in fact can be several orders of magnitude larger, (ii) thermal radiation at nanoscale is quasi monochromatic and (iii) it can be spatially strongly correlated despite the fact that thermal radiation is often taken as a textbook example for uncorrelated light, which is only true for distances larger than the thermal wavelength (Carminati and Greffet (1999); Polder and Van Hove (1971); Shchegrov et al (2000)). For some recent reviews see Refs. (Basu et al. (2009); Dorofeyev and Vinogradov (2011); Joulain et al. (2005); Vinogradov and Dorofeyev (2009); Volokitin and Persson (2007); Zhang (2007)).

Before we discuss possible applications exploiting the above mentioned thermal near-field properties, we first want to give a concise description of the physical origin of the electromagnetic fields radiated from the surface of a hot material within the framework of fluctutational electrodynamics. Based on this formal framework we derive the heat flux expression between two isotropic semi-infinite nonmagnetic media separated at a given distance by a vacuum gap. By means of this expression we discuss the modes which contribute to the heat flux in different distance regimes. In particular, we discuss the dominant contribution of the coupled surface modes at the nanoscale and illustrate the specific properties (i) and (ii) with some numerical results. Finally, we reformulate the heat flux expression in the same manner as it is done for the electronic transport at a mesoscopic scale (Datta (2002); Imry (2002)).

1.1 Fluctuating electrodynamics

Let's first consider a given medium at a fixed temperature T. We choose a volume $\mathcal{V}$ of this medium such that it is large compared to the size of the constituents of the material, i.e., the electrons, atoms or ions, but small on a macroscopic length scale as for example the size of the considered medium. Then the macroscopic electromagnetic fields **E**, **D**, **B** and **H** fullfilling the macroscopic Maxwell equations (Jackson (1998))

$$\nabla \cdot \mathbf{D}(\mathbf{r},t) = \rho^{e}(\mathbf{r},t) \quad \text{and} \quad \nabla \times \mathbf{E}(\mathbf{r},t) = -\frac{\partial \mathbf{H}(\mathbf{r},t)}{\partial t}, \tag{1}$$

$$\nabla \cdot \mathbf{B}(\mathbf{r},t) = 0 \quad \text{and} \quad \nabla \times \mathbf{H}(\mathbf{r},t) = \mathbf{j}^{e}(\mathbf{r},t) + \frac{\partial \mathbf{D}(\mathbf{r},t)}{\partial t} \tag{2}$$

can be regarded as the volume average over such a volume (Russakoff (1970)). Here ρ^{e} and $\mathbf{j}^{e}$ are external charges or currents, respectively. Within this macroscopic or continuum description, the material properties can be described by a permittivity tensor ϵ_{ij} and a permeability tensor μ_{ij} with $i, j = 1, 2, 3$ relating the fields **D** and **E** and **B** and **H**. When introducing the Fourier components as

$$\tilde{\mathbf{E}}(\mathbf{r},\omega) = \int_{-\infty}^{\infty} \mathrm{d}t\, \mathrm{e}^{\mathrm{i}\omega t}\mathbf{E}(\mathbf{r},t)\,, \qquad \tilde{\mathbf{H}}(\mathbf{r},\omega) = \int_{-\infty}^{\infty} \mathrm{d}t\, \mathrm{e}^{\mathrm{i}\omega t}\mathbf{H}(\mathbf{r},t)\,, \qquad \text{etc.,} \tag{3}$$

then we can write $\tilde{\mathbf{D}} = \epsilon_0 \boldsymbol{\epsilon} \cdot \tilde{\mathbf{E}}$ and $\tilde{\mathbf{B}} = \mu_0 \boldsymbol{\mu} \cdot \tilde{\mathbf{H}}$. Here, ϵ_0 and μ_0 are the permittivity and the permeability of the vacuum. In the following we are only interested in non-magnetic materials so that $\tilde{\mu}$ is given by the unit tensor, i.e., we have $\mathbf{B} = \mu_0 \mathbf{H}$. Note, that here we have already neglected any spatial dispersion of the permittivity, which can play an important role in the near-field of metals (Chapuis et al. (2008); Ford and Weber (1984); Joulain and Henkel (2006)).

Within a neutral material there are no external charges and currents, but the random thermal motion of the constituents of matter, i.e., of the electrons, atoms or ions, induces within the average volume $\mathcal{V}$ a macroscopic fluctuating charge density ρ^{f} and a current $\mathbf{j}^{f}$, which replace the external charges and currents in Maxwell's equations (1) and (2) and therefore generate fluctuating electromagnetic fields $\mathbf{E}^{f}$ and $\mathbf{H}^{f}$, which are now considered to be random processes as well as ρ^{f} and $\mathbf{j}^{f}$. Since the latter are the sum of many microscopic random charges and currents inside the average volume $\mathcal{V}$, we can apply the *central limit theorem* and infer that these fluctuating quantities are Gaussian distributed (Kubo et al. (1991)). That means that all higher moments of these quantities are determined by their mean value and variance. This statement is also true for the fluctuating fields, because there exists a linear relation between the electromagnetic fields and the generating currents which can be stated as

$$\tilde{\mathbf{E}}^{f}(\mathbf{r},\omega) = \mathrm{i}\omega\mu_0 \int_V \mathrm{d}\mathbf{r}''\, \mathbb{G}^{E}(\mathbf{r},\mathbf{r}'',\omega) \cdot \tilde{\mathbf{j}}^{f}(\mathbf{r}'',\omega), \tag{4}$$

$$\tilde{\mathbf{H}}^{f}(\mathbf{r},\omega) = \mathrm{i}\omega\mu_0 \int_V \mathrm{d}\mathbf{r}''\, \mathbb{G}^{H}(\mathbf{r},\mathbf{r}'',\omega) \cdot \tilde{\mathbf{j}}^{f}(\mathbf{r}'',\omega), \tag{5}$$

where the integrals are taken over the volume V which contains the fluctuating source currents; $\mathbb{G}^{E}$ and $\mathbb{G}^{H}$ are the classical dyadic electric and magnetic Green's functions (Chen-To Tai (1971)).

By assuming that due to the thermal motion no charges will be created or destroyed we have $\langle \rho^{f} \rangle = 0$, where the brackets symbolize the ensemble average. If we now further assume that the mean fluctuating currents are vanishing in average, i.e., $\langle \mathbf{j}^{\rm f} \rangle = \mathbf{0}$, then we find that $\langle \mathbf{E}^{\rm f} \rangle = \langle \mathbf{H}^{\rm f} \rangle = \mathbf{0}$ so that in average the fields do no work on external charges or currents. In order to complete the framework of fluctuating electrodynamics first developed by Rytov (Rytov et al. (1989)) we have to specify the second moment, i.e., the correlation function of the fluctuating currents or charges. In fluctuating electrodynamics this correlation function is specified by the *fluctuation dissipation theorem* and reads (Lifshitz and Pitaevskii (2002))

$$\langle \tilde{j}_i^{\rm f}(\mathbf{r}, \omega) \tilde{j}_j^{\rm f}(\mathbf{r}', \omega') \rangle = 2\pi\omega\Theta(\omega, T) \big[\tilde{\epsilon}_{ij}(\omega) - \tilde{\epsilon}_{ji}^*(\omega)\big] \delta(\omega - \omega') \delta(\mathbf{r} - \mathbf{r}'). \tag{6}$$

The delta-function $\delta(\mathbf{r} - \mathbf{r}')$ shows up because we have neglected spatial dispersion. The second delta function $\delta(\omega - \omega')$ reflects the fact that we have a stationary situation. Indeed, the fluctuation dissipation theorem is only valid in thermal equilibrium so that by applying this theorem we have assumed that the medium containing the fluctuating currents is in thermal equilibrium at temperature T. The function

$$\Theta(\omega, T) = \frac{\hbar\omega}{2} + \frac{\hbar\omega}{e^{\hbar\omega/(k_{\rm B}T)} - 1} \tag{7}$$

is the mean energy of a harmonic oscillator in thermal equilibrium and consists of the vacuum and the thermal part; $k_{\rm B}$ is Boltzmann's and $2\pi\hbar$ is Planck's constant. From the appearance of $\hbar$ in $\Theta(\omega, T)$ it becomes obvious that the fluctuation dissipation theorem is in principle a quantum mechanical relation. Hence, fluctuating electrodynamics combines the classical stochastic electromagnetic fields with the quantum mechanical fluctuation dissipation theorem and has therefore to be considered as a semi-classical approach (Rosa et al. (2010)).

Equipped with the correlation function for the source currents in Eq. (6) and the linear relations in Eqs. (4) and (5) we can now determine the correlation functions of the electromagnetic fields $\langle \tilde{E}_i^{\rm f}(\mathbf{r}, \omega) \tilde{E}_j^{\rm f}(\mathbf{r}', \omega') \rangle$, $\langle \tilde{H}_i^{\rm f}(\mathbf{r}, \omega) \tilde{H}_j^{\rm f}(\mathbf{r}', \omega') \rangle$, and $\langle \tilde{E}_i^{\rm f}(\mathbf{r}, \omega) \tilde{H}_j^{\rm f}(\mathbf{r}', \omega') \rangle$ in terms of the Green's functions. Hence, if we know the classical electromagnetic Green's functions $\mathbb{G}^{\rm E}$ and $\mathbb{G}^{\rm H}$ for a given geometry we can evaluate the correlation functions of the fields allowing for determining for example Casimir forces or heat fluxes. Although some purely quantum mechanical approaches exist (Agarwal (1975); Janowicz et al. (2003); Lifshitz and Pitaevskii (2002)) fluctuating electrodynamics has the advantage of being conceptionally simple while giving the correct results for the correlation functions of the fields.

1.2 Heat flux expression

Now we want to determine the heat flux between two semi-infinite media (see Fig. 1) which are at local thermal equilibrium and have the temperatures T_1 and T_2. We assume that both media are separated by a vacuum gap of thickness d. In order to determine the heat flux, we first consider $T_2 = 0$ so that we consider only fluctuating currents $\mathbf{j}_1^{\rm f}$ in medium 1. The fluctuating fields $\mathbf{E}_1^{\rm f}$ and $\mathbf{H}_1^{\rm f}$ inside the vacuum gap generated by the fluctuating sources in medium 1 can be expressed in terms of the relations (4) and (5). From these expressions one can determine the mean Poynting vector in z direction

$$\langle S_z^{1\to 2} \rangle = \langle \mathbf{E}_1^{\rm f} \times \mathbf{H}_1^{\rm f} \rangle \cdot \mathbf{e}_z \tag{8}$$

by means of the fluctuation dissipation theorem in Eq. (6). The resulting expression contains the dyadic Green's functions $\mathbb{G}^{\rm E}(\mathbf{r}, \mathbf{r}'', \omega)$ and $\mathbb{G}^{H}(\mathbf{r}, \mathbf{r}'', \omega)$ for that layered geometry with source points $\mathbf{r}''$ inside medium 1 and observation points $\mathbf{r}$ inside the vacuum gap. For the given layered geometry the Green's functions are well known and can for example be found in (Tsang et al. (2000)). For determining the net heat flux one has also to consider the opposite case with $T_1 = 0$ so that only fluctuating currents inside medium 2 are taken into account. Then the net heat flux inside the vacuum gap is given by the difference

$$\Phi = \langle S_z^{1\to 2}\rangle - \langle S_z^{2\to 1}\rangle. \tag{9}$$

For two isotropic media we find (Polder and Van Hove (1971))

$$\Phi = \int_0^\infty \frac{\mathrm{d}\omega}{2\pi}\,[\Theta(\omega,T_1) - \Theta(\omega,T_2)] \sum_{j=\{\mathrm{s,p}\}} \int \frac{\mathrm{d}^2\boldsymbol{\kappa}}{(2\pi)^2}\,\mathcal{T}_j(\omega,\boldsymbol{\kappa};d) \tag{10}$$

The second integral of the energy transmission coefficient $\mathcal{T}_j(\omega,\boldsymbol{\kappa};d)$ is carried out over all transverse wave vectors $\boldsymbol{\kappa} = (k_x,k_y)^{\mathrm{t}}$. This means it includes propagating modes as well as evanescent modes. The division into propagating and evanescent modes stems from the fact that the electromagnetic waves inside the vacuum gap region have a phase factor $\exp[\mathrm{i}(k_x x + k_y y + k_{z0} z) - \mathrm{i}\omega t]$ with $k_{z0} = \sqrt{\omega^2/c^2 - \kappa^2}$, where c is the velocity of light in vacuum. Hence, k_{z0} is purely real for all lateral wave vectors $\kappa < \omega/c$ and therefore the phase factor gives an oscillatory solution with respect to z, whereas k_{z0} is for all $\kappa > \omega/c$ purely imaginary so that the phase factor gives an exponential damping with respect to z. The latter modes are called evanescent modes, whereas modes with $\kappa < \omega/c$ are called propagating modes. Note, that the vacuum part in $\Theta(\omega,T)$ does not contribute to the flux Φ.

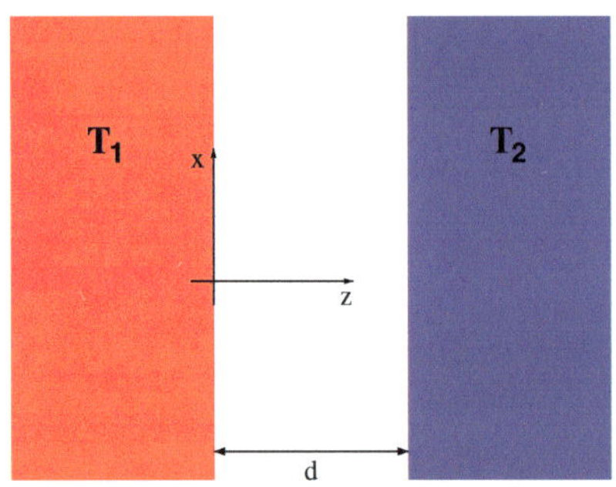

Fig. 1. Sketch of the considered geometry: Two semi-infinite materials at local thermal equilibrium with temperatures T_1 and T_2 are separated by a vacuum gap of thickness d.

The energy transmission coefficient $\mathcal{T}_j(\omega,\boldsymbol{\kappa};d)$ is different for propagating and evanescent modes and can be stated as (Polder and Van Hove (1971))

$$\mathcal{T}_j(\omega,\kappa;d) = \begin{cases} (1-|r_j^1|^2)(1-|r_j^2|^2)/|D_j^{12}|^2, & \kappa < \omega/c \\ 4\mathrm{Im}(r_j^1)\mathrm{Im}(r_j^2)\mathrm{e}^{-2|k_{z0}|d}/|D_j^{12}|^2, & \kappa > \omega/c \end{cases} \tag{11}$$

for $j = \{\mathrm{s,p}\}$ where r_j^1 and r_j^2 are the usual Fresnel coefficients

$$r_\mathrm{s}^i(\omega,\kappa) = \frac{k_{z0} - k_{zi}}{k_{z0} + k_{zi}} \quad \text{und} \quad r_\mathrm{p}^i(\omega,\kappa) = \frac{\epsilon_i(\omega)k_{z0} - k_{zi}}{\epsilon_i(\omega)k_{z0} + k_{zi}} \tag{12}$$

for s- and p-polarized light, where $k_{zi} = \sqrt{\epsilon_i(\omega)\omega^2/c^2 - \kappa^2}$. We have further introduced the Fabry-Pérot-like denominator D_j^{12}, defined by ($j = \{s, p\}$)

$$D_j^{12} = (1 - r_j^1 r_j^2 e^{2ik_{z0}d})^{-1} \tag{13}$$

which appears as a consequence of the multiple reflections inside the vacuum gap.

1.3 Nanoscale heat flux

The expression in Eq. (10) together with the energy transmission coefficient in Eq. (11) is very general and allows the determination of the heat flux between two arbitrary isotropic semi-infinite bodies kept at fixed temperatures T_1 and T_2 for any distance d. In particular this expression contains the Stefan-Boltzmann law for the heat flux between two *black bodies*. This can be seen as follows: a *black body* is a body which absorbs all incoming radiation. For a semi-infinite body this situation is realized, when the Fresnel reflection coefficients are exactly zero for both polarizations. Then all incoming radiation is transmitted and will be absorbed inside the semi-infinite medium. Hence, by assuming that the Fresnel coefficients are zero we obtain from Eq. (11) that the energy transmission coefficient $\mathcal{T}_j(\omega, \boldsymbol{\kappa}; d) = 1$ for s- and one for p-polarized light with $\kappa < \omega/c$ and $\mathcal{T}_j(\omega, \boldsymbol{\kappa}; d) = 0$ for $\kappa > \omega/c$. In other words, all propagating modes contribute with a maximal transmission of 1 to the heat flux. Then one can easily compute the heat flux from Eq. (10) yielding

$$\Phi_{\rm BB} = \int_0^\infty \left[\Theta(\omega, T_1) - \Theta(\omega, T_2)\right] \left(\frac{\omega^2}{c^3 \pi^2}\right) \frac{c}{4} = \sigma_{\rm BB}(T_1^4 - T_2^4) \tag{14}$$

which is the well-known Stefan-Boltzmann law for the heat flux between two black bodies with the Stefan-Boltzmann constant $\sigma_{\rm BB} = 5.67 \cdot 10^{-8}\,{\rm Wm^{-2}K^{-4}}$.

From this derivation of the Stefan-Boltzmann law we see that it can be a limit for the propagating modes only, since only for these modes ($\kappa < \omega/c$) on the left of the light line $\omega = c\kappa$ [see Fig. 2 (a)] the energy transmission coefficient has its maximum value and is zero for the evanescent modes ($\kappa > \omega/c$) on the right of the light line in Fig. 2 (a). This fact can also be formulated in terms of the number of contributing modes. To this end consider a quantisation box in x and y direction with a length $L_x = L_y = L$. For very large L the integral over the lateral wave vectors in Eq. (10) is equivalent to a sum over the modes $k_x = 2\pi n_x/L$ and $k_y = 2\pi n_y/L$ with $n_x, n_y \in \mathbb{N}$, i.e.,

$$\int \frac{d^2\kappa}{(2\pi)^2} = \frac{1}{L^2} \int \frac{d^2\kappa}{\left(\frac{2\pi}{L}\right)^2} \leftrightarrow \frac{1}{L^2} \sum_{n_x, n_y} 1. \tag{15}$$

Here $\sum_{n_x,n_y} 1$ is the number of contributing modes and $L^{-2} \sum_{n_x,n_y} 1$ the density of states. Hence, for a given frequency ω only the modes in a circle [see Fig. 2 (b)] with radius ω/c contribute, but with a transmission factor of one for each polarisation, this means the number of contributing modes is limited to the region $\kappa < \omega/c$.

For real materials, the number of contributing modes is not limited to the region $\kappa < \omega/c$. As was already put forward by (Cravalho et al., (1967)) total internal reflection modes become frustrated if the gap distance d is much smaller than the thermal wavelength $\lambda_{\rm th} = \hbar c/(k_B T)$ and can therefore tunnel through the vacuum gap and hence contribute to the heat flux. Since

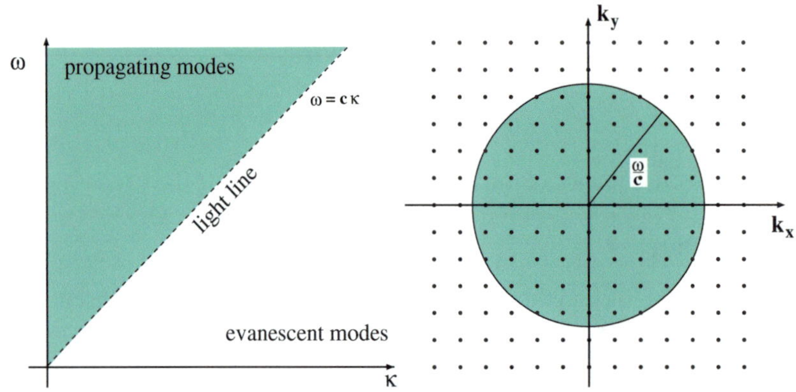

(a) ω-κ diagram (b) wave vector space

Fig. 2. Sketch of the contributing modes. (a) shows the ω-κ diagram. The light line at $\omega = c\kappa$ divides the ω-κ space into the propagating and evanescent part, i.e., the propagating modes inside the vacuum gap are on the left of the light line, whereas the evanescent modes are on the right of the light line. For a *black body* all propagating modes contribute with transmission 1 to the heat flux. (b) shows the space of lateral wave vectors for a fixed frequency ω. All modes inside the circle with radius ω/c are propagating modes, and all modes outside that circle are evanescent.

these modes are propagating inside the material but evanescent in the vacuum region they are determined by $\kappa > \omega/c$ and $\kappa < \sqrt{\epsilon_i(\omega)}\omega/c$. For a polar material as SiC, which can be described by the permittivity

$$\epsilon_1(\omega) = \epsilon_2(\omega) = \epsilon_\infty \left(\frac{\omega_{\rm L}^2 - \omega^2 - {\rm i}\gamma\omega}{\omega_{\rm T}^2 - \omega^2 - {\rm i}\gamma\omega} \right) \equiv \epsilon(\omega), \tag{16}$$

with the longitudinal phonon frequency $\omega_{\rm L} = 1.827 \cdot 10^{14}\,\text{rad/s}$, the transversal phonon frequency $\omega_{\rm T} = 1.495 \cdot 10^{14}\,\text{rad/s}$, the damping $\gamma = 0.9 \cdot 10^{12}\,\text{rad/s}$ and $\epsilon_\infty = 6.7$, we illustrate schematically in Fig. 3 the regions for which one can expect frustrated modes. For the sake of simplicity we neglect the damping for the discussion of the contributing modes and assume a real permittivity. We will later add the absorption again, since it is vital for the nanoscale heat transfer. Note in Fig. 3(a), that in the so called *reststrahlen region* $\omega_{\rm T} < \omega < \omega_{\rm L}$ no optical phonons can be excited. Within this frequency band the permittivity is negative so that the material behaves effectively like a metal, i.e., the reflectivity is close to one. From Fig. 3 (b) it is obvious that due to the frustrated internal reflection the number of contributing modes for the heat flux increases, but is still limited to $\kappa < \sqrt{\epsilon(\omega)}\omega/c$.

Before we can discuss the energy transmission coefficient and the heat flux, we need to discuss another kind of evanescent mode which is responsible for the tremendous increase of the heat flux at nanoscale, the so-called surface phonon polariton (Kliewer and Fuchs (1974)). This mode is characterized by the fact that the electromagnetic fields are evanescent inside and outside the medium so that these modes are confined to the boundary of the medium itself. Assuming an infinite large distance d between the halfspaces, then both can be considered as individual semi-infinite bodies with negligible coupling. For such isotropic nonmagnetic halfspaces the surface modes are purely p-polarized and fulfill the dispersion

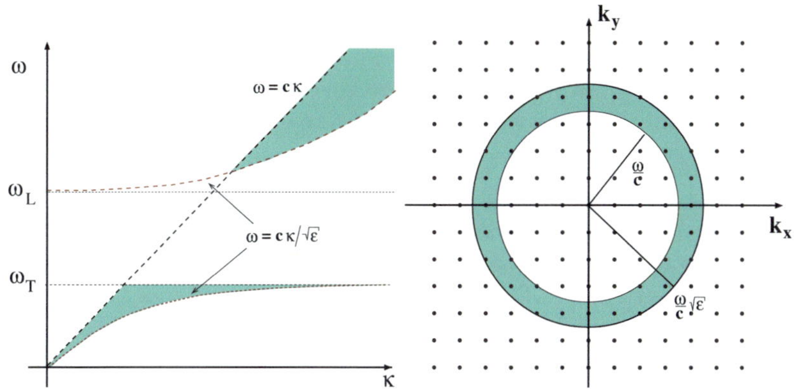

(a) ω-κ diagram (b) wave vector space

Fig. 3. Sketch of the frustrated modes. The modes which can propagate inside the dielectric are on the left of the polariton lines $\omega = c\kappa/\sqrt{\epsilon}$. The internal reflection modes are on the left of the polariton lines and on the right of the light line within the green region.

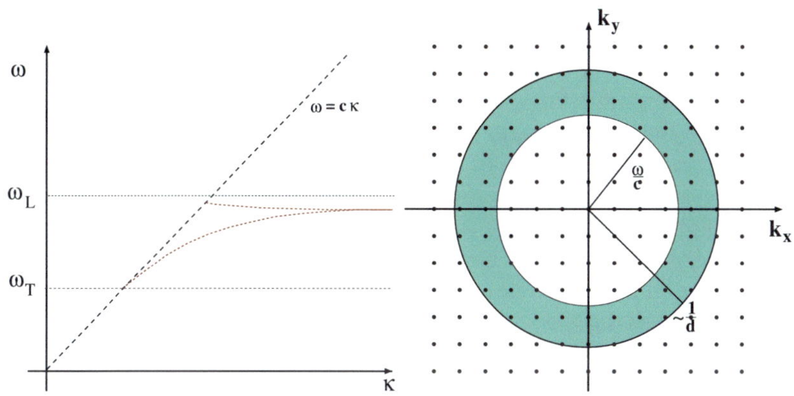

(a) ω-κ diagram (b) wave vector space

Fig. 4. Sketch of the surface phonon polariton modes for a given distance d.

relation (Kliewer and Fuchs (1974))

$$\kappa_{\text{SPhP}} = \sqrt{\frac{\epsilon(\omega)}{\epsilon(\omega)+1}}. \tag{17}$$

Furthermore, a necessary condition for having surface modes is that $\epsilon(\omega) < 0$, i.e., for polar materials like SiC separated by vacuum it can only exist within the reststrahlen band $\omega_{\text{T}} < \omega < \omega_{\text{L}}$. When the two semi-infinite material are placed at a distance d smaller than the penetration length of this surface mode in vacuum, i.e., $1/\text{Im}(\sqrt{\omega^2/c^2 - \kappa_{\text{SPhP}}^2})$, these modes will couple. This coupling removes the two-fold degeneracy and produces a splitting of the dispersion relation (Raether (1988)), which is determined by the relation

$$\left[-\text{Im}(r_{\text{p}})^2 + \text{Re}(r_{\text{p}})^2 + 2\text{i}\text{Im}(r_{\text{p}})\text{Re}(r_{\text{p}})\right]\text{e}^{-2\text{Im}(k_{z0}d)} = 1 \tag{18}$$

and schematically illustrated in Fig. 4 (a). Since we are interested in the transmission coefficient of such modes, we now consider a permittivity with absorption, or $\gamma \neq 0$. For small absorption, more precise for $\mathrm{Im}(r_\mathrm{p}) \ll \mathrm{Re}(r_\mathrm{p})$, the dispersion relation for the coupled surface modes coincides with the resonance condition of the energy transmission coefficient (Pendry (1999))

$$\big[-\mathrm{Im}(r_\mathrm{p})^2 + \mathrm{Re}(r_\mathrm{p})^2\big]\mathrm{e}^{-2\mathrm{Im}(k_{z0}d)} = 1 \tag{19}$$

for which the evanescent part of the energy transmission coefficient in Eq. (11) has its maximal value of one, i.e., the energy transmission coefficient is one for the surface phonon polaritons as long as $\mathrm{Im}(r_\mathrm{p}) \ll \mathrm{Re}(r_\mathrm{p})$ is fullfilled. Nonetheless, for very large $\kappa \gg d^{-1} \gg \omega/c$ the energy transmission coefficient in Eq. (11) is damped exponentially due to the exponential $\exp(-2\mathrm{Im}(k_{z0})d) \approx \exp(-2\kappa d)$. Here, the exact damping of the energy transmission coefficient is determined by the losses of the material (Biehs et al. (2010)). Hence, the coupled surface phonon polariton provides for distances smaller than $d \ll c/(\omega\sqrt{\epsilon(\omega)})$ a number of modes proportional to d^{-2} as illustrated in Fig. 4(b) contributing to the heat flux which eventually results in a larger contribution than that of the frustrated internal reflection modes.

Now we are in a good starting position to discuss the energy transmission coefficient between two semi-infinite SiC plates assuming that $T_1 = 300\,\mathrm{K}$ and $T_2 = 0$ so that $\lambda_{\mathrm{th}} = 7.6\,\mu\mathrm{m}$. For this purpose we plot in Fig. 5 the energy transmission coefficient $\mathcal{T}_\mathrm{p}(\omega, \kappa; d)$ in ω-κ space for distances (a) $d = 5\,\mu\mathrm{m}$, (b) $d = 500\,\mathrm{nm}$ and (c) $d = 100\,\mathrm{nm}$. In Fig. 5 (a) we observe that for a relatively large distance the transmission coefficient is dominated by the propagating modes on the left of the light line and is maximal for the Fabry-Pérot modes inside the gap. Nonetheless, the surface phonon polariton modes already contribute inside the reststrahlen region. One can observe that in this region the surface phonon mode dispersion is continued on the left of the light line. This mode is evanescent inside the medium, but propagating in the vacuum gap so that it can be considered as a wave guide mode. As for surface phonon polaritons these guided modes do not contribute to the energy flux if there is no absorption, i.e., if $\mathrm{Im}(\epsilon) = 0$, whereas the Fabry-Pérot modes and the frustrated modes do contribute. For smaller distances we can see in Fig 5(b) that the surface modes and frustrated modes come into play. For even smaller distances the energy transmission coefficient equals one for all modes which can exist inside the bulk SiC (on the left of the phonon polariton lines) and for the surface modes [see Fig. 5 (c)], which will give the main contribution to the heat flux, since the number of contributing modes is very large [see Fig. 5 (d)].

The resulting spectral heat flux Φ_ω is now plotted in Fig. 6(a). It can be observed that for very small distances the spectrum becomes quasi monochromatic around the frequency of the surface mode resonance $\omega_{\mathrm{SPhP}} = 1.787 \cdot 10^{14}\,\mathrm{rad/s}$ which is defined by the pole of the denominator in Eq. (17), i.e., through the implicit relations $\mathrm{Re}[\epsilon(\omega_{\mathrm{SPhP}})] = -1$ and $\mathrm{Im}[\epsilon(\omega_{\mathrm{SPhP}})] \ll 1$. The distance dependence is shown in Fig. 6(b) where the flux Φ is normalized to the heat flux between two black bodies $\Phi_{\mathrm{BB}} = 459.27\,\mathrm{Wm}^{-2}$. The contributions are divided into the propagating, the frustrated, and the surface phonon polariton part. One can clearly see that the heat flux rises for distances smaller than the thermal wavelength $\lambda_{\mathrm{th}} = 7.6\,\mu\mathrm{m}$ due to the frustrated modes and exceeds the black body limit at $d \approx 3\,\mu\mathrm{m}$. For even smaller distances ($d < 100\,\mathrm{nm}$) the surface modes start to dominate the heat flux completely and give a characteristic $1/d^2$ dependence, since the number of contributing modes is for these modes proportional to $1/d^2$. Note, that on the nanoscale at a distance of $d = 10\,\mathrm{nm}$ the heat flux exceeds the black body limit by a factor of 1000! For some asymptotic

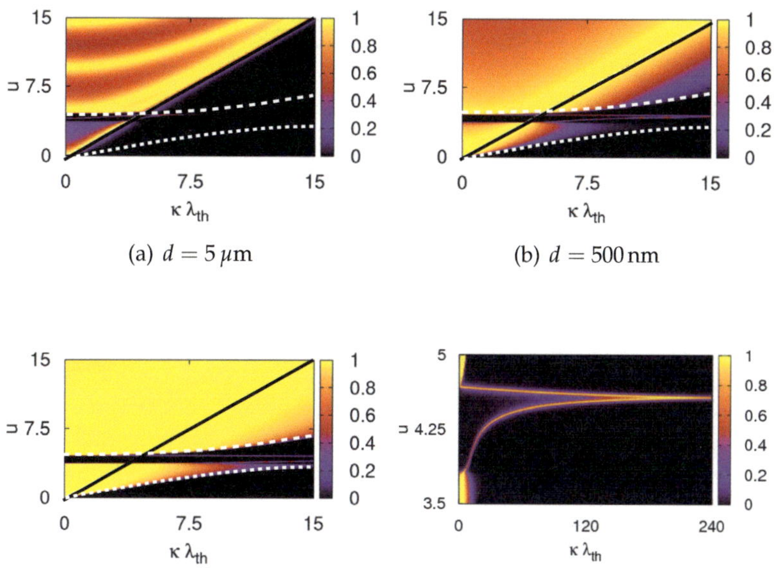

(a) $d = 5\,\mu$m (b) $d = 500\,$nm

(c) $d = 100\,nm$ (d) $d = 100\,nm$

Fig. 5. Transmission coefficient $\mathcal{T}_{\rm p}(\omega, \kappa; d)$ between two SiC plates for different distances in ω-κ space. Note that (d) is the same as (c) but for a large κ range, showing that the number of contributing modes for the coupled surface modes is much larger than for the frustrated modes. The dashed lines are the phonon polariton lines for SiC. Here, $u = \hbar\omega/(k_{\rm B}T)$ is a rescaled frequency so that for $T = 300\,$K we have $\omega = u \cdot 4.14 \cdot 10^{13}$rad/s.

expression concerning the heat flux in different distance regimes see (Rousseau et al. (2009b; 2010)).

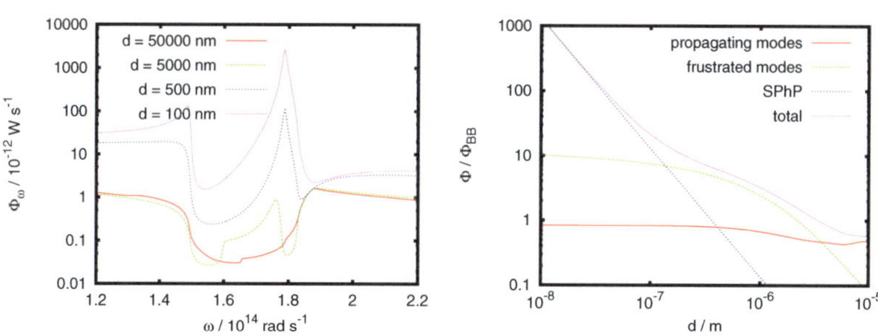

(a) spectral heat flux Φ_ω (b) total heat flux Φ

Fig. 6. (a) spectral heat flux Φ_ω between two SiC halfspaces at $T_1 = 300\,$K and $T_2 = 0\,$K for different distances. (b) total heat flux Φ over distance.

Finally, we want to express the formula for Φ in a way which highlights the number of modes contributing to the heat flux. To this end, we start with Eq. (10) assuming $T_1 = T$ and $T_2 = T + \Delta T$. For small temperature differences ΔT we can linearize Eq. (10) defining the radiative heat transfer coefficient h_{rad} through

$$\Phi = h_{\text{rad}}(T)\Delta T \equiv \frac{\partial \Phi}{\partial T}\Delta T. \tag{20}$$

By introducing the dimensionless variable $u = \hbar\omega/(k_{\text{B}}T)$ and the mean transmission coefficient

$$\overline{\mathcal{T}}_j = \frac{\int_0^\infty \mathrm{d}u\, f(u)\mathcal{T}_j(u,\kappa;d)}{\int_0^\infty \mathrm{d}u\, f(u)} \tag{21}$$

with $f(u) = u^2 \mathrm{e}^u/(\mathrm{e}^u - 1)^2$ we find a Landauer-like expression for the heat flux (Biehs et al. (2010))

$$\Phi = \frac{\pi^2}{3}\frac{k_{\text{B}}^2 T}{\text{h}}\Big(\sum_{j=\text{s,p}}\int \frac{\mathrm{d}^2\kappa}{(2\pi)^2}\overline{\mathcal{T}}_j\Big)\Delta T. \tag{22}$$

Here, $\pi^2 k_{\text{B}}^2 T/(3\text{h})$ is the universal quantum of thermal conductance (Pendry (1983); Rego and Kirczenow (1999)). Hence, each mode can at most contribute one quantum of thermal conductance, since the mean transmission coefficient $\overline{\mathcal{T}}_j \in [0,1]$. This representation allows for studying the tradeoff between the mean transmission coefficient and the number of modes. In Fig. 7 we show a plot of the mean transmission coefficient $\overline{\mathcal{T}}_{\text{p}}$ for two SiC slabs varying the distance. It can be seen that the mean transmission coefficient for the surface modes is extremely small. Nonetheless, the coupled surface modes give the dominant heat transfer mechanism for small distances. This is due to the number of modes which increases dramatically $\propto \kappa^2$ explaining the $1/d^2$ increase in heat flux due to the coupled surface modes. For polar materials there is a cutoff value for the spatial wave vectors of the phonons given by π/a, where a is the lattice constant. This sets an ultimate limit to the heat flux and removes the $1/d^2$ divergency. The limits of the heat flux in the near-field regime are for example disscussed in Refs. (Basu and Zhang (2009); Ben-Abdallah and Joulain (2010)).

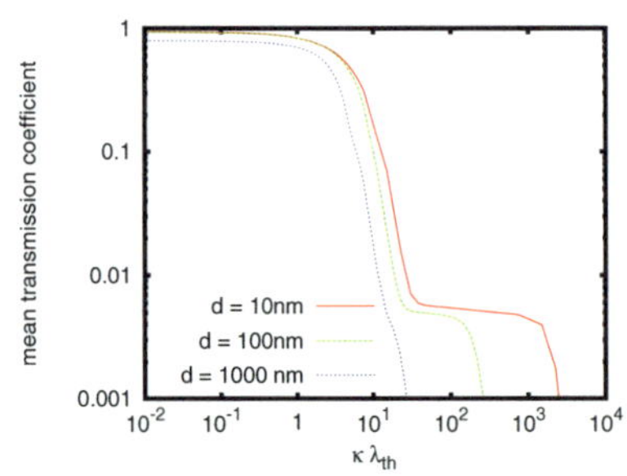

Fig. 7. Mean transmission coefficient $\overline{\mathcal{T}}_{\text{p}}$ for two SiC slabs with varying distances d.

2. Thermal imaging

Measurement and control of temperature at the nanoscale are important issues in nanotechnology. There are nowadays several possibilities for mapping the temperature above a surface or a nanostructure. For instance, the fluorescence polarization anisotropy of suspended molecules placed around a nanostructure can be used to map the local temperature of nanoscaled sources (Baffou et al. (2009)). On the other hand, the properties of the fluctuating electromagnetic fields can directly be used for a contact-free measurement of surface properties as local temperatures and local material properties. One step in this direction was done by De Wilde *et al.* (De Wilde et al. (2006)) [and recently Kajihara *et al.* (Kajihara et al. (2010))] who have developed a SNOM-based method in order to scatter the thermal near field into the far field and to measure the photonic local density of states of that surface. Very recently, a similar but promising imaging method was established which consist in measuring the thermal near-field spectra of surfaces in order to characterize their material properties as for instance the local free-carrier concentration and mobility (Huth et al. (2011)).

Here, we will review a method of thermal imaging called near-field scanning thermal microscopy (NSThM), which was developed in the Oldenburg group of Achim Kittel and Jürgen Parisi (Kittel et al (2005); Müller-Hirsch et al. (1999); Wischnath et al (2008)). It is based on an STM tip which is augmented by a thermocouple in the tip apex. With the help of the STM ability one can control the surface-tip distance, whereas the thermocouple allows for measuring the local temperature at the tip position, which can be varied in a distance range of 0.1 nm to about 100 nm. In contrast to usual thermal profilers used in scanning thermal microscopy (SThM), the NSThM probe operates at ultra high vacuum conditions rather than at ambient conditions (Majumdar (1999)). Hence, the energy or heat flow is not mediated by gas molecules, nor a liquid film of adsorbates, nor solid-solid conduction, but by the near field interaction between the tip and the sample mediated by the fluctating electromagnetic field. In other words, the NSThM exploits the enhanced radiative heat transfer at the nanoscale for surface imaging.

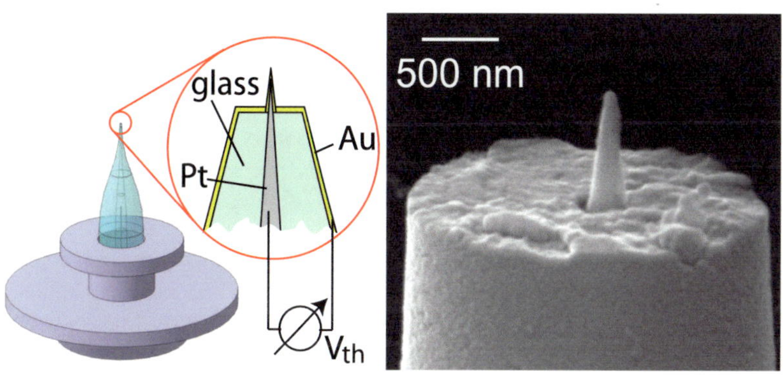

(a) Schematic drawing of the NSTHM tip. (b) SEM image of a typical NSTHM tip.

Fig. 8. Near field scanning thermal microscope developed in the group of Achim Kittel in Oldenburg (Kittel et al. (2008)). Reprinted with permission from Appl. Phys. Lett., Vol. **93**, 193109 (2008). Copyright 2008, American Institute of Physics.

As shown in Fig. 8 the tip consists of a platinum wire protruding about 500 nm from a glass capillary. A gold coating establishes the gold-platinum thermocouple in coaxial configuration at the very end of the tip. At the foremost part the tip radius is less than 50 nm allowing for a high lateral resolution when scanning a sample surface. When the probe is moving in proximity to a cold or hot sample surface the tip is slightly cooled down or heated up at the very end resulting in a temperature gradient within the tip. This temperature gradient is the source of a thermovoltage V_{th} which is the measured quantity. Now, the thermovoltage is directly proportional to the temperature difference ΔT_{tip} in the tip wich allows for determining the local temperature of the sample surface or the heat flow between the tip and the sample.

In particular, for the heat flux Φ one has (Wischnath et al (2008))

$$\Phi = V_{\text{th}} \frac{1}{S R_{\text{th}}^{\text{pr}}} \tag{23}$$

where S is the Seebeck coefficient of the probe's thermocouple and $R_{\text{th}}^{\text{pr}}$ is its thermal resistance, which can in principle be determined experimentally. Hence, the heat flow is also directly proportional to V_{th} times a tip-dependent factor of proportionality.

In order to provide an interpretation of the data measured with an NSThM tip one can model it as a simple dipole associated with a given temperature T_1 situated at $\mathbf{r}_{\text{tip}}$ above the sample surface with a temperature T_2 as sketched in Fig. 9. Within such a simple dipole model it can be shown that the heat flux is given by (Chapuis et al. (2008b); Dedkov and Kyasov (2007); Dorofeyev (1998); Mulet et al. (2001); Pendry (1999))

$$\Phi = \sum_{i=\text{E,M}} \int_0^\infty \mathrm{d}\omega\, 2\omega \text{Im}[\alpha_i(\omega)] \left[\Theta(\omega, T_1) - \Theta(\omega, T_2)\right] D^i(\omega, \mathbf{r}_{\text{tip}}) \tag{24}$$

where $\alpha_{\text{E/M}}$ is the electric/magnetic polarizability of the tip apex and $D^{\text{E/M}}(\omega, \mathbf{r}_{\text{tip}})$ is the electric/magnetic local densitiy of states (LDOS) above the sample surface (Joulain et al. (2003)). Here, the spectral power absorbed by the tip apex is given by $\text{Im}[\alpha_i(\omega)]D^i(\omega, \mathbf{r}_{\text{tip}})\Theta(\omega, T_2)$, i.e., it is proportional to the imaginary part of the polarizability of the tip and proportional to the energy density above the surface which is given by the product $D^i(\omega, \mathbf{r}_{\text{tip}})\Theta(\omega, T_2)$. On the other hand the power emitted by the tip and absorbed within the bulk medium is proportional to $\text{Im}[\alpha_i(\omega)]D^i(\omega, \mathbf{r}_{\text{tip}})\Theta(\omega, T_1)$. In fact, when considering the flux between two metals not supporting surface plasmons for $T_1 = 300\,\text{K}$ and $T_2 \ll T_1$ this expression simplifies to (Biehs et al. (2008); Rüting et al. (2010))

$$\Phi \propto \text{Im}[\alpha_M(\omega_{\text{th}})] D^M(\omega_{\text{th}}, \mathbf{r}_{\text{tip}}). \tag{25}$$

This means, the heat flux is directly proportional to the magnetic LDOS above the sample evaluated at the tip position $\mathbf{r}_{\text{tip}}$ and the thermal frequency $\omega_{\text{th}} \approx 2.82 k_{\text{B}} T/\hbar$. Hence, roughly speaking by measuring the thermovoltage the NSThM measures the LDOS of the sample surface. Note, that this expression is strictly valid for surface tip distances much larger than the tip radius only assuming a spherical metallic sensor tip. Indeed the value of the heat flux as well as the thermal near-field image of a structured surface depend on the shape and the material properties of the tip apex as was shown for ellipsoidal sensor tips (dielectric and metallic) in (Biehs et al. (2010b); Huth et al. (2010)). Hence, for a more refined model it is important to account for the sensor shape and to include the contributions of higher multipoles.

For structured as well as for rough surfaces (Biehs et al. (2010c;d; 2008); Rüting et al. (2010)) the LDOS can be calculated pertubatively by using for example the perturbation approach of (Greffet (1988)) if the height differences of the surface profile are the smallest length scales and in particular smaller than the thermal wavelength (Henkel and Sandoghdar (1998)). This allows for comparision of the NSThM data with theory, i.e., with the numerically evaluated LDOS $D^M(\omega_{\text{th}}, \mathbf{r}_{\text{tip}})$. To this end, one can use the STM ability of the NSThM probe to obtain the topographical information of the sample surface. Using this data for the theoretical calculation one can compare the theoretical results for the LDOS with the measured thermovoltage V_{th}.

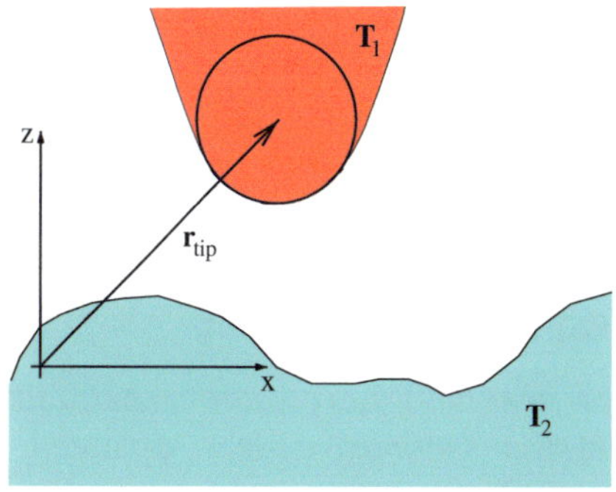

Fig. 9. Schematic of the tip-sample geometry. The sensor tip is assumed to have a spherical tip apex so that it can be modeled by a simple dipole placed in the center of the tip apex at $\mathbf{r}_{\text{tip}}$.

Such a comparision is shown in Fig. 10 for the scan of a 100 nm × 100 nm gold surface. During the measurement the gold surface is cooled down to about 110 K, whereas the tip is kept at 293 K. The tip-surface distance is kept constant and is smaller than 1 nm. Due to this small distance the dipole model together with the first-order perturbation theory is strictly speaking not valid anymore. Nonetheless, the data fit very well with the LDOS calculated for a constant distance of 9 nm above the surface showing that the measured signal follows qualitatively the LDOS of the thermal electromagnetic field above the surface profile evaluated at the dominant thermal frequency $\omega_{\text{th}} \approx 10^{14}$ rad/s. Further quantitative comparisions with the predictions of a refined model for different samples and scan modi are desirable for exploring the possibilities opened up by the NSThM.

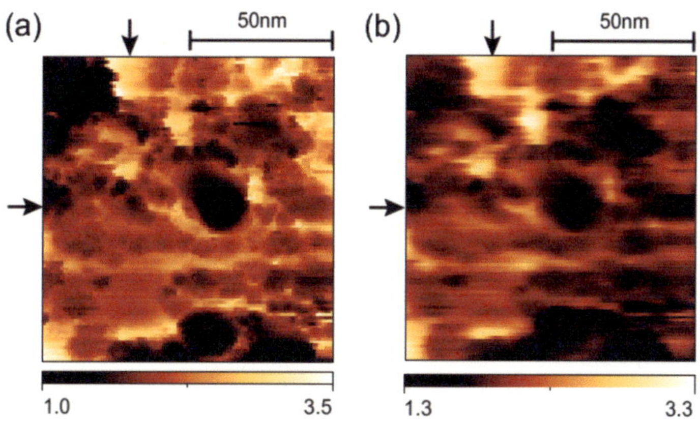

Fig. 10. (a) Numerically calculated LDOS in 10^6 m^{-3} s at a constant distance of 9 nm above the two-dimensional topography directly extracted from the STM data (b) A plot of the thermovoltage data which corresponds to the temperature gradient in the tip and varies as the temperature in the tip apex varies (in arbitrary units). Reprinted with permission from Appl. Phys. Lett., Vol. **93**, 193109 (2008). Copyright 2008, American Institute of Physics.

In summary, the NSThM provides the possibility for a contact-free measurement of surface properties by exploiting the enhanced radiative heat flux at the nanoscale. In particular, it allows for measuring the heat flux between the nanometer-sized tip and a surface in a distance regime of a few nanometers above the sample surface. The measured thermovoltage signal is not only proportional to the heat flux between the tip and the sample surface, but also facilitates the measurement of the local surface temperature. On the other hand, since the heat flux depends on the material properties of the sample, the measured signal could also be used to access local material properties, while a spectral measurement as in (Huth et al. (2011)) is more suitable for that purpose.

3. Thermal management with anisotropic media

A proper understanding of near-field heat transfer naturally gave rise to new ideas on how to control the heat flux between closely separated structures, i.e., on *thermal management* at the micro/nanoscale. Such a control can be achieved, for instance, by thermal rectifiers (Basu and Francoeur (2011); Otey et al (2010)), thermal transistors (Ojanen and Jauho (2008)) and thermal modulators (Biehs et al. (2011b); van Zwol et al (2010)) for thermal photons. Here we review a very interesting approach to heat-flux modulation that consists in actively changing the relative orientation of electrically anisotropic materials, while keeping a fixed (small) distance between them. However, before going into the specifics of that subject we would like to briefly comment another approach to heat flux modulation, namely, with the use of phase change materials.

In our context, a phase change material (PCM) can be defined as a medium that shows two distinct solid phases, one amorphous and the other crystalline, and that can be switched from one to the other in a sufficiently short time (Wuttig and Yamada (2007)). The switching typically goes through the liquid phase as well, and can be summarized in a series of three steps (Wuttig and Yamada (2007)). First we take a PCM in the crystalline phase and heat it up quickly with an intense short pulse. The subsequent cooling is thus also very fast and leads to a quenching process, trapping the material in an amorphous state. The return to the crystalline state is performed by a weaker and longer pulse, that heats up the medium just enough to allow the transition. The considerable difference in optical and electrical properties between the amorphous and crystalline states of some PCMs opens the door to several potentially interesting applications. Among them, we find the possibility of actively controlling the heat flux by switching the PCM back and forth among its two phases, which can be done as fast as 100 ns (Wuttig and Yamada (2007)). Not only the modulation is quick, but it was also shown (van Zwol et al (2010; 2011b)) that for certain distances the switching changes the heat flux by one order of magnitude and that the cycle is fairly repeatable ($10^7 - 10^{12}$ times), making it a good candidate for possible applications in thermal management.

3.1 The heat transfer between planar anisotropic materials

In order to fix ideas, let us consider the situation depicted in Fig. 1,where we have two linear semi-infinite media at different temperatures, but at this point not necessarily homogeneous or isotropic. The expression for the transfered heat is given by

$$\mathrm{H_F}(T_1, T_2, a) = \int_A \mathbf{dA} \cdot \langle \mathbf{S}^{1\to 2} - \mathbf{S}^{2\to 1} \rangle = \int_{z=0} \mathrm{d}^2\mathbf{r}_{\|} \langle S_z^{1\to 2} - S_z^{2\to 1} \rangle, \tag{26}$$

where $\mathbf{r}_{\|} = (x, y)$ and $S_z^{1\to 2}$ is given by (8) and the integration can be over any surface A that completely separates the bodies, that for convenience (and with no loss of generality) we took as the plane $z = z_0$. By using the Fourier expansions (3) and the Green's dyad introduced in

(4, 5), we can recast the integrand of the previous expression into

$$\langle S_z \rangle = \int_0^\infty \frac{d\omega}{2\pi} \left[\Theta(\omega, T_1) - \Theta(\omega, T_2) \right] \langle S_\omega \rangle, \tag{27}$$

where (Volokitin and Persson (2007))

$$\langle S_\omega \rangle = 2\,\mathrm{Re}\,\mathrm{Tr} \int d\mathbf{r}'_{\parallel} \left(\mathbb{G}(\mathbf{r}, \mathbf{r}') \partial_z \partial'_z \mathbb{G}^\dagger(\mathbf{r}, \mathbf{r}') - \partial_z \mathbb{G}^\dagger(\mathbf{r}, \mathbf{r}') \partial'_z \mathbb{G}(\mathbf{r}, \mathbf{r}') \right) \Bigg|_{z'=z=z_0} . \tag{28}$$

and $\Theta(\omega, T_i)$ was defined in (10).

The conclusion that we draw from Eqs. (26)-(28) is that, in order to evaluate the heat transfer for a given geometry we have to determine the Green's dyadic inside the gap region. In most cases this is surely a formidable task, but for planar homogeneous media, even if anisotropic, it is possible to simplify things enough so semi-analytic expressions are obtainable. This is not to say that everything was made easy - in fact even in this simplified case the calculations are fairly long (Chew (1995); Tomaš (2002)) [or requires some indirect arguments, see (Philbin and Leonhardt (2008))], so we shall just quote the final result for the Green tensor

$$\begin{aligned} \mathbb{G}(\mathbf{r}, \mathbf{r}') &= \frac{i}{2} \int d^2\boldsymbol{\kappa}\, \frac{e^{i\boldsymbol{\kappa}\cdot(\mathbf{r}_\parallel - \mathbf{r}'_\parallel)}}{k_{z0}} \Bigg[\mathbb{D}_{12} \left(\mathbb{1} e^{ik_{z0}(z-z')} + \mathbb{R}_1 e^{ik_{z0}(z+z')} \right) \\ &+ \mathbb{D}_{21} \left(\mathbb{R}_2 \mathbb{R}_1 e^{ik_{z0}(z'-z)} e^{2ik_{z0}d} + \mathbb{R}_2 e^{2ik_{z0}d} e^{-ik_{z0}(z+z')} \right) \Bigg], \end{aligned} \tag{29}$$

where $\mathbb{R}_i$ $(i = 1, 2)$ are the 2×2 reflection matrices characterizing interfaces (to be extensively discussed in the next section) and $\mathbb{D}_{ij}$ are defined by

$$\mathbb{D}_{ij} = (\mathbb{1} - \mathbb{R}_i \mathbb{R}_j e^{2ik_{z0}d})^{-1}. \tag{30}$$

When inserting Eq. (29) into the heat flux formula we find the analogue of (10) for anisotropic media, which reads (Biehs et al. (2011))

$$\langle S_\omega \rangle = \int \frac{d^2\boldsymbol{\kappa}}{(2\pi)^2} \mathcal{T}_{\mathrm{A}}(\omega, \boldsymbol{\kappa}, d), \tag{31}$$

where

$$\mathcal{T}_{\mathrm{A}}(\omega, \boldsymbol{\kappa}, d) = \begin{cases} \mathrm{Tr}\left[(\mathbb{1} - \mathbb{R}_2^\dagger \mathbb{R}_2) \mathbb{D}_{12} (\mathbb{1} - \mathbb{R}_1^\dagger \mathbb{R}_1) \mathbb{D}_{12}{}^\dagger \right], & \kappa < \omega/c \\ \mathrm{Tr}\left[(\mathbb{R}_2^\dagger - \mathbb{R}_2) \mathbb{D}_{12} (\mathbb{R}_1 - \mathbb{R}_1^\dagger) \mathbb{D}_{12}{}^\dagger \right] e^{-2|k_{z0}|d}, & \kappa > \omega/c \end{cases} \tag{32}$$

where Tr stands for the two-dimensional trace. From the previous equation we see that the whole problem is now reduced essentially to the calculation of the reflection matrices $\mathbb{R}_1, \mathbb{R}_2$, meaning that the problem has become essentially classical: the reflection coefficients can be found by considering a classical plane wave impinging on a vacuum/magnetodielectric interface, with no fluctuating fields involved. Since this is a somewhat long exercise, we give an outline for it in the next section.

3.2 Reflection coefficients for anisotropic materials

Let us consider the situation depicted on Fig. 11, that shows an incoming plane wave being reflected by an anisotropic (homogeneous) half-space. In the orthonormal coordinate system defined by the incident plane the incident fields are

$$\mathbf{E}_{\text{in}} = \left[e^{\text{s}}_{\text{in}}\hat{\mathbf{y}}' + e^{\text{p}}_{\text{in}}\frac{c}{\omega}(q_{\text{in}}\hat{\mathbf{x}}' - k_{x'}\hat{\mathbf{z}}')\right] e^{i(k_{x'}x' + q_{\text{in}}z' - \omega t)}, \tag{33}$$

$$\mathbf{H}_{\text{in}} = \left[e^{\text{p}}_{\text{in}}\hat{\mathbf{y}}' - e^{\text{s}}_{\text{in}}\frac{c}{\omega}(q_{\text{in}}\hat{\mathbf{x}}' - k_{x'}\hat{\mathbf{z}}')\right] e^{i(k_{x'}x' + q_{\text{in}}z' - \omega t)}, \tag{34}$$

where e^{s}_{in}, e^{p}_{in} are respectively the transverse electric (TE) and transverse magnetic (TM) incoming amplitudes, and we defined $k_{x'} = (\omega/c)\sin\theta_{\text{in}}$ and $q_{\text{in}} = (\omega/c)\cos\theta_{\text{in}}$. The reflected wave has a similar expression

$$\mathbf{E}_{\text{ref}} = \left[e^{\text{s}}_{\text{ref}}\hat{\mathbf{y}}' - e^{\text{p}}_{\text{ref}}\frac{c}{\omega}(q_{\text{in}}\hat{\mathbf{x}}' + k_{x'}\hat{\mathbf{z}}')\right] e^{i(k_{x'}x' - q_{\text{in}}z' - \omega t)}, \tag{35}$$

$$\mathbf{H}_{\text{ref}} = \left[e^{\text{p}}_{\text{ref}}\hat{\mathbf{y}}' + e^{\text{s}}_{\text{ref}}\frac{c}{\omega}(q_{\text{in}}\hat{\mathbf{x}}' + k_{x'}\hat{\mathbf{z}}')\right] e^{i(k_{x'}x' - q_{\text{in}}z' - \omega t)}, \tag{36}$$

where we have used $q_{\text{ref}} = -q_{\text{in}}$. Our problem now consists in finding the amplitudes $e^{\text{s}}_{\text{ref}}$, $e^{\text{p}}_{\text{ref}}$, so we can construct the reflection matrix given by

$$\mathbb{R}_j = \begin{bmatrix} r_j^{\text{s,s}}(\omega,\boldsymbol{\kappa}) & r_j^{\text{s,p}}(\omega,\boldsymbol{\kappa}) \\ r_j^{\text{p,s}}(\omega,\boldsymbol{\kappa}) & r_j^{\text{p,p}}(\omega,\boldsymbol{\kappa}) \end{bmatrix}, \tag{37}$$

where, by definition

$$r_j^{\text{s,s}}(\omega,\boldsymbol{\kappa}) = \frac{e^{\text{s}}_{\text{ref}}}{e^{\text{s}}_{\text{in}}} \quad , \quad r_j^{\text{p,s}}(\omega,\boldsymbol{\kappa}) = \frac{e^{\text{p}}_{\text{ref}}}{e^{\text{s}}_{\text{in}}}$$
$$r_j^{\text{s,p}}(\omega,\boldsymbol{\kappa}) = \frac{e^{\text{s}}_{\text{ref}}}{e^{\text{p}}_{\text{in}}} \quad , \quad r_j^{\text{p,p}}(\omega,\boldsymbol{\kappa}) = \frac{e^{\text{p}}_{\text{ref}}}{e^{\text{p}}_{\text{in}}} \tag{38}$$

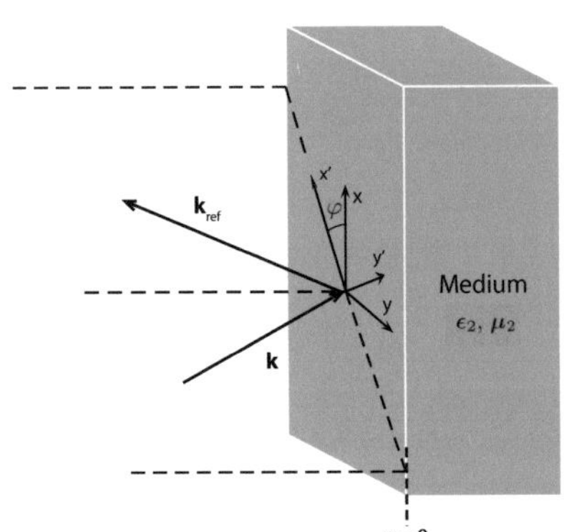

Fig. 11. An incident plane wave impinging on an anisotropic material.

The determination of such amplitudes is carried out by solving Maxwell's equations and imposing the proper boundary conditions on the interface (and on infinity). That means that we have to find the transmitted amplitudes as well, which in turn requires that we state the constitutive relations for the materials involved. For anisotropic magnetodielectric media we have

$$\mathbf{D} = \boldsymbol{\epsilon} \cdot \mathbf{E} = \begin{bmatrix} \epsilon_{xx} & \epsilon_{xy} & \epsilon_{xz} \\ \epsilon_{yx} & \epsilon_{yy} & \epsilon_{yz} \\ \epsilon_{zx} & \epsilon_{zy} & \epsilon_{zz} \end{bmatrix} \cdot \begin{bmatrix} E_x \\ E_y \\ E_z \end{bmatrix} \quad \text{and} \quad \mathbf{B} = \boldsymbol{\mu} \cdot \mathbf{H} = \begin{bmatrix} \mu_{xx} & \mu_{xy} & \mu_{xz} \\ \mu_{yx} & \mu_{yy} & \mu_{yz} \\ \mu_{zx} & \mu_{zy} & \mu_{zz} \end{bmatrix} \cdot \begin{bmatrix} H_x \\ H_y \\ H_z \end{bmatrix} \tag{39}$$

where it is assumed that (i) the material tensors $\boldsymbol{\epsilon}$ and $\boldsymbol{\mu}$ are functions of frequency only, so no spatial dispersion is present, and that (ii) the materials involved do not present what is called bi-anisotropy (Tsang et al. (2000)), which manifests itself in non-vanishing cross couplings between **D** and **H** (and also **B** and **E**), and finally that (iii) the whole system is time-reversible, implying on $\epsilon_{xy} = \epsilon_{yx}$, $\mu_{xy} = \mu_{yx}$ (Landau and Lifshitz (2007)).

The degree of anisotropy of a material is roughly governed by the eigenvalues and eigenvectors of $\boldsymbol{\epsilon}$ and $\boldsymbol{\mu}$, which are in turn connected to the crystallographic structure of the material (Landau and Lifshitz (2007)). In the simplest case we have a cubic lattice, which has completely degenerate eigenvalues and is therefore not different from an isotropic medium. In the next level we have the trigonal, tetragonal and hexagonal lattices (Kittel (1962)), all characterized by two degenerate eigenvalues, or, in other words, by a preferred axis. Increasing the complexity a bit more we get to the orthorhombic lattice (Kittel (1962)), which presents 3 different eigenvalues but still has the eigenvectors crystallographic fixed (and orthogonal to each other). Finally, in the top of the list are the monoclinic and triclinic lattices (Kittel (1962)), which have no eigenvalue degeneracy and show also the so-called dispersion of axes (Landau and Lifshitz (2007)), meaning that the direction of the eigenvectors depend upon frequency.

Substituting the constitutive relations into Maxwell's equations (1)-(2), we get

$$\nabla \cdot (\boldsymbol{\epsilon} \cdot \mathbf{E}) = 0\,, \qquad \nabla \cdot \mathbf{B} = 0 \tag{40}$$

$$\nabla \times \mathbf{E} = -\frac{1}{c}\frac{\partial \mathbf{B}}{\partial t}\,, \qquad \nabla \times (\boldsymbol{\mu}^{-1} \cdot \mathbf{B}) = \frac{1}{c}\boldsymbol{\epsilon} \cdot \frac{\partial \mathbf{E}}{\partial t}, \tag{41}$$

again reminding that we are now solving a classical reflection/transmission problem, so $\rho(\mathbf{r}, t) = \mathbf{j}(\mathbf{r}, t) = 0$. By assuming plane waves as solutions inside the material as well, we get

$$\begin{aligned} \mathbf{E} &= \mathbf{e}(z')e^{i(k_{x'}x' - \omega t)}\,, & \mathbf{e} &= (e_{x'}, e_{y'}, e_{z'}), \\ \mathbf{H} &= \mathbf{h}(z')e^{i(k_{x'}x' - \omega t)}\,, & \mathbf{h} &= (h_{x'}, h_{y'}, h_{z'}), \end{aligned} \tag{42}$$

and using that k_x is conserved across the interface, we see that the z' components can be eliminated as

$$e_{z'} = -ck_{x'}h_{y'}/\omega\epsilon_{z'z'}\,, \qquad h_{z'} = ck_{x'}e_{y'}/\omega\mu_{z'z'}, \tag{43}$$

leaving a total of 4 linearly independent solutions for a given $k_{x'}$ and ω (Chew (1995)). In order to determine the remaining x' and y' components of **e** and **h** it is convenient to introduce a vector $\mathbf{u}$ with components $u_1 = e_{x'}$, $u_2 = e_{y'}$, $u_3 = h_{x'}$ and $u_4 = h_{y'}$. With the ansatz $u_j = u_j(0)e^{iqz'}$ we can transform (40)-(41) into an algebraic linear system of equations

$$\mathbf{L} \cdot \mathbf{u} = -\frac{c}{\omega} q \, \mathbf{u}, \tag{44}$$

where $\mathbf{L}$ is a 4×4 matrix and the possible q's are determined by

$$\det \left(\mathbf{L} + \frac{\omega q}{c} \mathbf{I} \right) = 0. \tag{45}$$

The analytical solutions of (44) and (45) for an arbitrary anisotropic magnetoelectric behavior are certainly very cumbersome, and to best of our knowledge they were never written down explicitly. The general case for electric anisotropy only ($\mu_{ij} = \delta_{ij}$) was dealt in (Teitler et al (1970)), while the magnetoelectric orthorhombic case was treated in (Rosa et al. (2008)). Due to the size and scope of this work it is not possible to reproduce the details here, so the interested reader is kindly referred to the references just mentioned in order to find the explicit solutions not only to (44) and (45) but also to the reflection coefficients themselves.

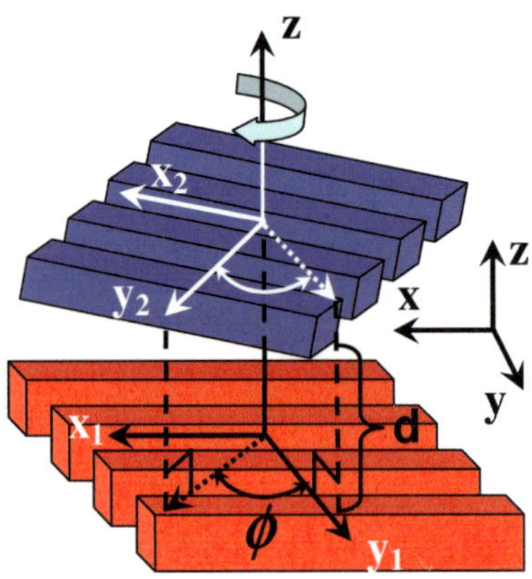

Fig. 12. Two gratings at different temperatures twisted with respect to each other. Reprinted with permission from Appl. Phys. Lett., Vol. **98**, 243102 (2011). Copyright 2008, American Institute of Physics.

3.3 Results

With the explicit expressions for the reflection matrices, we can calculate the transmission factor in (32) and therefore the heat transfer (27) with (31). In order to have a concrete situation in mind, let us imagine that we have the situation depicted in the Fig. 12, where two grating structures are facing each other at an arbitrary twisting angle (Biehs et al. (2011b)). In the effective medium approximation, those gratings may be described as anisotropic media with different dielectric/conduction properties in y and x, z directions. Assuming a simple Maxwell-Garnett model (Tao et al (1990)) for the respective permittivities, we get

$$\epsilon^i_{xx}(\omega) = \epsilon^i_{zz}(\omega) = \epsilon_{\mathrm{h}_i}(\omega)(1 - f_i) + f_i \quad , \quad \epsilon^i_{yy}(\omega) = \frac{\epsilon_{\mathrm{h}_i}(\omega)}{(1 - f_i) + f_i \epsilon_{\mathrm{h}_i}(\omega)}, \tag{46}$$

where ϵ_{h_i} is the permittivity of the i-th host medium, and f_i is the filling factor of the air inclusions in the i-th grating.

Substituting expressions (46) into (35)-(40) of Ref. (Rosa et al. (2008)) and then into (31) we get the heat transfer between the two gratings in the effective medium approximation. In Fig. 13(a) we plot the heat flux between two gold gratings as a function of the relative angle of twist between them, for fixed distances. We see that the flux is dramatically reduced as we twist the gratings, up to almost 80% at $\phi = \pi/2$ for distances as large as 1 μm. Unfortunately there is no simple physical picture that allows us to understand such effect, but it clearly indicates that symmetric configurations transmit heat more efficiently that asymmetric ones. This is further supported by Fig. 13(b), where the heat flux between two SiC gratings is shown. The reduction in the flux is less impressive in this case (although still quite significative), but the upside is that here we have more direct interpretation: for SiC gratings the surface modes give an important contribution to the flux, so it is intuitive that mismatching surface mode dispersion relations (for twisted structures) couple less effectively than matching ones (for parallel gratings) and will therefore give rise to a smaller transmission factor, and that is indeed what is observed.

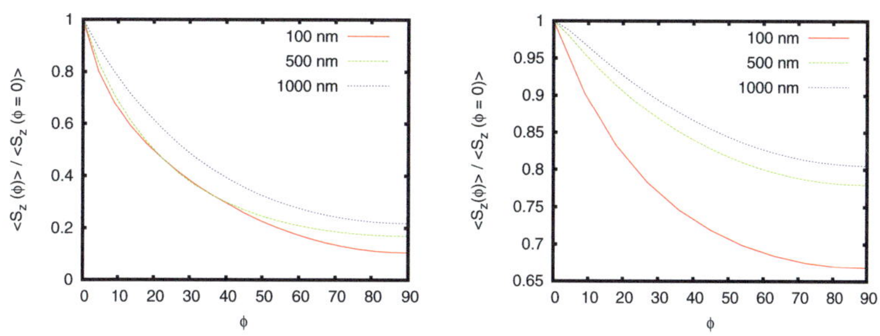

(a) Au (b) SiC

Fig. 13. The heat flux $\langle S_z \rangle(\phi)$ between two (a) Au and (b) SiC gratings, normalized by the flux $\langle S_z \rangle(0^\circ)$ when the gratings are aligned. The angle ϕ measures the relative twisting between the gratings, and the filling factor is fixed at $f = 0.3$. Reprinted with permission from Appl. Phys. Lett., Vol. **98**, 243102 (2011). Copyright 2008, American Institute of Physics.

Going back to the Au gratings, we see that the large suppression obtained by just rotating the structures with respect to one another suggests that such a setup could be used as a thermal modulator controlled by the twisting angle: in the parallel position there would be a heat flux (position "on"), in the orthogonal one there would not (position "off"). The on/off switching could as fast as several tens of kilohertz, and it would be extremely robust as the relative rotation does not wear off the material. Such thermal modulators can for example be interesting for fast heat flux modulation and thermal management of nano-electromechanical devices (Biehs et al. (2011b)).

4. Near-field thermophotovoltaics

Thermophotovoltaic (TPV) devices (Coutts (1999)) are energy conversion systems that generate electric power directly from thermal radiation. The basic principle (see Fig. 14) is similar to the classical photovoltaic conversion. A source of photons radiates in the direction of a p-n junction which converts the photons which have a sufficient energy into electron-hole pairs which, in turn, can be used to generate electricity. However contrary to classical systems, TPV devices operates in the near-infrared and not in the visible range. The efficiency of a photovoltaic cell is defined as the ratio $\eta = P_{\rm el}/P_{\rm rad}$ of the electric power $P_{\rm el}$ produced by the photovoltaic cell and the net radiative power P_{rad} exchanged between the hot source and the p-n junction.

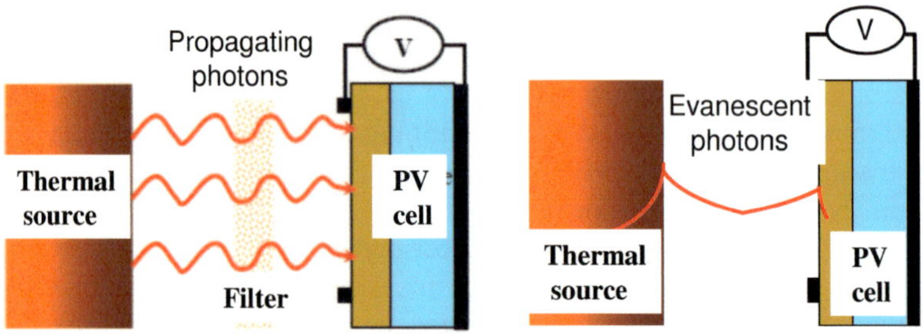

Fig. 14. Principle of thermophotovoltaic energy conversion devices. (a) In far field, the photovotlaic (PV) cell is located at long distance (compared to the thermal wavelength) from a thermal source. Propagating photons only reach the cell. A filter can eventually select the photons with an energy higher than that of the energy gap of the cell. (b) In near-field TPV the cell is located at subwavelength distance from the source. Evanescent photons are the main contributors to the radiative power transfered from the source to the cell.

In far field, this efficiency is in principle limited by the thermodynamic Schockley-Queisser limit (Shockley and Queisser (1961)) wich corresponds to the case where the source is a perfect black body and is typically about 33%. This limit could be easily overcome with a monochromatic source when the frequency of emission coincides with the gap energy of the semiconductor. In this case η would be equal to one. However, first it is difficult to have natural materials with a monochromatic emission so that some photons are generally dissipated incide the cell without participating to the conversion. Second, the production of electricity depends directly on the magnitude of radiative flux received by the cell. But, in the far field, the heat flux cannot exceed that of black body. On the other hand, in the near-field the heat flux can be several orders of magnitude larger than that of a black body, so that near-field TPV conversion (Basu et al. (2009); Laroche et al. (2006); Narayanaswamy and Chen (2003); Pan et al. (2000); Park et al. (2007)) seems to be a promising technology for an intensive production of electricity.

Generaly speaking, in (far or near-field) TPV devices, the maximal power which can be extracted from the cell reads (Laroche et al. (2006))

$$P_{el} = F_{\text{fill}} I_{\text{ph}} V_{\text{oc}}, \tag{47}$$

where I_{ph} is the photogeneration current (which corresponds to photons that are effectively converted), V_{oc} is the open-circuit voltage (which correspond to a vanishing current into the diode). The factor F_{fill} is called fill factor and depends on I_{ph} and on the saturation current I_0 of the diode. When we assume that each absorbed photon with an energy higher than the gap energy E_g produces an electron-hole pair, the photogeneration current reads (Laroche et al. (2006))

$$I_{\text{ph}} = e \int_{E_g/\hbar}^{\infty} \mathrm{d}\omega \, \frac{P_{\text{rad}}(\omega)}{\hbar\omega}. \tag{48}$$

It immediately follows from this equation that an increase in the radiative power exchanged between the source and the cell leads to an enhancement of the photogeneration current. On the other hand, the fill factor is given by (Laroche et al. (2006))

$$F_{\text{fill}} = \left[1 - \frac{1}{\ln(I_{\text{ph}}/I_0)}\right]\left[1 - \frac{\ln(\ln(I_{\text{ph}}/I_0))}{\ln(I_{\text{ph}}/I_0)}\right], \tag{49}$$

with the dark current (Ashcroft and Mermin (1976))

$$I_0 = e\left(\frac{n_i^2 D_{\text{h}}}{N_{\text{D}}\tau_{\text{h}}^{1/2}} + \frac{n_i^2 D_{\text{e}}}{N_{\text{A}}\tau_{\text{e}}^{1/2}}\right). \tag{50}$$

In Eq.(50) n_i denotes the intrinsic carrier concentration, N_D (N_A) the donor (acceptor) concentration, D_{e} (D_{h}) the diffusion constant of electrons (holes) and τ_{e} and τ_{h} represent the electron-hole pair lifetime in the p-doped and n-doped domains of the cell. In Fig. 15 we see that for a plane tungsten thermal source in front of a GaSb cell (see Palik (1998) for optical properties) the radiative power exchanged increases dramaticaly at subwavength distances compared to what we observe in far field. As direct consequence, the photocurrent generated in the GaSb cell follows an analog behavior.

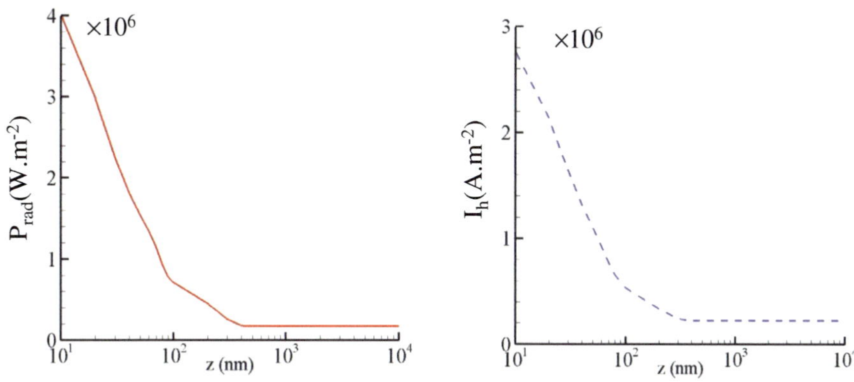

Fig. 15. (a) Radiative power exchanged between a tungsten source at 2000K and a GaSb cell at 300K. (b) Photocurrent in the GaSb cell with respect to the separation distance z of the thermal source. $N_{\text{A}} = N_{\text{D}} = 10^{-17}\text{cm}^{-3}$; $n_i = 4.3 \times 10^{1}2\text{cm}^{-3}$. Physical properties are taken from (Rosencher and Vinter (2002))

Once the photocurrent and the dark current are known, the electric power [see Eq. (47)] can be calculated using the open circuit voltage (Laroche et al. (2006))

$$V_{\text{oc}} = \frac{k_{\text{B}}T}{e}\log\left(\frac{I_{\text{h}}}{I_0}\right). \tag{51}$$

Fig. 16 clearly shows that the near-field TPV device produces much more electricity than a classical TPV conversion system. At a distance between the thermal source and the cell of $z = 100nm$ the production is approximatly enhanced by a factor of 5. At 10 nm this factor reaches a value of about 50 times the far-field value. These results show that the near-field TPV conversion is a promising technology that could offer new solutions for energy production in the next decades.

P. B.-A. and F.S.S. R. acknowledge the support of the Agence Nationale de la Recherche through the Source-TPV project ANR 2010 BLANC 0928 01. This research was partially supported by Triangle de la Physique, under the contract 2010-037T-EIEM.

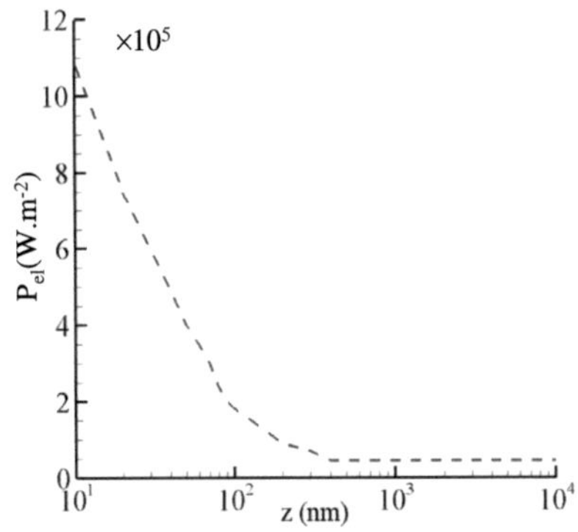

Fig. 16. Electric power generated by a Tungsten-GaSb cell with respect to the separation distance cell-source (same parameters as in (Laroche et al. (2006))).

5. References

G. S. Agarwal (1975), Quantum electrodynamics in the presence of dielectrics and conductors. I. Electromagnetic-field response functions and black-body fluctuations in finite geometries, *Physical Review A*, Vol. 11 (No. 1): 230-242.

N. Ashcroft and N. D. Mermin (1976), *Solid-State Physics*, Harcourt College Publishers, Philadelphia.

G. Baffou, M. P. Kreuzer, and R. Quidant (2009), Temperature mapping near plasmonic nanostructures using fluorescence polarization anisotropy, *Optics Express*, Vol. 17 (No. 5), 3291-3298.

S. Basu and M. Francoeur (2011), Near-field radiative transfer based thermal rectification using doped silicon, *Applied Physical Letters*, Vol. 98: 113106.

S. Basu, Z. M. Zhang, and C. J. Fu (2009), Review of near-field thermal radiation and its application to energy conversion,*International Journal of Energy Research*, Vol. 33 (No. 13), 1203-1232.

S. Basu and Z. M. Zhang (2009), Maximum energy transfer in near-field thermal radiation at nanometer distances, *Journal of Applied Physics*, Vol. 105: 093535.

P. Ben-Abdallah and K. Joulain (2010), Fundamental limits for noncontact transfers between two bodies,*Physical Review B(R)*, Vol. 82 (No. 12): 121419.

S.-A. Biehs, E. Rousseau, and J.-J. Greffet (2010), A mesoscopic description of radiative heat transfer at the nanoscale, *Physical Review Letters*, Vol. 105(No. 23): 234301.

S.-A. Biehs, O. Huth, F. Rüting, M. Holthaus (2010b), Spheroidal nanoparticles as thermal near-field sensors, *Journal of Applied Physics*, Vol. 108 (No. 1): 014312.

S.-A. Biehs and J.-J. Greffet (2010c), Influence of roughness on near-field heat transfer between two plates, *Physical Review B*, Vol. 82 (No. 24): 245410.

S.-A. Biehs and J.-J. Greffet (2010d), Near-field heat transfer between a nanoparticle and a rough surface, *Physical Review B*, Vol. 81 (No. 24): 245414.

S.-A. Biehs, O. Huth, and F. Rüting (2008), Near-field radiative heat transfer for structured surfaces, *Physical Review B*, Vol. 78 (No. 8): 085414.

S.-A. Biehs, F. S. S. Rosa, P. Ben-Abdallah, K. Joulain, and J.-J. Greffet (2011), Nanoscale heat flux between nanoporous materials, *Optics Express*, Vol. 19 (No. S5): A1088-A1103.

S.-A. Biehs, F. S. S. Rosa, and P. Ben-Abdallah (2011b), Modulation of near-field heat transfer between two gratings, *Applied Physical Letters*, Vol. 98: 243102.

R. Carminati and J.-J. Greffet (1999), Near-field effects in spatial coherence of thermal sources, *Physical Review Letters*, Vol. 82 (No. 8): 1660-1663.

P.-O. Chapuis, S. Volz, C. Henkel, K. Joulain, and J.-J. Greffet (2008), Effects of spatial dispersion in near-field radiative heat transfer between two parallel metallic surfaces,*Physical Review B*, Vol. 77: 035431.

P.-O. Chapuis, M. Laroche, S. Volz, and J.-J. Greffet (2008b), Near-field induction heating of metallic nanoparticles due to infrared magnetic dipole contribution, *Physical Review B*, Vol. 77 (No. 12): 125402.

W. C. Chew (1995), *Waves and Fields in Inhomogeneous Media*, IEEE Press.

T. J. Coutts (1999), A review of progress in thermophotovoltaic generation of electricity,*Renewable and Sustainable Energy Reviews*, Vol. 3: 77-184.

E. Cravalho, C. Tien, and R. Caren (1967), Effect of small spacings on radiative transfer between 2 dielectrics, *J. Heat Transfer*, Vol. 89 (No. 4): 351.

S. Datta (2002), *Electronic Transport in Mesoscopic Systems*, Cambridge University Press.

Y. De Wilde, F. Formanek, R. Carminati, B. Gralak, P.A. Lemoine, K. Joulain, J.P. Mulet, Y. Chen and J.J. Greffet (2006), Thermal radiation scanning tunnelling microscopy, Nature (London), Vol. 444: 740-743.

G. V. Dedkov and A. A. Kyasov (2007), Thermal radiation of nanoparticles occurring at a heated flat surface in vacuum, *Technical Physics Letters*, Vol. 33 (No. 4), 305-308.

I. A. Dorofeyev (1998), Energy dissipation rate of a sample-induced thermal fluctuating field in the tip of a probe microscope,*Journal of Physics D: Applied Physics*, Vol. 31 (No. 6): 600-601.

I. A. Dorofeyev and E. A. Vinogradov (2011), Fluctuating electromagnetic fields of solids, *Physics Reports*, Vol. 504 (No. 1-2): 75.

G. W. Ford and W. H. Weber (1984), Electromagnetic interactions of molecules with metal surfaces, *Physics Reports*, Vol. 113 (No. 4): 195-287.

J.-J. Greffet (1988), Scattering of electromagnetic waves by rough dielectric surfaces, *Physical Review B*, Vol. 37 (No. 11): 6436-6441.

C. Henkel and V. Sandoghdar (1998), Single-molecule spectroscopy near structured dielectrics, Optics Communications, Vol. 158: 250-262.

L. Hu, A. Narayanaswamy, X. Chen, and G. Chen (2008), Near-field thermal radiation between two closely spaced glass plates exceeding Planck's blackbody radiation law, *Applied Physics Letters*, Vol. 92 (No. 13), 133106.

O. Huth, F. Rüting, S.-A. Biehs, M. Holthaus (2010), Shape-dependence of near-field heat transfer between a spheroidal nanoparticle and a flat surface, *The European Physical Journal Applied Physics*, Vol. 50 (No. 1): 10603.

F. Huth, M. Schnell, J. Wittborn, N. Ocelic and R. Hillenbrand (2011), Infrared-spectroscopic nanoimaging with a thermal source, *Nature Materials*, Vol. 10: 352-356.

Y. Imry (2002), *Introduction to Mesoscopic Physics*, Oxford University Press.

J. D. Jackson (1998), *Classical Electrodynamics*, 3rd ed., John Wiley, New York.

M. Janowicz, D. Reddig, and M. Holthaus (2003), Quantum approach to electromagnetic energy transfer between two dielectric bodies, *Physical Review A*, Vol. 68 (No. 4): 043823.

K. Joulain, R. Carminati, J.-P. Mulet, and J.-J. Greffet (2003), Definition and measurement of the local density of electromagnetic states close to an interface, *Physical Review B*, Vol. 68 (No. 68), 245405 (2003).

K. Joulain, J.-P. Mulet, F. Marquier, R. Carminati, and J.-J. Greffet (2005), Surface electromagnetic waves thermally excited: Radiative heat transfer, coherence properties and Casimir forces revisited in the near field, *Surface Science Report*, Vol. 57 (No. 3-4): 59-112.

C. Henkel and K. Joulain (2006), Electromagnetic field correlations near a surface with a nonlocal optical response,*Applied Physics B: Lasers and Optics*, Vol. 84 (No. 1-2): 61-68.

Y. Kajihara, K. Kosaka, S. Komiyama (2010), A sensitive near-field microscope for thermal radiation, *Review of Scientific Instruments*, Vol. 81 (No. 3): 033706.

A. Kittel, W. Müller-Hirsch, J. Parisi, S.-A. Biehs, D. Reddig, and M. Holthaus (2005), Near-field heat transfer in a scanning thermal microscope, *Physical Review Letters*, Vol. 95 (No. 22), 224301. 301 (2005).

A. Kittel, U. F. Wischnath, J. Welker, O. Huth, F. Rüting, and S.-A. Biehs (2008), Near-field thermal imaging of nanostructured surfaces, *Applied Physics Letters*, Vol. 93 (No. 19): 193109.

C. Kittel (1962), *Introduction to Solid State Physics*, 2nd ed. John Wiley and Sons, New York.

K. L. Kliewer and R. Fuchs (1974), Theory of dynamical properties of dielectric surfaces, *Advances in Chemical Physics*, Vol. 27: 355-541.

J.A. Kong (2007), *Electromagnetic Wave Theory*, 2nd ed., Wiley.

R. Kubo, M. Toda, and N. Hashitsume (1991),*Statistical Physics II*, 2nd ed., Springer, Berlin Heidelberg.

L.D. Landau, E.M. Lifshitz, and L.P. Pitaevskii (2007), *Electrodynamics of Continuous Media*, 2nd ed. Elsevier, Oxford.

M. Laroche, R. Carminati, and J. J. Greffet (2006), Near-field thermophotovoltaic energy conversion, *Journal of Applied Physics*, Vol. 100: 063704.

E. M. Lifshitz, L. P. Pitaevskii, *Statistical Physics, Part 2*, Butterworth-Heinemann, Oxford.

A. Majumdar, 1999, Scanning thermal microscopy, *Annual Review of Materials Science*, Vol. 29: 505-585.

J.-P. Mulet, K. Joulain, R. Carminati, and J.-J. Greffet (2001), Nanoscale radiative heat transfer between a small particle and a plane surface, *Applied Physics Letters*, Vol. 78 (No. 19): 2931-2933.

W. Müller-Hirsch, A. Kraft, M. T. Hirsch, J. Parisi, and A. Kittel (1999), Heat transfer in ultrahigh vacuum scanning thermal microscopy, *Journal of Vacuum Science & Technology A*, Vol. 17 (No. 4): 1205-1211.

A. Narayanaswamy and G. Chen (2003), Surface modes for near field thermophotovoltaics, *Applied Physics Letters*, Vol. 82: 3544-3546.

A. Narayanaswamy, S. Shen, and G. Chen (2008), Near-field radiative heat transfer between a sphere and a substrate ,*Physical Review B*, Vol. 78 (No. 11): 115303.

T. Ojanen and A.-P. Jauho (2008), Mesoscopic photon heat transistor, *Physical Review Letters*, Vol. 100: 155902 (2008).

C. R. Otey, W. T. Lau, and S. Fan (2010), Thermal rectification through Vacuum, *Physical Review Letters*, Vol. 104: 154301.

R. Ottens, V. Quetschke, S. Wise, A. Alemi, R. Lundock, G. Mueller, D. H. Reitze, D. B. Tanner, B. F. Whiting (2011), Near-Field Radiative Heat Transfer between Macroscopic Planar Surfaces, *Physical Review Letters*, Vol. 107 (No. 1): 014301.

E.D. Palik (1998), *Handbook of Optical Constants of Solids*, Academic Press, London.

J. L. Pan, H. K. H. Choy, C. G. Fonstad (2000), Very large radiative transfer over small distances from a black body for thermophotovoltaic applications, *IEEE Transactions on Electron Devices*, Vol. 47(No. 1): 241 - 249.

K. Park, S. Basu, W. P. King (2007), Z. M. Zhang, Performance analysis of near-field thermophotovoltaic devices considering absorption distribution, *Journal of Quantitative Spectroscopy & Radiative Transfer*, Vol. 109: 305-316.

J. B. Pendry (1983), Quantum limits to the flow of information and entropy, *Journal of Physics A: Mathematical and General*, Vol. 16(No. 10), 2161-2171.

J. B. Pendry (1999), Radiative exchange of heat between nanostructures,*Journal of Physics: Condensed Matter*, Vol. 11 (No. 35): 6621-6634.

T.G. Philbin and U. Leonhardt (2008), Alternative calculation of the Casimir forces between birefringent plates, *Physical Review A*, Vol. 78: 042107

D. Polder and M. Van Hove (1971), Theory of radiative heat transfer between closely spaced bodies, *Physical Review B*, Vol. 4(No. 4): 3303-3314.

H. Raether (1988),*Surface Plasmons on Smooth and Rough Surfaces and on Gratings*, Springer, Heidelberg.

L. G. C. Rego and G. Kirczenow (1999), Fractional exclusion statistics and the universal quantum of thermal conductance: A unifying approach, *Physical Review B*, Vol. 59(No. 20): 13080-13086.

F. S. S. Rosa, D. A. R. Dalvit, and P. W. Milonni (2008), Casimir interactions for anisotropic magnetodielectric metamaterials, *Physical Review A*, Vol. 78: 032117.

F. S. S. Rosa, D. A. R. Dalvit, and P. W. Milonni (2010), Electromagnetic energy, absorption, and Casimir forces: Uniform dielectric media in thermal equilibrium, *Physical Review A*, Vol. 81 (No. 3): 033812.

E. Rosencher and B. Vinter (2002), *Optoélectronique*, Dunod, Paris.

E. Rousseau, A. Siria, G. Jourdan, S. Volz, F. Comin, J. Chevrier, and J.-J. Greffet (2009), Radiative heat transfer at the nanoscale, *Nature Photonics*, Vol. 3: 514-517.

E. Rousseau, M. Laroche, and J.-J. Greffet (2009b), Radiative heat transfer at nanoscale mediated by surface plasmons for highly doped silicon, *Applied Physics Letters*, Vol. 95 (No. 23): 231913.

E. Rousseau, M. Laroche, and J.-J. Greffet (2010), Radiative heat transfer at nanoscale: Closed-form expression for silicon at different doping levels, *Journal of Quantitative Spectroscopy & and Radiative Transfer*, Vol. 111(No. 7-8): 1005-1014.

G. Russakoff (1970), A derivation of the macroscopic Maxwell equations, *American Journal of Physics*, Vol. 38 (No. 10): 1188-1195.

F. Rüting, S.-A. Biehs, O. Huth, and M. Holthaus (2010), Second-order calculation of the local density of states above a nanostructured surface, *Physical Review B*, Vol. 82 (No. 11): 115443.

S. M. Rytov, Y. A. Kravtsov, and V. I. Tatarskii (1989), *Principles of Statistical Radiophysics*, Vol. 3, Springer, New York.

W. Shockley and H. Queisser (1961), Detailed balance limit of efficiency of pĆÄên junction solar cells, *Journal of Applied Physics*, Vol. 32: 510.

A. V. Shchegrov, K. Joulain, R. Carminati, and J.-J. Greffet, 2000, Near-field spectral effects due to electromagnetic surface excitations, *Physical Review Letters*, Vol. 87 (No. 7): 1548-1551.

S. Shen, A. Narayanaswamy, and G. Chen (2009), Surface phonon polaritons mediated energy transfer between nanoscale gaps, *Nano Letters*, Vol. 9 (No. 8), 2909-2913.

C.-T. Tai (1971), *Dyadic Green's Functions in Electromagnetic Theory*, Intext Educational Publishers, Scranton.

R. Tao, Z. Chen, and P. Sheng (1990), First-principles Fourier approach for the calculation of the effective dielectric constant of periodic composites, *Physical Review B* Vol. 41: 2417.

M. S. Tomaš (2002), Casimir force in absorbing multilayers, *Physical Review A*, Vol. 66: 052103.

S. Teitler and B.W. Henvis (1970), Refraction in stratified, anisotropic media, *Journal of the Optical Society of America*, Vol. 60: 830.

L. Tsang, J. A. Kong, K.-H. Ding (2000), *Scattering of Electromagnetic Waves*, John Whiley, New York.

P. J. van Zwol, K. Joulain, P. Ben-Abdallah, J.-J. Greffet, J. Chevrier (2011), Fast heat flux modulation at the nanoscale, *Physical Review B* Vol. 83: 201404(R).

P. J. van Zwol, K. Joulain, P. Ben-Abdallah, J. Chevrier (2011), Phonon polaritons enhance near-field thermal transfer across the phase transition of VO2, *Physical Review B* Vol. 84: 161413(R).

E. A. Vinogradov and I. A. Dorofeyev, Thermally stimulated electromagnetic fields of solids, *Physics Uspekhi*, Vol. 52 (No. 1), 425-459.

A. I. Volokitin and B. N. J. Persson (2007), Near-field radiative heat transfer and noncontact friction, *Review of Modern Physics*, Vol. 79 (No. 4): 1291-1329.

U. F. Wischnath, J. Welker, M. Munzel, and A. Kittel (2008), The near-field scanning thermal microscope, *Review of scientific instruments*, Vol. 79 (No. 7): 073708.

M. Wuttig and N. Yamada (2007), Phase-change materials for rewritable data storage, *Nat. Mat.*, Vol. 6: 824.

Z. M. Zhang (2007), *Nano/Microscale Heat Transfer*, McGraw-Hill, New York.

2

IR Spectroscopy as a Possible Method of Analysing Fibre Structures and Their Changes Under Various Impacts

Barbara Lipp-Symonowicz,
Sławomir Sztajnowski and Anna Kułak
Technical University of Lodz
Department of Material and Commodity Sciences and Textile Metrology
Poland

1. Introduction

Infrared absorption spectroscopy, broadly applied to analyse polymer structures (within the spectral wavenumber range of 400-4000 cm^{-1}), is also employed as a method of studying fibre structures and their changes. This is one of the instrumental methods commonly applied in Fibre Physics for the purpose of qualitative and quantitative analyses of fibre orientation, studies of fibre crystalline structure, and selective evaluation of the structure of fibre surface layers, as well as the effects of superficial and volumetric modification of fibre structures.

The experimental basis in the studies are the absorption spectra of IR radiation through adequately prepared fibre samples.

Studies of the share of crystalline material in fibres (crystallinity) employ so-called fibre pellet preparations, with fibres homogenously dispersed in KBr, and the IR radiation technique. For quantitative evaluation of crystallinity, the ratio taken is that of the intensity of the so-called "crystalline" or "amorphous" absorption band to the intensity of the so-called "internal standard" absorption band.

2. Studies of fiber surface layer molecular structure using ATR-IR technique

The Attenuated Total Reflectance method (ATR - *Attenuated Total Reflection*), has found application in spectrophotometry, for studies of the surfaces of materials, including fibers[1].

In ATR-IR spectroscopy, infrared radiation passes through a crystal characterized by high refraction index, which makes it possible for the rays to be reflected many times in its interior (figure 1).

Close contact of the studied sample with the crystal surface is a prerequisite for correct measurement.

To obtain reflection inside the crystal, the angle of incidence of the rays must exceed the so-called „critical" angle θ_c. This angle is a function of radiation beam, both in the sample and in the crystal, formula 1:

$$\theta = \arcsin\left(\frac{n_2}{n_1}\right) \quad (1)$$

where:

- n_2 is the radiation beam refraction index for the sample,
- n_1 is the radiation beam refraction index for the crystal.

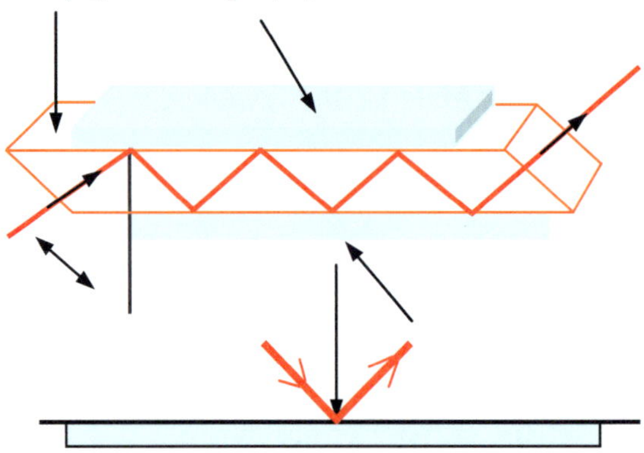

Fig. 1. The Schematic of a path radiation in repeated reflection in the crystal [2].

The radiation entering the crystal may, depending on the type of the investigated sample, penetrate the sample to the depth of to 2μm [3,4] (fig.1). The penetration depth of the „vanishing" wave d_p in defined as the distance between the crystal and the sample, formula 2 [4].

$$d_p = \frac{\lambda}{2\pi n_1\left[\sin^2\theta - \left(\frac{n_2}{n_1}\right)^2\right]^{1/2}} \quad (2)$$

where: - λ is the wavelength of IR radiation.

The measurements are performed by means of a FTIR spectrophotometer, equipped with an ATR sampling accessory.

The conditions most commonly applied for the measurements are as follows:

- the range of IR radiation used in the studies for wool, natural silk and polyamide fibers is 2000÷600 cm^{-1}, for polyacrylnitrile fibers - 2500÷600 cm^{-1},
- the mirror movement speed - 5 mm/s,
- the radiation wavenumber resolution recorded by the detector - 4 cm^{-1},
- as a rule, at least 32 scans of each sample with IR radiation beam are performed to obtain the absorption spectra.

For the investigated fiber samples, IR absorption spectra are plotted in the A=f(1/λ), or T=f(1/λ) system. They provide the basis for interpretation of changes in the molecular and supramolecular structures of the surface layer. An example of a spectrum has been presented in figure 2. The absorption bands are corrected by determination of the baseline, and then their absorbance is determined.

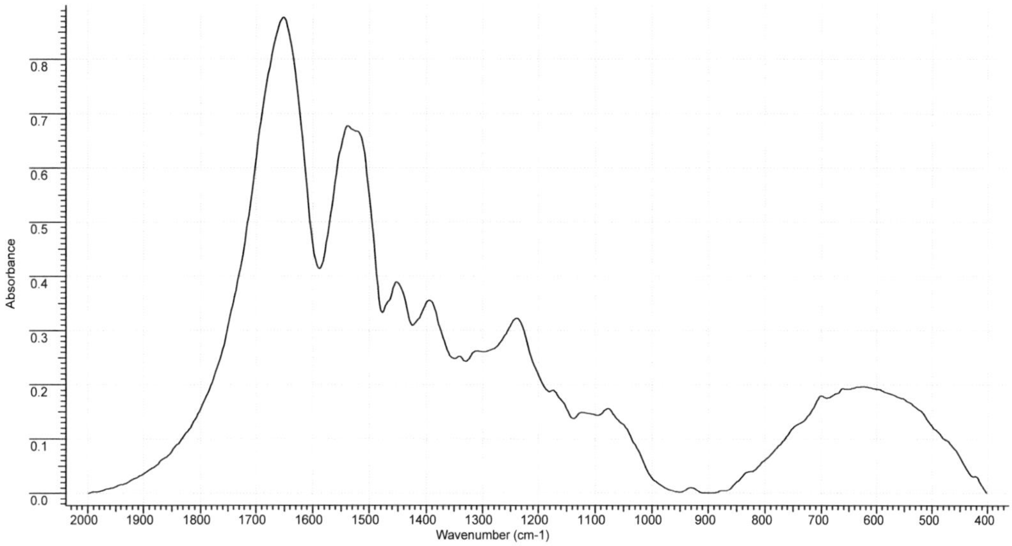

Fig. 2. FT-IR spectrum of wool fibre.

Table 1 presents the correlation between the location of the absorption bands and the type of the chemical groups for wool fibers.

Wavenumber cm^{1}	**Absorbing group and type of vibration**
2950	v_a (CH_2), v_a (CH_3)
1658	NH_2
1539	amide II
1393	COO^-
1233	CNH
1074	CC,CN_3
932	CH_3, CONH

Table 1. The correlation between the location of the absorption bands and the type of the chemical groups for wool fibers under Rau, Urbańczyk [2,5].

3. Studies of fiber structure with IR absorption spectroscopy using the transilluminating technique

3.1 Studies of overall fiber structure

Spectroscopic IR transilluminating technique is used for qualitative studies of changes in the overall structure of fibers. Qualitative analysis of the absorption spectra makes it possible to assess transformations of chemical groups and macromolecules as a result of interactions with external factors.

An IR absorption spectrum represents the changes of radiation intensity, or a value proportional to it, as a function of wavelength λ or another value associated with λ. Absorption of energy portions by intra- and intermolecular bonds, dependent on the frequency of normal vibrations of the chemical group, determines the location and intensity of the corresponding absorption band in the spectrum. Correlation between the location of the band in the absorption spectrum (value of the wavenumber for the band maximum), and the type of the absorbing chemical group is determined experimentally. The dependences of location of the absorption bands on the type of the absorbing groups for fibers are included in the correlation tables and provide the basis for quantitative analysis of the investigated chemical compounds. The correlation of band location with the type of the absorbing chemical group for PAN fibers is presented in table 2 as an example.

The changes in molecular structure manifest themselves by appearance of new absorption bands correlated with the newly formed chemical groups, or by changes in intensity of the absorption bands correlated with the existing chemical groups.

The measurements are performed using a FTIR spectrophotometer.

Wavenumber , cm^{-1}	**Absorbing group and type of vibration**
2950	ν_a (CH_2)
2930	ν (CH)
2870	ν_a (CH_2)
2237	ν (CN)
1447	δ (CH_2)
1362	δ (CH)
1335	γ_w (CH_2) + ν_a (CC)
1310	γ_w (CH_2) - δ (CH)
1247	γ_w (CH)+ γ_w (CH_2) - ν_a $(C\text{-}C)$
1115	ν_s $(C\text{-}C)$ - δ (CH)
1073	ν_s $(C\text{-}C)$, γ_r (CH_2) - δ $(C\text{-}C\text{-}CN)$
865	γ_r (CH_2)
778	ν $(C\text{-}CN)$+ γ_t (CH_2), ν $(C\text{-}CN)$ - γ_r (CH_2)
570	δ $(C\text{-}C\text{-}CN)$ - δ $(C\text{-}C\text{-}N)$
537	δ $(C\text{-}C\text{-}CN)$ - δ $(C\text{-}C\text{-}N)$

Table 2. The IR absorption bands for poliacrylnitryle fibres according Rau, Yamader, Urbańczyk [2,5,6].

The experiments make use of tablet preparations with 1% mass/weight content of the investigated fiber. The tablets are obtained from powdered fiber (segments of length equal to fiber thickness), dispersed in dried potassium bromide with a homogenizer. The tablets are prepared by hydraulic press moulding. Standard conditions are applied for tablet preparation, i.e.:

- homogenization time - 2 minutes,
- moulding time - 10 minutes,
- forcing pressure - 10 MPa.

Standard conditions for measurements:

- to obtain the IR absorption spectra, transmission technique is used,
- the range of IR radiation used in the studies e.g. for wool, natural silk and polyamide fibers is 2000÷400 cm^{-1}, for polyacrylnitrile fibers - 2500÷400 cm^{-1},
- the mirror movement speed - 5 mm/s,
- the radiation wavenumber resolution recorded by the detector - 4 cm^{-1},
- as a rule, at least 32 scans of each sample with IR radiation beam are performed to obtain the absorption spectra.

The spectra obtained in the $A = f(1/\lambda)$ system provide the basis for determination of absorbance A (figure 3), the values of which are adopted for the final interpretation of changes in the molecular structure of the studied fibers. An example of a spectrum is presented in figure 4.

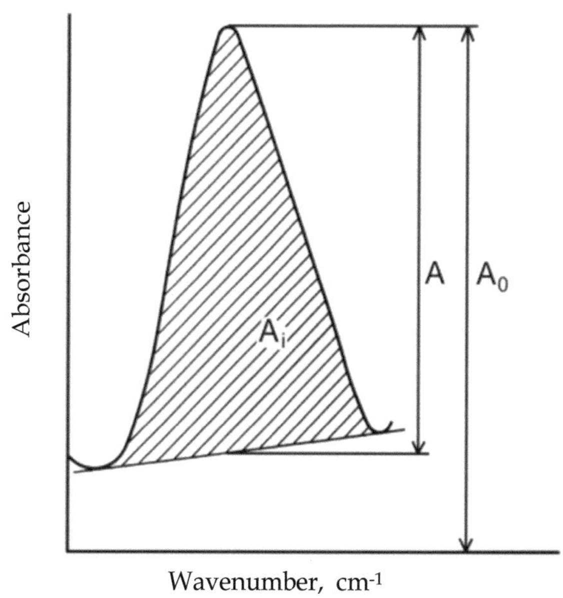

Fig. 3. The method of determining the volume of absorbance A [4].

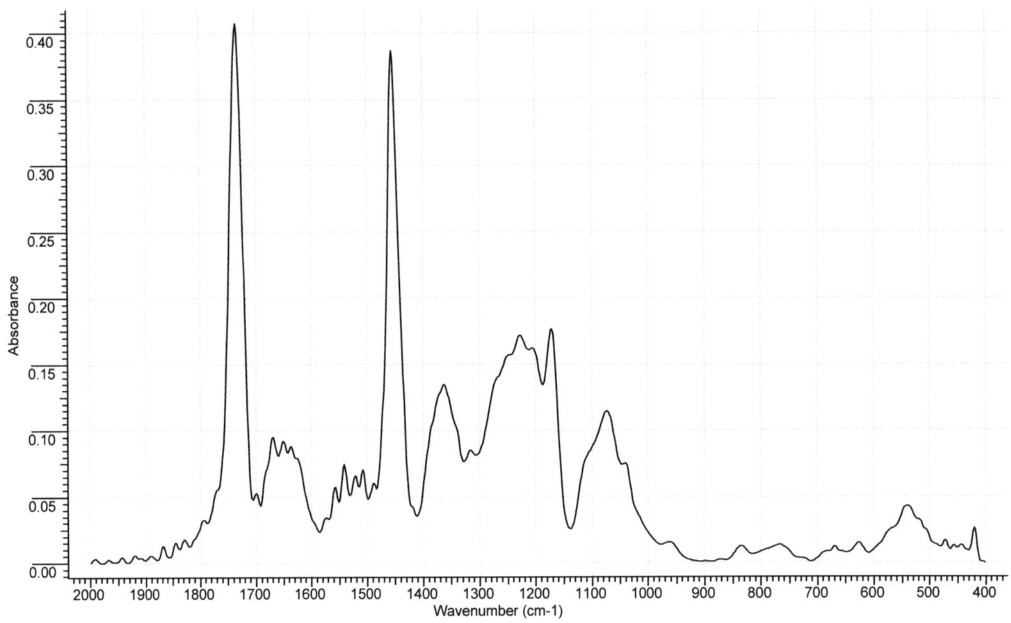

Fig. 4. FT-IR spectrum of PAN fibre „Nitron semi-dull".

3.2 Studies of fiber crystallization grade on the example of PA6 fibers

The assessment of changes in the polyamide fiber structure due to UV irradiation was made[7].

IR analysis was carried out by the transilluminating technique using tablet specimens containing 1% of powdered fiber. IR absorption spectra were recorded within the wavelength range from 4000 cm^{-1} to 400 cm^{-1}, in the following systems: $T = f(1/\lambda)$, $A = f(1/\lambda)$.

The changes in the crystallinity index values were determined from so-called crystalline absorption bands and the bands of internal standard. The values of fiber crystallinity index $x_{IR\alpha}$ were calculated using Dechant's formulas [4], namely for modification α $x_{IR\alpha} = A_{i1030}/A_{i1074}$, and for modification γ $x_{IR\gamma} = A_{i976}/A_{i1074}$ where: $x_{IR\alpha}$, $x_{IR\gamma}$ - examined indices of crystallinity for crystallographic forms α and γ and A_{i1030}, A_{i976}, A_{i1074} - integral absorptions of the bands at wave number values 1030 cm^{-1}, 976 cm^{-1} and 1074 cm^{-1}, respectively.

Table shows the results of assessment of the changes in supermolecular structure of the fiber-forming polyamides based on the changes in crystallinity indices $x_{IR\alpha}$ and $x_{IR\gamma}$. These data characterize the fractions of crystallographic modifications α and γ in polyamide fibers versus exposure time. So, one can conclude that recrystallisation process takes place in the fibers under the influence of UV radiation, namely so-called crystallographic transformation or the reconstruction of crystallographic lattice from modification alfa (α) into gamma (γ) type modification. In the case of polyamide fiber round "semi-dull", the fraction of crystallographic form γ increases during the initial stage of exposure, while the fraction of α form is decreased. This indicates the crystallographic transformation of more ordered modification α into modification γ with a lower degree of order. When the exposure time is prolonged the fraction of crystallographic form γ clearly decreases, while the fraction of modification α decreases only to a slight extent. The most serious decrease in the total crystallinity index x_{IR} is observed for polyamide fibers round "bright" and round "semi-dull". Clearly lower changes of this index are observed in mat fibers of round cross-section. In these fibers one can observe the clearest process of transformation of crystallographic modification α into modification γ. The fibers showing triangular cross-sections do not show such drastic changes in crystallinity due to the exposure to UV radiation.

Fiber type	Exposure time (number of summer seasons)	Modification α [-]	Modification γ [-]
Polyamide fiber round-"bright"	0 1 2 5	0,27 0,28 0,26 0,25	0,42 0,33 0,37 0,33
Polyamide fiber round-"semi-dull"	0 1 2 5	0,28 0,24 0,27 0,25	0,41 0,47 0,34 0,20
Polyamide fibers round-"dull"	0 1 2 5	0,26 0,22 0,17 0,19	0,32 0,26 0,27 0,35
Polyamide fiber triangle-"bright"	0 1 2 5	0,23 0,24 0,25 0,27	0,29 0,27 0,28 0,28

Table 3. Fractions of crystallographic modifications α- and γ- in PA6 fibers.

3.3 Studies of fiber internal orientation on the example of PA6 fibers

In fibre orientation studies the experimental basis are the absorption spectra of linearly polarised IR radiation applied parallel and perpendicular to the axis of the irradiated fibre. This takes to develop an adequate so-called web preparation being a monofibrous layer of compact and parallel fibres. The study produces two IR absorption spectra with discernible absorption bands correlated with the relevant relative ordered states of macromolecules, so-called "crystalline", "amorphous", and "independent" bands [1,8].

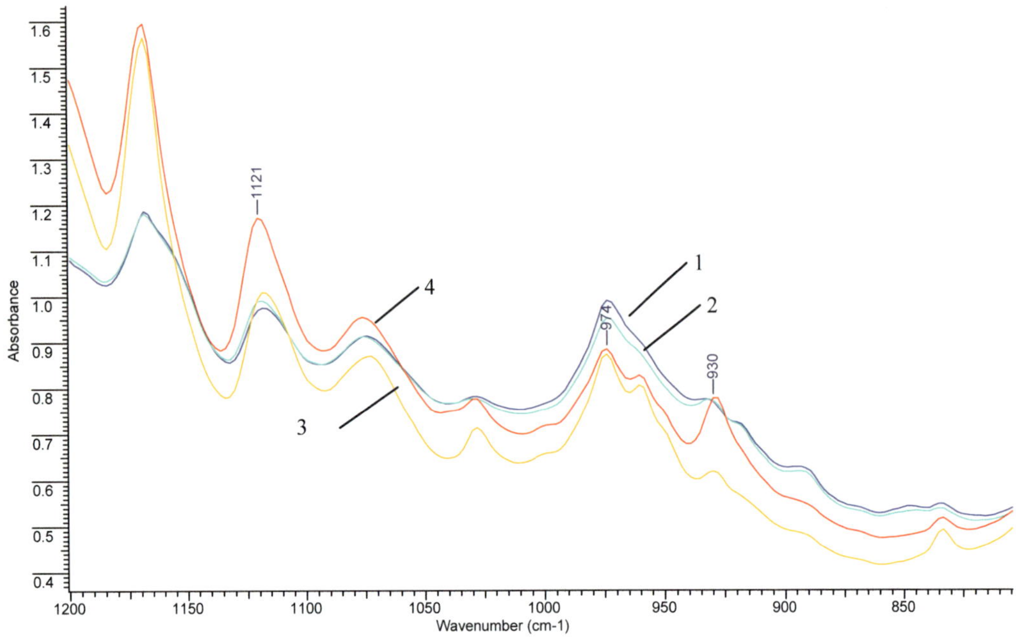

Fig. 5. FT-IR spectra of PA6 fibre: 1) draw ratio R=1x, parallel direction of IR radiation, 2) draw ratio R=1x, perpendicular direction of IR radiation, 3) draw ratio R=4,7x, parallel direction of IR radiation, 4) draw ratio R=4,7x, perpendicular direction of IR radiation.

This provides the basis for fibre orientation analysis in terms of the ordered state separately for: crystallites, macromolecules in non-crystalline material, and the resultant – as total (fig.5, table 4, 5). The indicator in the quantitative evaluation of fibre orientation is the value of the dichroism quotient R which expresses the intensity ratio of the dichroic absorption band appropriate for the parallel irradiation of the fibre to the intensity of that absorption band found in the spectrogram for the perpendicular irradiation of the sample. The more the R value diverges from 1, the higher is the fibre orientation level. The quantitative fibre orientation indicators [1,8] are determined based on certain experimental relationships[1,8].

Wavenumber cm^{-1}	**Absorbing group, type of vibration**	**The type of band**	**The type of dichroism**	**Dichroic ratio, R**	**Orientation index f_{IR}**
1119	ν (C-C)	o	π	1,073	0,024
974	ν_0 , amide IV	a	σ	0,892	0,075
932	ν_0 (0, π)	c	π	2,364	0,313

Table 4. The data for selected dichroic absorption bands: “crystalline”, “amorphous”, and “independent” for initial fibre PA6 draw ratio R=1x.

Wavenumber cm^{-1}	Absorbing group, type of vibration	The type of band	The type of dichroism	Dichroic ratio, R	Orientation index, f_{IR}
1119	ν (C-C)	o	π	1,313	0,095
974	ν_0 amide IV	a	σ	0,692	0,229
932	ν_0 (0, π)	c	π	4,889	0,565

Table 5. The data for selected dichroic absorption bands: "crystalline", "amorphous", and "independent" for initial fibre PA6 draw ratio R=4,7x.

4. Studies of changes in fiber structure due to external factors

The IR absorption spectroscopy method used to assess changes in fiber structure is aimed to detect changes in molecular structure and crystalline structure. The measurements may be carried out using the ATR technique, or transilluminating technique. For this purpose, a fibrous sample appropriately positioned in relation to a special prism, or a tablet preparation is used. Both the changes in structure of the fiber volume (transilluminating technique) and the changes of its surface characteristics (ATR technique) are assessed.

4.1 Changes in structure of wool and polyamide fibres

The changes in fiber structure may take place as a result of interactions with many external factors, chemicals, as well as intentional processing aimed at modification of the fiber surface.

For example, assessment may concern the character of fiber-forming material remodeling due to such factors as: exposure to sunlight, heat and humidity, skin secretions - sweat and sebum [8], including:

- changes in fiber-forming material structure in the aspect of its negative effect on the skin due to remodeling of the fiber material, so-called „aging" as a result of various factors, or due to development of mites, for which it constitutes a favorable medium;
- changes of morphological and macroscopic structure - as a mechanical factor causing skin irritation.

The fibers are studied for changes in their structure due to the effect of external factors using IR absorption spectroscopy techniques - ATR and transilluminating technique.

4.1.1 Molecular structure of the wool fibre surface layer and its changes under the influence of the external factors used (example)

The test results of three types of wool fibres with various thicknesses show a clearly different molecular structure of their surface layer. Thinner fibres with a thickness of 15,5 μm or 19,5 μm show similar structures, which results from the comparable absorbance values of absorption bands correlated with the peptide group being characteristic of keratin (fig 6.). A considerably lower absorbance of the band correlated with the peptide group of the fibre with a higher thickness (25,5 μm) indicates a lower keratin content in its surface layer. This seems to result from a more developed cuticule layer (greater and thick scales), the structure of which contains a higher quantities of other chemical substances then keratin, e.g. lipids.

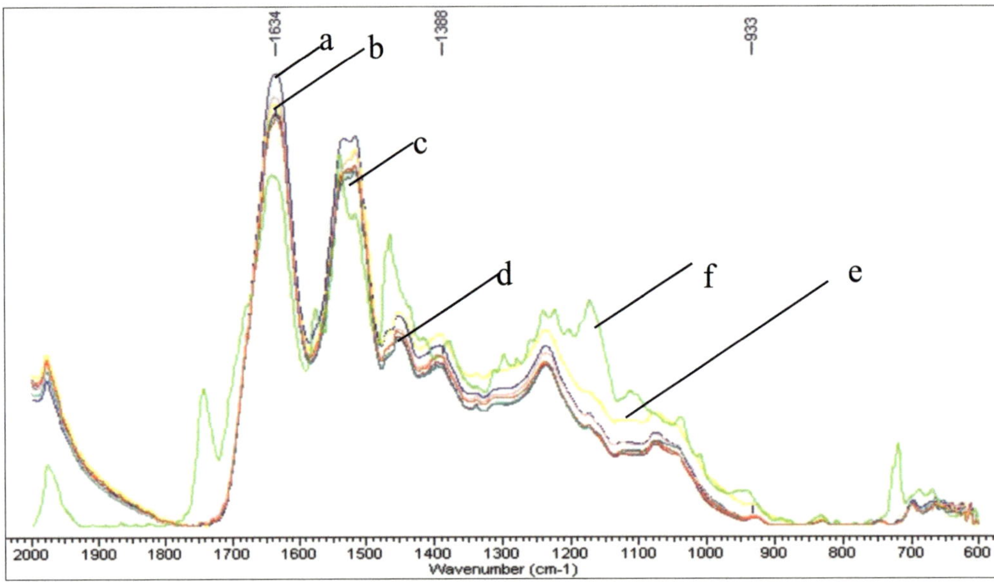

Fig. 6. FT-IR spectra of wool fibre after exposure to external factors: a) initial fibre, b) fibre and alkaline sweat, c) fibre and acid sweat, d) fibre and temp. 37,5, humidity 65%, e) fibre and temp. 37,5, humidity 25%, f) fibre and sebum, g) fibre and UV radiation, 2 seasons

The effect of the external factor used on the keratin depolymerisation is minimal as shown by a low differentiation of value absorbance A in relation to the absorbance value of the absorption band of –CO-NH- group in the initial fibre. The strongest effect on the keratin molecular structure is exerted by acidic perspiration and heat at RH 65%. In the case of thinner fibres, the decrease in the absorbance value of the absorption band correlated with peptide group is not always accompanied by the increase in the number of -COOH and NH_2 groups. Thus the molecular restructuring process seems to be more complex and concerns first of all the groups occurring in the substituents of α-amino acids radicals of a acidic and basic character probably resulting from the change in their energetic state as a result of changes in intermolecular interactions due to e.g. the decomposition or formation of new intermolecular bonds as well as the accompanying conformational changes and even the electric state transition (change in the range of isoelectric point) [9].

In the case of thicker fibres, the number of acid and base groups is clearly increased with a slight increase in the number of -NH-CO- groups as indicated by the increased intensity of the corresponding absorption band (increase in the value of absorbance A). The impossibility of the unmistakable interpretation of changes in the molecular structure of wool fibres may however allow one to conclude that these changes take place. Thus it can be assumed that regardless of their character they can influence the interactions between fibre and skin considering e.g. the change in the range of isoelectric point being characteristic of polypeptide fibres, thus the appearance of a more or less specified beneficial interaction resulting from the amphoteric character of keratin.

4.1.2 Molecular structure of the polyamide fibre surface layer and its changes caused by the external factors used

Changes in the polyamide fibre surface layer take place in a different way than those in the remaining fibres. The interpretation principles assumed disappoint in the case of these fibres despite the relatively uncomplicated chemical structure of polyamide (fig. 7).

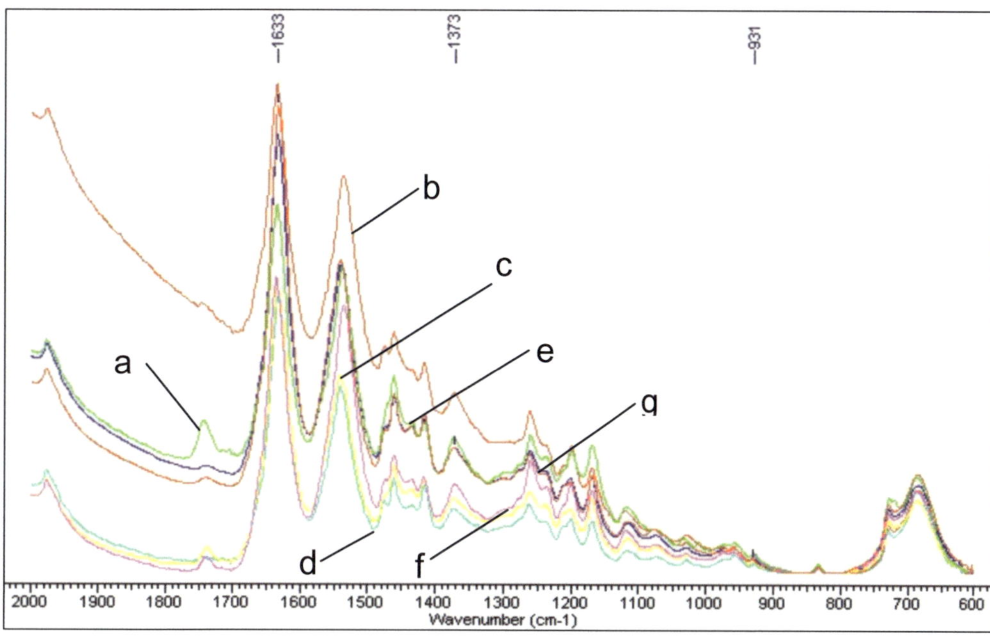

Fig. 7. FT-IR spectra of polyamide fibre after exposure to external factors: a) initial fibre, b) fibre and acid sweat, c) fibre and alkaline sweat, d) fibre and temp. 37,5, humidity 25%, e) fibre and temp. 37,5, humidity 65%, f) fibre and sebum, g) fibre and UV radiation, 2 seasons.

This different behaviour of polyamide fibres under the influence of the factors used seems to result from the occurrence and superimposition of the parameters of the fibre physical microstructure, especially the high degree of crystallisation on the fibre molecular structure. The effects of interaction will be different in relation to the fibre amorphous and crystalline materials. The decrease in the absorbance values of all the absorption bands tested in relation to bands A of the initial fibre in the case of the action of alkaline perspiration can be connected with a strong "etching" of the amorphous surface layer by alkaline perspiration. Acidic perspiration showing swelling effects on the polyamide material [11] will cause the restructuring of the crystalline material into a non-crystalline matter, releasing characteristic groups from intermolecular interactions, which will be seen in the increase in the intensity of bands correlated with these groups. The assumed mechanism of the effect on restructuring the PA6 fibre surface layer concerns heat and moisture. One may assume that when moisture is lower the effect resolves itself into the removal of low-molecular fractions from the fibre surface, while within the range of high moisture the crystalline matter is restructurised with releasing the groups from intermolecular interactions.

4.2 Assessment of surface wool fibres as a result of a biopolymer deposition

Considering the possibility of skin irritation [12, 13] due to a physical mechanism (friction of the fibers against the skin surface), the fibers were subjected to surface modification using a bioactive polymer - chitosan - as a modifier. The modification of wool fibers was carried out by subjecting them to water bath with 5% chitosan content, pH = 5,0 and processing temperature of 80°C for differentiated periods of time of 10 min and 30 min. The studies were conducted on intact fibers, and fibers which had undergone preliminary enzymatic processing with two different enzyme concentrations - 1% and 3% [12].

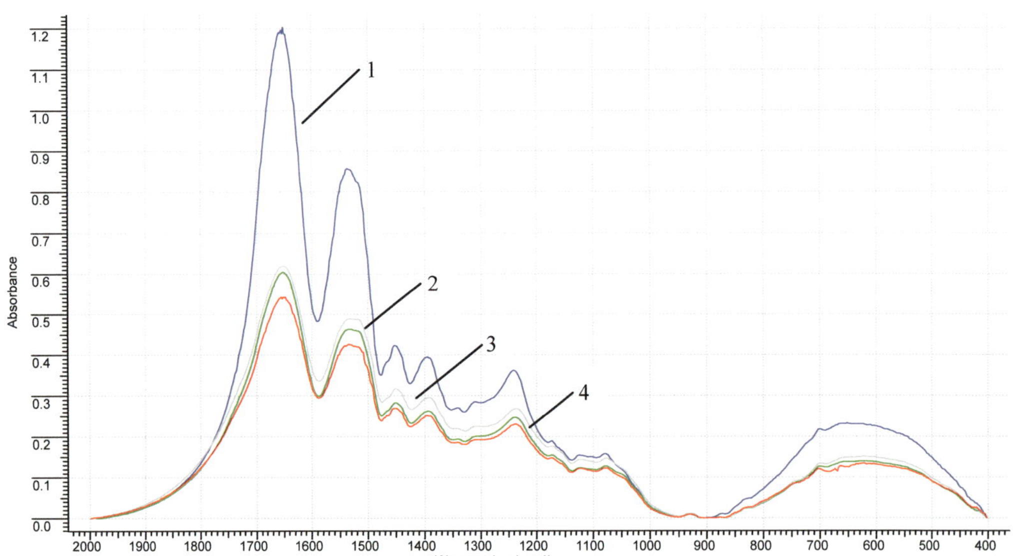

Fig. 8. FT-IR spectra of wool fibre- 25,5μm: 1- initial fibre, 2- chitosane (M_2) treated fibre, 3- fibre after enzyme (3%) and chitosane (M_2) modification, 4- fibre after enzyme (1%) and chitosane (M_2) modification, where:

(M_1) chitosane preparation – molecular mass $\overline{M}$ =360 thousand, deacetylation degree (DD)= 76.3 %

(M_2) chitosane preparation –molecular mass $\overline{M}$ =229 thousand, deacetylation degree (DD) = 97.0 %

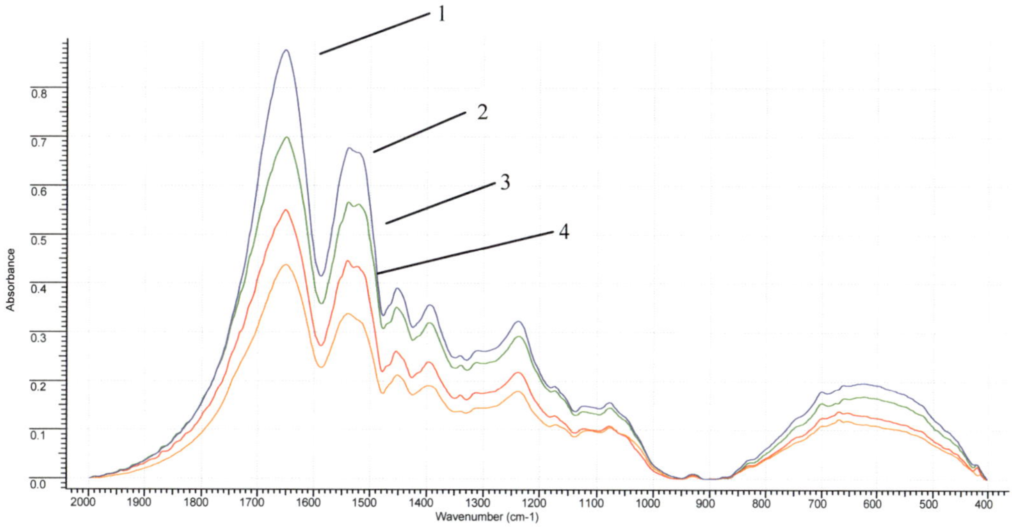

Fig. 9. FT-IR spectra of wool fibre- 19,5μm: 1- initial fibre, 2- chitosane (M_2) treated fibre, 3- fibre after enzyme- and chitosane (M_1) modification, 4- fibre after enzyme - and chitosane (M_2) modification, where:
(M_1) chitosane preparation – molecular mass $\overline{M}$ =360 thousand, deacetylation degree (DD)= 76.3 %
(M_2) chitosane preparation –molecular mass $\overline{M}$ =229 thousand, deacetylation degree (DD) = 97.0 %

The absorption spectra obtained with IR - spectroscopy demonstrate post-processing changes in fibre structure of wool, and polyacrylonitrile fibres. On the basis of the test results, the effect of enzymatic processing on fibre surface, confirmed by fibre mass decrease, can be observed and, in case of chitosan-treated fibres, the presence of bands specific for chitosan, which indicate binding of the polymer to the fibres.

5. Conclusion

The possibility of applying IR- spectroscopy analysis in the fibres structure and properties investigations is very broad. On the basis of examples shown it can be concluded, that the IR spectroscopic methods can be used:

- for identification of fibres polymer matter (kind of fibre) [14, 15],
- for investigation of fibres structure parameters as: crystallinity degree [7,16] , inner orientation (molecular orientation in noncrystalline part of fibres matter, crystallites orientation and total orientation- separately),
- for assessment of fibres ageing effects[7],
- for assessment of fibres surface changes under the different action-physical, chemical and biochemical treatments,
- for analysis the kind of substances deposited on the fibres surface during their modification processes.

6. References

[1] Urbańczyk G. 1986. *Mikrostruktura Włókna - Orientacja Wewnętrzna*, WNT, ISBN 83-204-0745-1, Warszawa

[2] Urbańczyk G.W. 1988. *Mikrostruktura Włókna - Badanie Struktury Krystalicznej i Budowy Morfologicznej*, WNT, ISBN 83-204-1014-2, Warszawa

[3] Material Thermo Scientific Smart iTR, *Available from http://www.thermoscientific.com*

[4] Dechant J. 1972. *Ultrarotspektroskopische Untersuchungen an Polymeren*, Berlin

[5] Reu J.H. Melliand Textilberichte, 44/1963, s.1102, 1197, p. 1320

[6] Yamader R. Journal Chemisty Physics, 15/1974, p. 749

[7] Bieniek A., Lipp-Symonowicz B., Sztajnowski S. 2009. *Influence of the structures of polyamide 6 fibers on their ageing under intensive insolation conditions*, Polimery, no. 11-12, p. 54

[8] Czubak W, 2004. Master`s thesis,Technical Univerity of Lodz, Fibre Physics Department

[9] Kułak A., 2010. *The investigation of fibres and textiles in the aspect of allergic properties,* Scientific Papers Technical Univerity of Lodz, abridged dissertation, No. 1092/2011,p.37-65

[10] Urbańczyk G. 1986. Nauka o Włóknie, WNT, ISBN 83-204-0632-3, Warszawa

[11] Peters R.H. 1963.*Textile chemistry. Volume I. The chemistry of fibres*, Elsevier Publishing Company, Amsterdam/ London/ Neu York

[12] Kułak A, B. Lipp-Symonowicz, E. Lis, 2008. *Modification of fibres and textiles in aspect of skin irritation prophylaxis*, 4th International Textile, Clothing & Design Conference - Croatia, ISBN 978-953-7105-26-6, Dubrovnik 2008, p.100-105

[13] Klaschka F. 1994. *Textiles and human skin fact and faction- a description of the situation from a dermatological of view*, Melliand Textilberichte No. 3, pp.193-202, ISSN 0341-0781

[14] Wojciechowska D., A., Lipp-Symonowicz B., Sztajnowski S., 2011. *New Commercial Fibres Called 'Bamboo Fibres – Their Structure and Properties*, FIibres & Textiles in Eastern Europe, Vol. 19, No. 1 (84) pp. 18-23

[15] Tszydel M., Sztajnowski S., Michalak M., Wrzosek H., Kowalska S., Krucińska I., Lipp-Symonowicz B. 2009. *Structure and Physical and Chemical Properties of Fibres from the Fifth Larval Instar of Caddis-Flies of the Species Hydropsyche Angustipennis*. FIibres & Textiles in Eastern Europe, Vol. 17, No. 6 (77) pp. 7-12.

[16] Kardas I., Lipp-Symonowicz B., Sztajnowski S. 2009. *Comparison of the Effect of PET Fibre Surface Modification Using Enzymes and Chemical Substances with Respect to Changes in Mechanical Properties.* FIibres & Textiles in Eastern Europe, Vol.17, No. 4 (75) p. 93-97.

3

The Relation of FTIR Signature of Natural Colorless Quartz to Color Development After Irradiation and Heating

Fernando Soares Lameiras
Nuclear Technology Development Center, National Nuclear Energy Commission
Brazil

1. Introduction

The infrared spectroscopy is a powerful technique to identify colorless quartz with potential for color development by irradiation and heating for jewelry. It is being routinely used in Brazil since 2005. In this chapter the use of FTIR for this purpose is described in detail.

Amethyst, prasiolite, citrine, and morion are gemstones widely used to make jewelry, to compose mineral collections of museums or private persons, or to decorate homes and offices. They are all alpha quartz (Dana et al., 1985) and the differences between them are very small. The natural quartz always has trace elements in its crystal structure, such as aluminum, iron, hydrogen, lithium, sodium, and potassium. The content of these trace elements is important for color development. But without exposure to ionizing radiation, these trace elements are not able to produce colors in quartz.

Since the discovery of radioactivity, it has become clear that the color of several minerals, including the quartz, can be modified by their exposure to radiation emitted by radioactive substances (Nassau, 1978 and Nassau, 1980). An amethyst is colorless quartz with small contents of aluminum, iron, hydrogen, sodium, lithium, and potassium. It was formed as a result of a hydrothermal process similar to the one used to produce cultured quartz for the electronic industry. The hydrothermal solution and the environment of the amethyst (e.g., feldspar) may be rich in potassium. Potassium is a natural radioactive element due do its ^{40}K isotope, which emits gamma rays with an energy of 1.46 MeV and has a half life of 1,260,000,000 years. Other natural radioactive elements (thorium and uranium) can also provide the radiation. Such an exposure during a geologic time span can accumulate irradiation doses high enough to produce the violet color of amethyst. However, if the natural irradiation was weak, the amethyst remains colorless. It can be exposed to gamma rays of a man-made ^{60}Co-source to develop intense violet colors. The same is true for prasiolites. Other colors of quartz can be formed by a similar process.

Jewelry demands tons of colored quartz that cannot be found in nature. It is necessary to extracted colorless quartz from nature and submit it to irradiation and heating. The problem is separating the quartz that can develop colors of commercial value from the quartz which cannot. Usually this separation is done through irradiation tests of representative samples.

But these tests can take weeks, because the samples need to be sent to an irradiation facility to be irradiated to varying doses. The irradiated samples then need to be sent back to the owner, who must perform the heating (in some cases) and evaluate the quality of color. Another problem is that some irradiators do not inform the applied doses.

The FTIR signature of the colorless quartz can be very valuable in indicating whether or not it can develop colors. Through a previous correlation with the results of irradiation tests (Nunes et al. 2009), it is possible to forecast the color that will be achieved after an irradiation dose. This analysis takes only a few minutes and can be performed close to the mining site if a FTIR spectrometer is available.

Infrared measurements were widely reported in electrodiffusion studies of quartz with alkalis (Martin, 1988). The purpose of these studies was to understand the electronic performance of cultured quartz for the electronic industry. There are few studies concerning color formation. The region of interest is from 2400 cm^{-1} to 3650 cm^{-1}. Table 1 shows the bands that can be observed in colorless quartz at room temperature.

Band number	Wavenumber (cm^{-1})	Remark
1	2499	Strong, always present
2	2600	Strong, always present
3	2677	Strong, always present
4	2771	Strong, always present
5	2935	Small band, always present
6	3063	Small band, always present
7	3202	Small band, always present, Si-O overtone
8	3303*	Small band, always present, Si-O overtone
9	3381*	Al-OH related
10	3433*	Al-OH/Na^+ related
11	3483*	Al-OH/Li^+ related
12	3404*-510* doublet	Not assigned
13	3441* and 3585 doublet	3441 cm^{-1} is a broad band
14	3441* and 3585 doublet	3441 cm^{-1} is a broad band with strong absorption above 3000 cm^{-1}
15	3595	Not assigned

(*) The position of these bands may vary within ± 10 cm^{-1}

Table 1. FTIR bands observed in the spectrum of colorless quartz at room temperature

2. Acquisition of the FTIR spectra for determination of the potential for color development

The spectrum should be acquired in crystals, because some bands in the 2400 to 3650 cm^{-1} region are too weak to be resolved in pulverized samples. A thick fragment of a quartz crystal (1 to 5 cm) can be usually used to perform this measurement. The exception to this is prasiolite, which has a strong absorption. In this case, a thin fragment (less than 1 mm) should be used. The fragments can be obtained by a crude and rapid process, like beating the sample with a hammer. It is not necessary to orient the fragments of the crystals. One should protect the eyes with glass and the hands with gloves when beating the samples, because the quartz fragments are very sharp.

The resolution of the FTIR measurement should be at least 4 cm^{-1} with a minimum of 16 scans. In case a high noise is observed, a relocation of the sample in the measurement compartment is required.

Since there is no control of the fragment thickness, the absolute absorbance values are meaningless. All the analysis should be performed on normalized spectra. The normalization is carried out relative to the bands at 2499 to 2771 cm^{-1}, because they are related to the Si-O bond, and are not affected by the irradiation or heating, and are present in all samples. The height of one of these bands is taken equal to 1 (in the spectra of these Chapter, the band at 2677 cm^{-1} was chosen) and the absorption at all wave numbers is proportionally corrected.

A background of infrared absorption that rises, falls, or is constant for wave numbers above 3000 cm^{-1} can be observed (see Figure 1). This background is used to calculate the spectra baselines by fitting a third-order polynomial in the range of 4100 to 5300 cm^{-1}, where no bands are observed, and then subtracted from the normalized spectra.

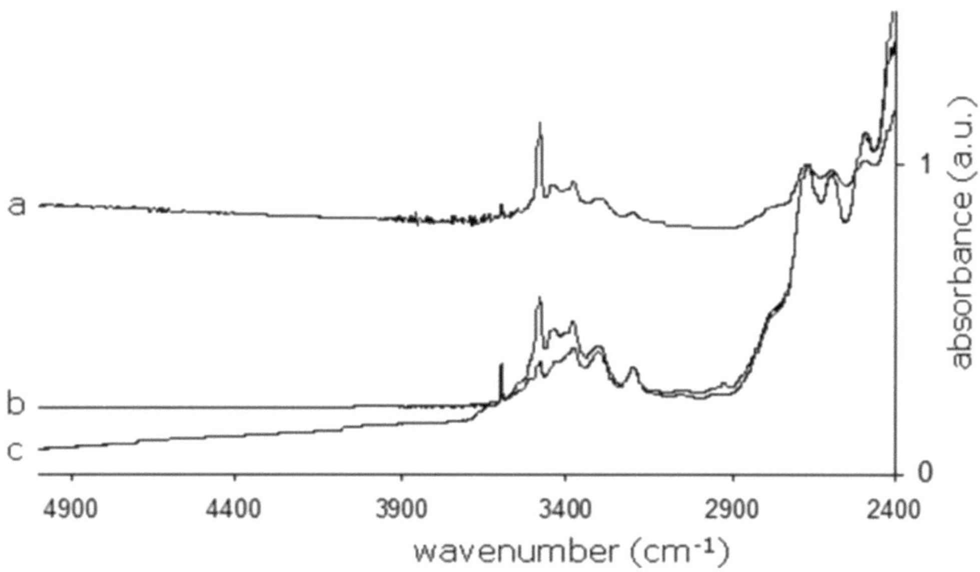

Fig. 1. Examples of colorless samples of natural quartz that show a background of infrared absorption that rises (line a), is constant (line b), or falls (line c) for wavenumbers above 3000 cm^{-1}. The typical noise between 3600 and 4000 cm^{-1} is observed in a and b. (Nunes et al., 2009).

The rising background is attributed to internal turbidity of the samples or to micro-inclusions (Hebert and Rossman, 2008). Many samples show a noise in the range of 3600 to 4000 cm^{-1} that also is attributed to micro inclusions.

3. The FTIR spectrum relationship to color formation

3.1 Smoky, morion, green gold, and brown quartz

The FTIR signature of this kind of quartz is shown in Figure 2. The band at 3483 cm^{-1} is related to Al-OH/Li^+ and plays an import role in color formation. It appears prominent in the samples that develop colors of commercial value after irradiation and heating.

After irradiation, these samples become smoky or black (morion). The band at 3483 cm^{-1} decreases and the bands at 3389 and 3313 cm^{-1} increase (see Figure 3). The band at 3483 cm^{-1} partially recovers after heating. These samples become green gold, yellow, or brown after heating, depending on the irradiation doses. A new band appeared at 3539-3451 cm^{-1} in colored samples. If the band at 3483 cm^{-1} is weak in the colorless sample, it may become colorless or weak in color after irradiation and heating.

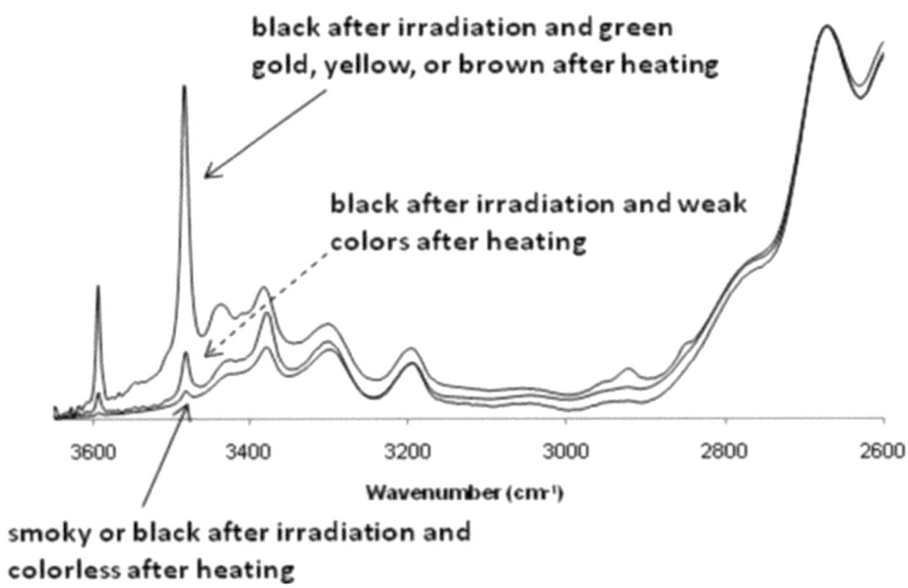

Fig. 2. The FTIR signature of quartz that become smoky or black after irradiation and green gold, yellow, or brown after heating.

One observes that the ultraviolet irradiation can partially recover the spectrum before radiation and bleach the smoky or black color developed by the γ-irradiation (see Figure 4). The colors developed after heating are resistant to ultraviolet rays.

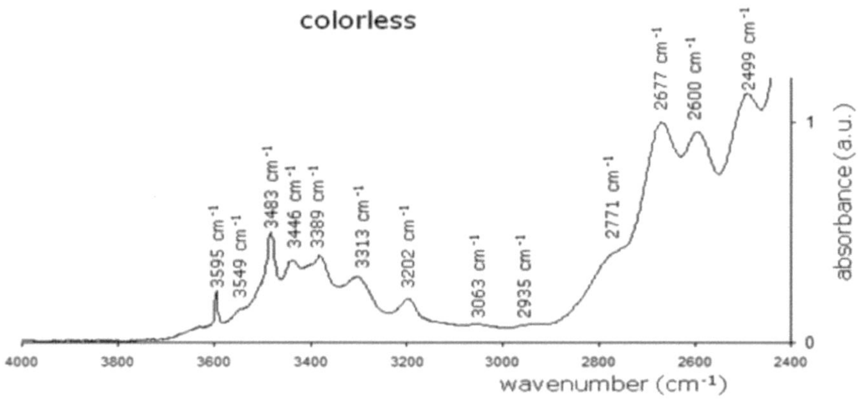

(a)

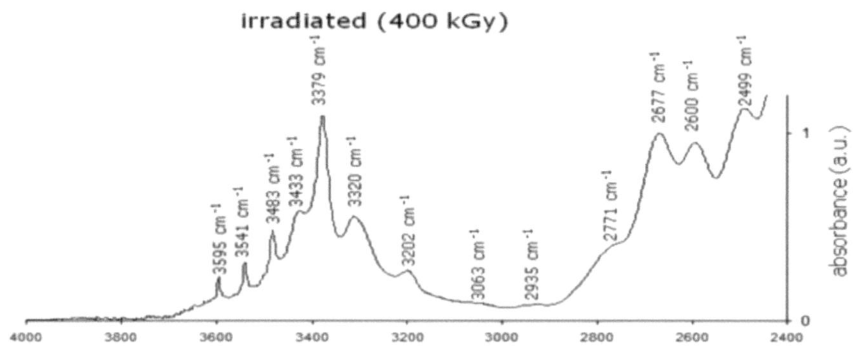

(b)

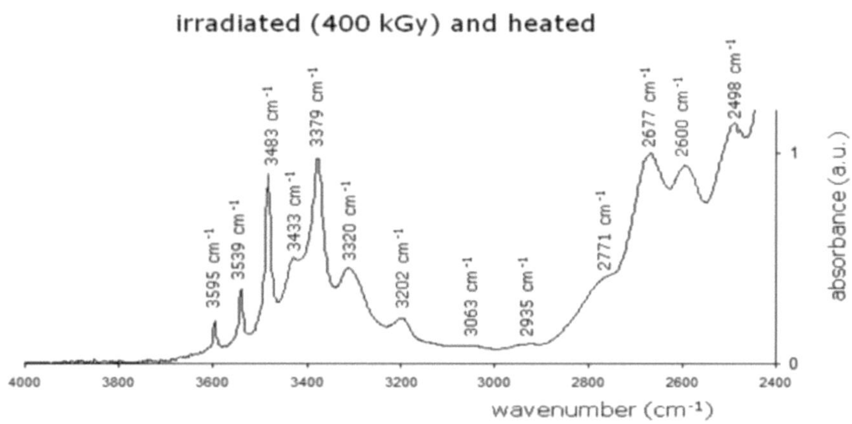

(c)

Fig. 3. The influence of irradiation and heating: (a) colorless, before γ-irradiation; (b) black, after γ-irradiation; (c) colored (greenish, greenish yellow, yellow, or brown), after irradiation and additional heating (Nunes et al., 2009).

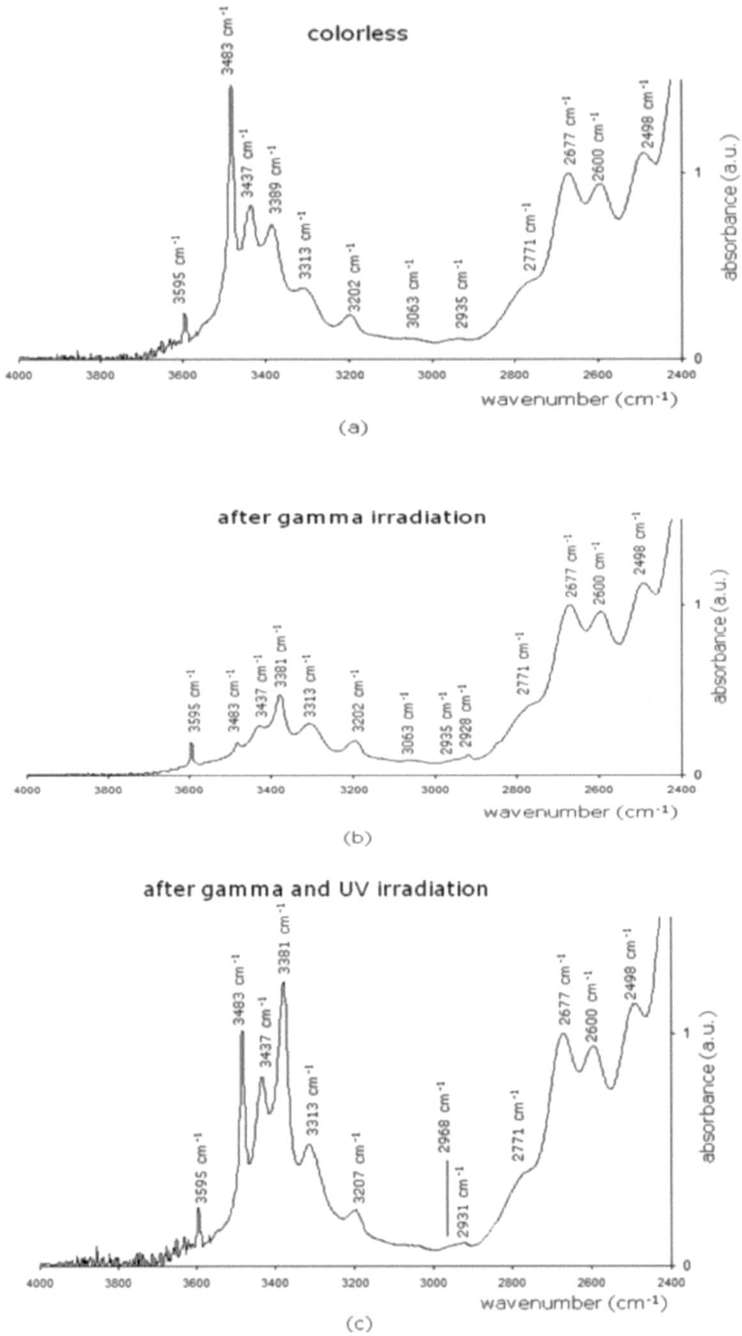

Fig. 4. The influence of ultraviolet irradiation: (a) colorless, before γ-irradiation; (b) black, after γ-irradiation; (c) lightly smoky, after γ-irradiation and UV irradiation (exposure to a 5 kW iron iodide lamp for 60 minutes, with the sample temperature kept below 333 K by air ventilation). (Nunes et al., 2009).

3.2 Amethyst and prasiolite

The signature of colorless quartz that becomes amethyst or prasiolite after irradiation is shown in Figure 5. Both have the 3441 cm^{-1} and 3585 cm^{-1} doublet. The band at 3441 cm^{-1} is very broad. The difference between these gems is their absorption in this region. The prasiolite has a stronger absorption in the 2700 cm^{-1} to 3650 cm^{-1} range. There is a transition zone (the spectrum in the middle in Figure 5), where the quartz is neither amethyst nor prasiolite. In this zone the colorless quartz cannot develop colors of commercial value.

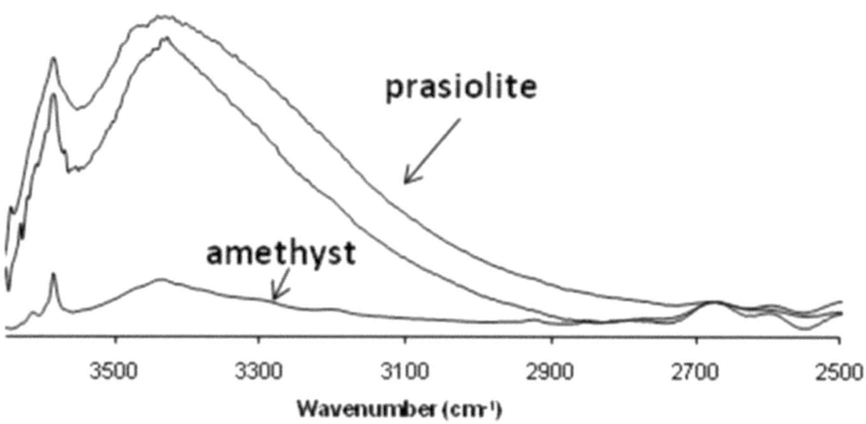

Fig. 5. The FTIR signature of colorless quartz that become amethyst or prasiolite after irradiation.

Figure 6 shows spectra at room temperature and at 93 K of a colorless sample of quartz that develops the violet color after irradiation (amethyst). At low temperature the bands appear higher, narrower, and shifted up to 10 cm^{-1} (Fritzgerald et al., 2006). The broad band at 3441 cm^{-1} at room temperature is decomposed into several bands at 93 K. The absorbance between 3450 and 3575 cm^{-1} is lower at 93 K, indicating that there are either no bands or only bands of very low intensity in this region. Figure 6c shows the spectrum at 93 K after irradiation (violet). With respect to the colorless spectrum, the irradiated one shows a band at 3306 cm^{-1}, a remarkable growth of the band at 3365 cm^{-1}, a slight decrease of the band at 3580 cm^{-1}, and the growth of a small band at 3549 cm^{-1}.

Ultraviolet rays can bleach the colors of amethyst and prasiolite developed after irradiation. Prasiolites become grayish-green after irradiation. Through a short exposition to ultraviolet rays, the grayish component can be bleached out.

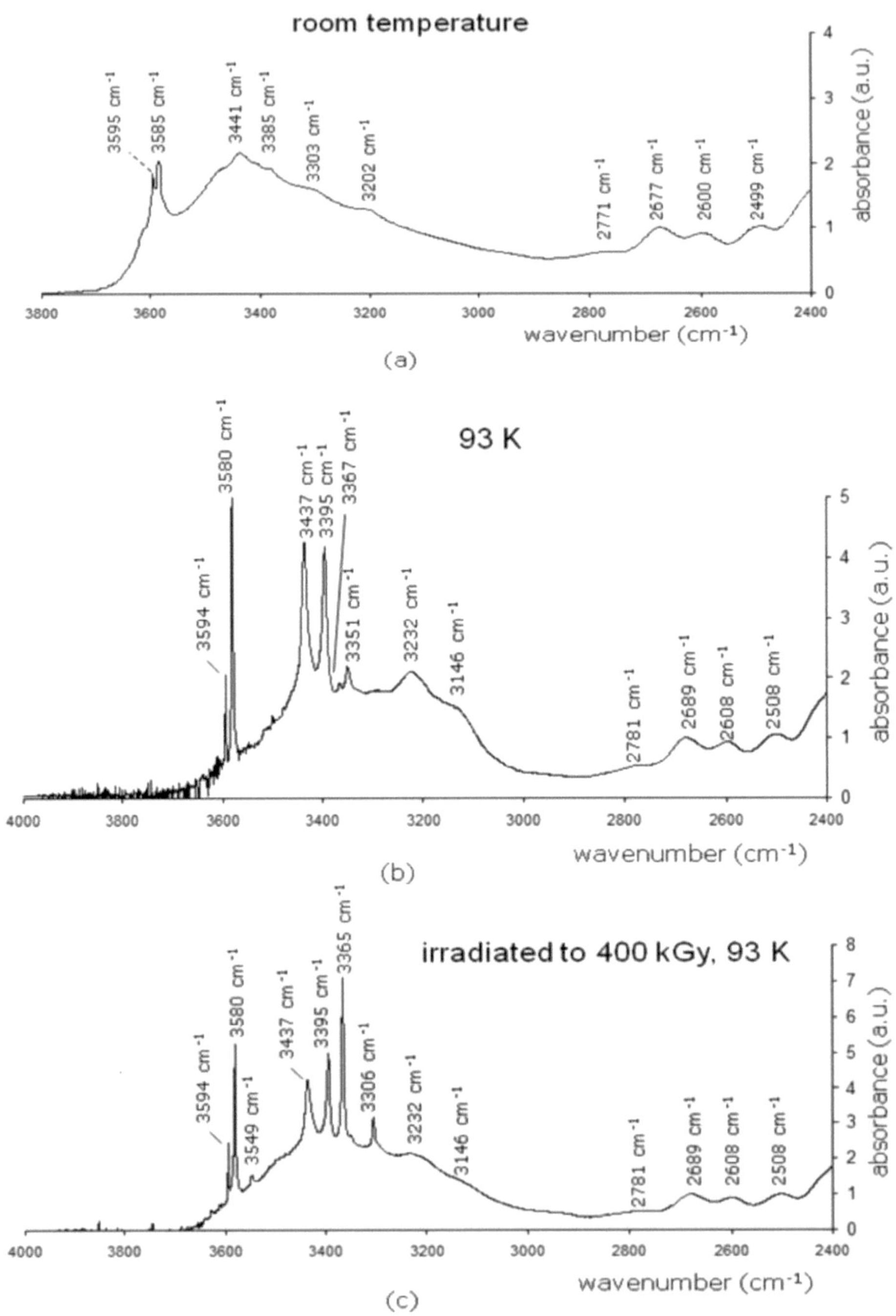

Fig. 6. Infrared spectra of a colorless sample of quartz that becomes violet (amethyst) after γ-irradiation: (a) at room temperature, before γ-irradiation (colorless); (b) at 93 K, before γ-irradiation (colorless); (c) at 93 K, after γ-irradiation (violet). (Nunes et al., 2009).

3.3 Olive green quartz

There is a kind of quartz that becomes grayish-olive green after irradiation. The grayish component can be bleached out through a brief period of heating. Its FTIR signature is shown in Figure 7. It shows a pair of bands at 3404 and 3510 cm^{-1}. Ultraviolet rays can bleach the color developed by these samples.

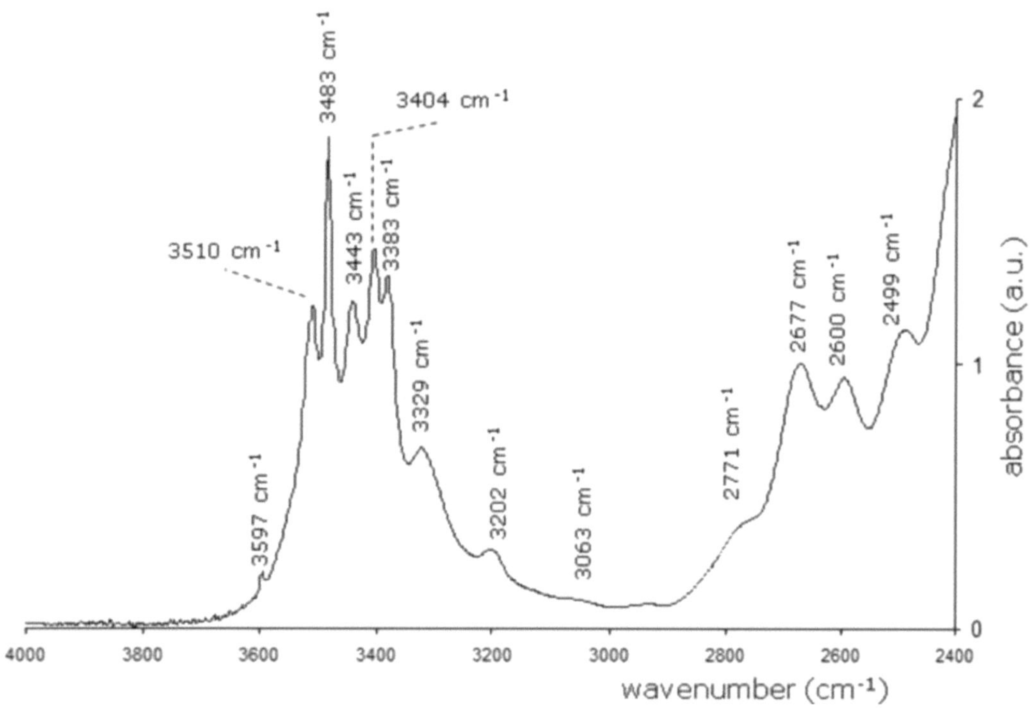

Fig. 7. The spectrum at room temperature of colorless quartz that becomes grayish-olive green after γ-irradiation and olive green after additional short heating. (Nunes et al., 2009).

4. Quantitative analysis of the FTIR spectra

4.1 The lithium factor

The lithium factor can be calculated from the spectra shown in Figure 2. The bands at 3483, 3446, and 3339 cm^{-1} are, respectively, related to lithium, sodium, and hydrogen. Figure 8 shows how the lithium factor, f_{Li}, is calculated. One observes that samples with $f_{Li} > 2.0$ develop intense colors. If $1.0 \leq f_{Li} \leq 2.0$, the samples can develop intense colors, depending on the area below the spectrum between 3000 cm^{-1} and 3600 cm^{-1}. If $f_{Li} < 1.0$, the samples develop weak colors with no commercial value.

The area is calculated according to the following formula:

$$\mathbf{area} = \int_{3000\,\mathrm{cm}^{-1}}^{3600\,\mathrm{cm}^{-1}} \mathbf{h(w) \cdot dw} \tag{1}$$

where h(w) is the high in a.u. at the wave number w. The area gives a measured of the content of trace elements in the sample (see Figure 9), the greater the area, the higher the content of trace elements.

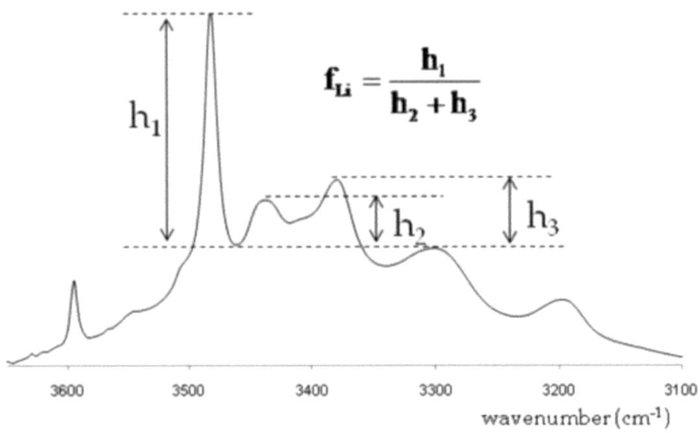

Fig. 8. Calculation of the lithium factor, f_{Li}. h_i are the heights measured in a.u. in normalized spectra after subtraction of the base line.

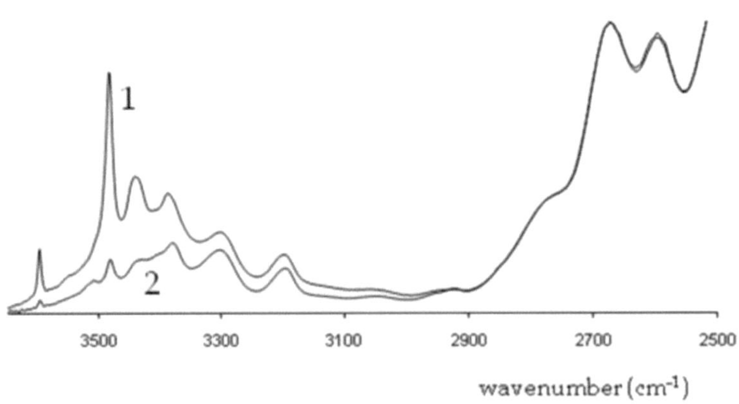

Fig. 9. In sample 1, f_{Li} = 2.1 and area = 126 u.a.cm^{-1} and develops intense colors. In sample 2, f_{Li} = 0,3 and area = 69 u.a.cm^{-1} and develops weak colors.

Table 2 summarizes the potential for color development according to the lithium factor, area, and doses of γ-ray exposure.

f_{Li}	area (u.a.cm^{-1})	dose (kGy)	color development
< 1.0	independent	< 100	grayish after irradiation and weak colors after heating
		> 100	black after irradiation and weak colors after heating
$1.0 \leq f_{Li} \leq 2.0$	< 100	< 100	grayish after irradiation and weak colors after heating
		> 100	black after irradiation and weak colors after heating
	≥ 100	60 - 100	black after irradiation and greenish-yellow after heating
		100 – 200	black after irradiation and yellow after heating
		> 200	black after irradiation and brown after heating
> 2.0	independent	60 - 100	black after irradiation an greenish-yellow after heating
		100 – 200	black after irradiation and yellow after heating
		> 200	black after irradiation and brown after heating

Table 2. Potential of color development according to the lithium factor and area.

4.2 The amethyst factor

The amethyst factor can be calculated from the spectra shown in Figure 5. Figure 10 shows how the amethyst factor, f_a, is calculated. h_1 is the height of the band at 3441 cm^{-1}, and h_2 is the height of the minimum between 3441 and 3585 cm^{-1}. One observes that samples in which $f_a > 3.3$ develop violet colors. If $2.7 \leq f_a \leq 3.3$, the samples can develop colors between violet and green with no commercial value. If $f_a < 2.7$, the sample develops the green color of prasiolite. However, it is also important to observe the area, as shown in Table 3.

5. Mechanism of color formation

Gamma rays can remove electrons from their positions within the quartz crystal lattice, creating electron-electron hole pairs. The removed electrons are trapped in interstitial positions and the electron holes have a high mobility within the crystal lattice. The localized electric charge unbalances created by the electron holes together with the trapped electrons contribute to their rapid recombination. For this reason, pure quartz remains unchanged after γ-ray irradiation.

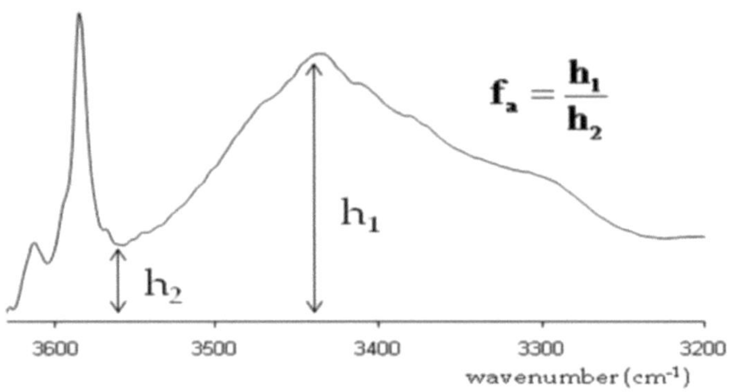

Fig. 10. Calculation of the lithium factor, f_{Li}. h_i are the heights measured in a.u. in normalized spectra after subtraction of the base line.

f_a	area (u.a.cm^{-1})	dose (kGy)	color development
< 2.7	< 900	> 600	grayish-green after irradiation and weak green (prasiolite) after exposure to ultraviolet rays and colorless or weak yellow after irradiation and heating
	> 900		grayish-green after irradiation and green (prasiolite) after exposure to ultraviolet rays and colorless or yellow (citrine) after irradiation and heating
$2.7 \leq f_a \leq 3.3$	independent	independent	no definition between violet and green
> 2.7	< 200	> 200	weak violet after irradiation and colorless or weak yellow (citrine) after irradiation and heating
	> 200		violet after irradiation and colorless or yellow (citrine) after irradiation and heating

Table 3. Potential of color development according to the amethyst factor and area.

But if Al^{3+} replaces Si^{4+} in certain positions within the quartz lattice (Al_{Si}), the situation is quite different, because there is one unpaired electron in the $Al_{Si}O_4^-$ tetrahedron. One electron should be donated by a hydrogen or alkaline atom. The positive hydrogen or alkaline ion should remain in the vicinity bounded by the electron attraction of the $[Al_{Si}O_4]^-$ tetrahedron. The OH^- vibrations modes in quartz are influenced by these bounded positive ions and are observed in the infrared spectrum (see Table 1). If an electron is removed by a gamma photon from the $[Al_{Si}O_4]^-$ tetrahedron, the positive ion becomes unbounded and free to diffuse in the crystalline lattice, according to

$$[Al_{Si}O_4^- / M^+] + \gamma \rightarrow [Al_{Si}O_4 / h^+]^0 + M^+ + e^- \tag{2}$$

where M^+ is the positive ion (H^+, Li^+, or Na^+), h^+ is an electron hole, and e^- is an electron. Hydrogen and alkali ions linked to Al-associated defects are known to reside in the c-axis channels of quartz crystal lattice, which contribute to their mobility (Walsby et al., 2003). The $[Al_{Si}O_4/h^+]^0$ is the $[Al_{Si}O_4]^-$ after the removal of one electron. It is called a color center and is related to the absorption of light in the visible spectrum, producing the colors gray to black after irradiation. The electron hole remains trapped in this center, but it can move within the oxygen electron clouds around the Al_{Si}.

The bands at 3303 and 3381 cm^{-1} are related to aluminum and hydrogen. The H^+ should be bounded to oxygen in the vicinity of aluminum due to the electrostatic attraction, forming a hydroxyl (Bahadur, 1995). It has a very high mobility in the quartz lattice once it is free to diffuse, due to the formation of the $[Al_{Si}O_4/h^+]^0$ center. It can combine with an interstitial electron and form a hydrogen atom, which also has a high mobility in the lattice. The hydrogen may eventually approach the vicinity of another $[Al_{Si}O_4/h^+]^0$ center, donate its electron, and become attached to an oxygen. These processes should occur at room temperature, because the bands at 3303 ant 3381 cm^{-1} are not affected or can even grow after irradiation if hydrogen atoms were interstitially present in the lattice before the irradiation.

The lithium is a light and small ion. Once it is free to diffuse due to the formation of a $[Al_{Si}O_4/h^+]^0$ center, it can diffuse within the quartz lattice with moderate mobility. The correspondent infrared band at 3483 cm^{-1} should decrease after irradiation (see Figure 4). Upon heating, the Li^+ can move farther from the vicinity of the aluminum atom. At the same time, the removed electrons can return to the aluminum vicinity and combine with $[Al_{Si}O_4/h^+]^0$ center, forming an $[Al_{Si}O_4]^-$ center, which is probably related to the band at 3539 - 3451 cm^{-1} observed in greenish-yellow, yellow, or brown samples (see Figure 4c). This band was also reported in citrine (obtained by the heating of amethyst) (Maschmeyer et al., 1980). With prolonged heating, the color is bleached because Li^+ can approach the $[Al_{Si}O_4]^-$ center and form the configuration it had before the irradiation, $[Al_{Si}O_4/Li^+]^0$.

The sodium is a heavy and large ion. Once it is free to diffuse, it cannot move away from the vicinity of an aluminum atom. It attracts removed electrons to its vicinity, that can combine with the $[Al_{Si}O_4]^-$ center. The infrared band related to sodium should remain nearly unaffected by the irradiation. This is the band at 3437 cm^{-1} at room temperature. Upon heating, the electrons which remained at interstitial positions after the irradiation are attracted by the Na^+ ions and combine with the $[Al_{Si}O_4/h^+]^0$ center. This ion cannot contribute to the formation of color after heating.

The potassium ion is very large and heavy ion. It probably cannot exist in the vicinity of an aluminum atom because great distortions of the lattice would be created. Figure 11 depicts the mechanism of color formation in quartz.

The ultraviolet irradiation can move the trapped interstitial electrons that were displaced by the gamma rays. They may eventually combine with the $[Al_{Si}O_4/h^+]^0$ center, forming an $[Al_{Si}O_4]^-$ center, which in turn combines with the positive ion within its vicinity. The ultraviolet irradiation can recover the infrared spectrum produced by the gamma irradiation (see Figure 4) and bleach the resultant colors. The colors produced by heating cannot be bleached by ultraviolet irradiation because Li^+ cannot be brought back to the vicinity of aluminum at temperatures below 453 K.

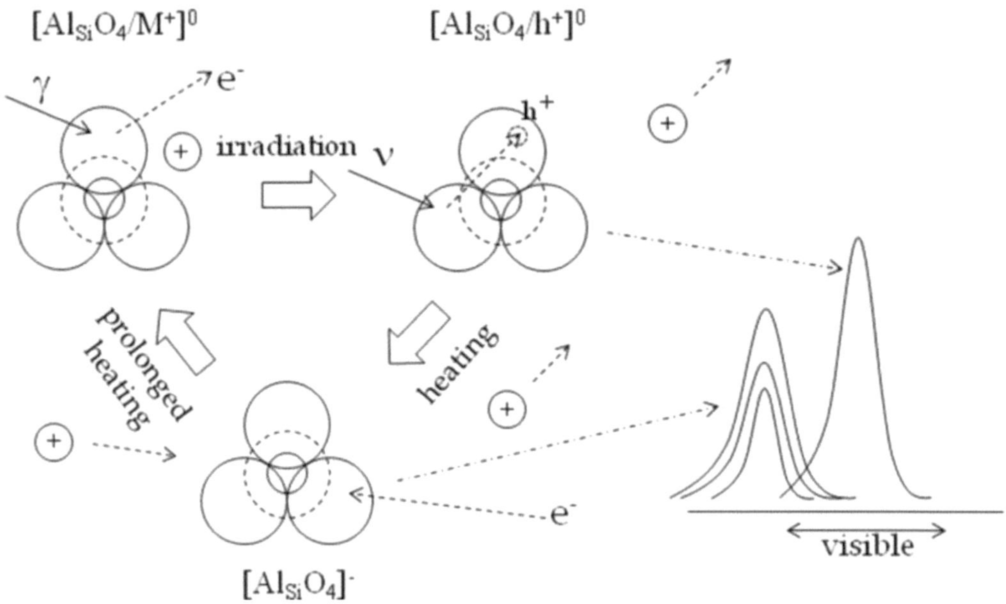

Fig. 11. Mechanism of color formation in quartz. The large circles represent oxygen atoms; the small solid circles represent aluminum atoms; the small dashed circle represents an electron hole, h^+; the circles with + are the charge compensator ions (H^+, Li^+, or Na^+); γ and ν are, respectively, a gamma photon and a photon within the visible spectrum. The $[Al_{Si}O_4/h^+]^0$ center absorbs light in the visible region; the $[Al_{Si}O_4]^-$ center has an absorption band in the ultraviolet with a tail in the visible region, which is responsible for color formation.

If Fe^{3+} replaces Al^{3+}, Fe_{Si}, the situation is more complex. The ionic radius of Fe^{3+} is large enough to cause substantial distortions in the tetrahedral configuration of the quartz lattice. Al^{3+} has an electronic configuration similar to Si^{4+} (both have the [Ne] configuration), whereas Fe^{3+} has a very different one, $[Ar]3d^5$. This difference causes additional distortions in the lattice in the vicinity of an iron atom. These distortions caused by the iron substitution for silicon provides the condition for K^+ to be accommodated in its vicinity. Potassium was reported in prasiolite and amethyst (Hebert and Rossman, 2008). The bands that grew after irradiation of amethyst (at 3306 and 3365 cm^{-1}, see Figure 6) are related to aluminum and hydrogen. They are probably also related to the formation of $[Al_{Si}O_4/h^+]^0$ centers. The lattice distortions may cause a shift in the absorption spectrum of the $[Al_{Si}O_4/h^+]^0$ center to larger wavelengths and produce the violet and green colors after γ-irradiation of colorless quartz (see Figure 12). If the concentration of aluminum and lithium are high, the heating of irradiated violet or green quartz can produce yellow colors by a mechanism like the one previously described.

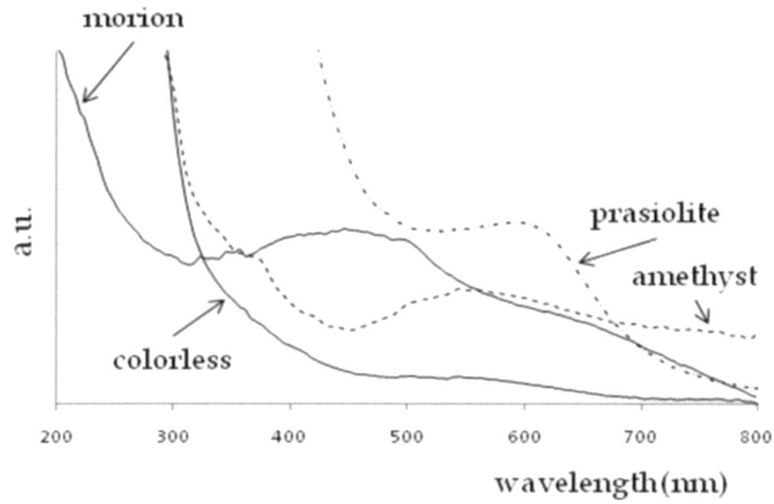

Fig. 12. Absorption spectra of fragments of colorless quartz, morion (black), amethyst (violet), and prasiolite (green) in the visible and ultraviolet range. The maximum absorption in the visible range is shift to larger wavelengths from morion (450 nm) to amethyst (467 nm) and to prasiolite (610 nm).

The nature of the bands at room temperature at 3404, 3510, 3585, and 3595 cm^{-1} is not yet assigned. The band at 3585 cm^{-1} is probably related to iron, because this is a trace element in amethyst and prasiolite.

6. Final remarks

The infrared spectrum at all colorless samples of natural alpha quartz show bands at 2499, 2600, 2677, and 2771 cm^{-1}. Two small bands are also observed at 2935 and 3063 cm^{-1}. The samples that do not develop colors after irradiation show bands at 3202 and 3304 cm^{-1} (overtones of the Si-O bond). The samples that become grayish to black after irradiation show three more bands at 3381, 3433, and 3483 cm^{-1}. This last band is related to the development of the colors greenish-yellow, yellow, or brown after heating. The intensity of these colors is proportional to the intensity of this band. The samples that become grayish-olive green after irradiation show, in addition to the former bands, a pair of bands at 3404 and 3510 cm^{-1}. The samples that become violet (amethyst) after irradiation show a broad band at 3441 cm^{-1} and a band at 3585 cm^{-1}. The samples that become grayish-green after irradiation and green (prasiolite) after short exposure to ultraviolet rays also show the band at 3585 cm^{-1}, but the broad band at 3433 cm^{-1} is more intense. These samples show strong absorption of infrared radiation in the range above 2700 cm^{-1}. Finally, some samples of quartz show a band at 3595 cm^{-1} that does not seem related to the formation of any color.

All colors developed after irradiation can be bleached by ultraviolet irradiation can be bleached by ultraviolet irradiation because they are due to the dislocation of electrons in the quartz lattice. The colors developed after irradiation and heating cannot be bleached by ultraviolet irradiation because the heating causes the diffusion of lithium in quartz lattice. The band at 3539 - 3451 cm^{-1} is probably related to the colors obtained after irradiation and heating. A quantitative analysis can be performed on the FTIR spectrum of natural quartz in order to evaluate the potential of color development after irradiation and heating. A similar analysis can also be performed in other gemstones whose colors need to be enhanced by irradiation, such as topaz, beryl, spodumene, and tourmaline and should be the subject of future researches.

7. References

Bahadur, H. (1995). Sweeping investigations on as grown Al-Li^+ and Al-OH centers in natural crystalline quartz. *IEEE Transactions on Ultrasonics, Ferroelectrics, and Frequency Control*, Vol. 41, No. 2, pp. 153-158, ISSN 0885-3010.

Dana, J. D., Hurlbut, C. S., and Klein, D. (1985). *Manual of Mineralogy* (20 ed.), John Wiley & Sons, ISBN 0-471-80580-7, New York.

Fritzgerakd, S. A., Churchill, H. O. H., Korngut, P. M., Simmons, C. B., and Strangas, Y. E. (2006). Low-temperature infrared spectroscopy of H_2 in crystalline C_{60}.*Physical Review B*, Vol. 73, pp. 155409, ISSN 1098-0121.

Hebert, L. B. and Rossman, G. R. (2008). Greenish quartz from the Thunder Bay Amethyst Mine Oanorama, Thunder Bay, Ontario, Canada. *Canadian Mineralogist*, Vol. 46, pp. 111 - 124, ISSN

Nassau, K. (1978). The origins of color in minerals. *American Mineralogist*, Vol. 63, pp. 219-229, ISSN 0003-004X.

Martin, J. J. (1988) Electrodiffusion (Sweeping) of ions in quartz - A Review. *IEEE Transactions on Ultrasonics, Ferroelectrics, and Frequency Control*, Vol. 35, No. 3, pp. 288-296, ISSN 0885-3010.

Maschmeyer, D., Niemann, H., Hake, H., Lehmann, G., and Räuber, A. (1980). Two modified smoky quartz centers in natural citrine. *Physics and Chemistry of Minerals*, Vol. 6, No. 2, pp. 145 - 156, ISSN 0342-1791.

Nassau, K. (1978). The origins of color in minerals. *American Mineralogist*, Vol. 63, pp. 219-229, ISSN 0003-004X.

Nassau, K. (1980) The causes of color. *Scientific American*, Vol. 243, No. 4, Oct. 1980, pp. 124 - 154, ISSN 0036-8733.

Nunes, E. H., Melo, V., Lameiras, F., Liz, O., Pinheiro, A., Machado, G. and Vasconcelos, W. (2009). Determination of the potential for extrinsic color development in natural colorless quartz. *American Mineralogist*, Vol. 94, pp. 935-941, ISSN 0003-004X.

Walsby, C. J., Lees, N. S., Claridge, R. F. C., and Weil, J. A. (2003). The magnetic properties of oxygen-hole aluminum centres in crystalline SiO_2. VI: A stable AlO_4/Li centre. *Canadian Journal of Physics*, Vol. 81, pp. 583 - 598, ISSN 0008-4204.

4

Transmission, Reflection and Thermal Radiation of a Magneto-Optical Fabry-Perot Resonator in Magnetic Field: Investigations and Applications

Anatoliy Liptuga, Vasyl Morozhenko and Victor Pipa
V. Lashkaryov Institute of Semiconductor Physics, Kyiv, Ukraine

1. Introduction

Recent years, considerable attention is paid to the study of the optical properties of plane-parallel mono- and multilayer resonator structures based on dielectric, semiconductor and metallic media. Interference effects in such structures set conditions for their selective properties with respect to wavelength, direction of propagation and polarization of light. These effects result in modification of the spectral and angular characteristics of the intensity of transmitted, reflected, and self-emitted (e.g., thermal emission) light. Application of magnetic materials as components of the structures has opened up quite a number of possibilities, which have both scientific and applied importance. First, the magneto-optical methods are very effective for investigation of the parameters of materials and characteristic properties of the structures. Second, the synthesis of new magnetic materials, selection of their dimensions and location will make possible to create a new generation of optical devices controlled by a magnetic field (H): displays and data-transmission systems, sources and sensors of light, magneto-optical modulators and shutters etc. In this case, it concerns not only visible but also infrared (IR) light.

To use the multiple reflections for enhancement of the magneto-optical rotation was proposed by Faraday itself. As a result of the fact that the direction of the rotation does not depend on direction of a magnetic field, he achieved an increase of the light path length in the sample by silvering its surfaces. In those years, samples were sufficiently thick and less than perfect and they did not take into account the interference effects.

The first one who has denoted on a necessity to consider interference effects in the measurements of the Faraday rotation angle was Voigt (Voigt, 1904), but only in the second half of the 20th century with the appearance of the plane-parallel samples with sufficient quality, the study of peculiarities of the Faraday effect in the presence of interference has found its experimental and theoretical advancement. In one of the first papers (Rosenberg et al., 1964) it was shown that the Faraday rotation can be resonantly enhanced by a Fabry-Perot resonator. The importance of considering the effects of internal multiple reflections in the measurements of the Faraday rotation angle was noted by the authors (Piller 1966; Rheinhlander et al, 1975; Stramska et al. 1968; Srivastava et al, 1975; Vorobev et al, 1972).

In the subsequent studies (Jacob et al, 1995; Ling, 1994; Wallenhorst et al 1995) attention was paid to the theoretical and experimental studies of not only the Faraday rotation angle, but also to the studies of a state of light polarization and transmission function of the simple resonance objects such as Fabry-Perot resonators. Based on the obtained results authors have concluded that the application of the Fabry-Perot resonator is an effective method for a measuring of the Faraday rotation, especially in the inefficient media. It was also mentioned that this effect can be used in the spectroscopic devices.

As an impact to active study of composite resonant magneto-optical structures it was the paper (Inoue & Fujii, 1997), in which the magneto-optical properties of the Bi:YIG films with random multilayer structures were researched. The authors have found a large enhancement in Faraday and Kerr effects in the structures. In the some subsequent papers such multilayer structures were named the magnetophotonic crystals (MPCs). Detailed review of these objects, their peculiarities and applications is presented in (Inoue et al. 2008).

Recently, the Faraday rotation was investigated in different types of structures including 2D and 3D MPCs (Dokukin et al. 2009; Fujikawa et al. 2009) and optical Tam structures (Goto et al., 2009a, 2009b). Moreover, the studies were done not only for the visible and near infrared light, but also in the far-infrared range (Zhu et al. 2011).

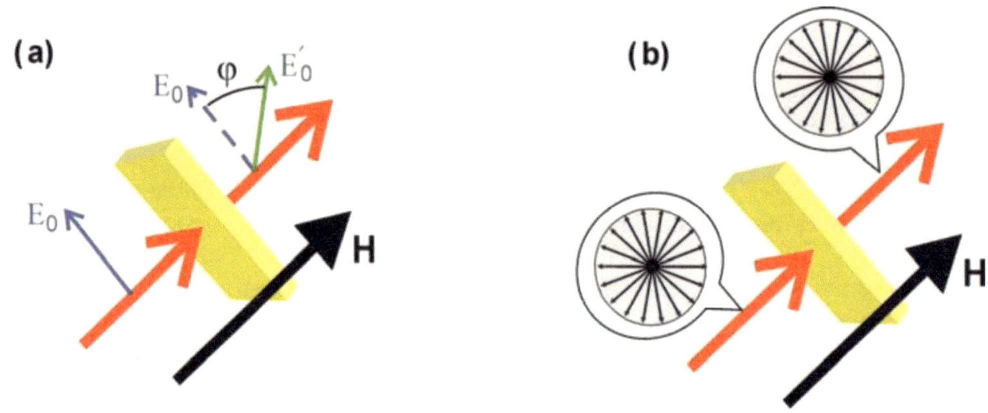

Fig. 1. (a) Demonstration of the Faraday rotation of linearly polarized light. $\mathbf{E}_0$ is a vector of electric field of the incident wave; $\mathbf{E}_0'$ is a vector of electric field of the transmitted wave, φ is the Faraday rotation angle.
(b) Demonstration of the Faraday rotation using unpolarized light. Balloons show the azimuth distribution of the linearly polarized components.

The investigations (Kollyukh et al., 2005; Morozhenko & Kollyukh, 2009) have shown that the influence of a magnetic field appears not only with respect to external linearly polarized light, but also with respect to their own thermal radiation (TR) of structure. This puts a question: does the Faraday rotation exist in unpolarized light? By definition, the Faraday effect is a rotation of the plane of polarization in a parallel to direction of propagation magnetic field, as it is shown in Fig.1a. Based on this, the obvious answer is no, because in the absence of the selected plane of polarization the rotation can not be fixed. This is clearly seen in Fig.1b. For this reason all studies of magneto-optical rotation are carried out with the linearly polarized light.

In this paper, based on the example of a magneto-optical Fabry-Perot resonator (MOR), it is shown that the uniqueness of the resonant magneto-optical structures is also that they are applicable for an unpolarized light. A change of their reflective and transmitting characteristics in a magnetic field using an unpolarized light is so effective as polarized.

The authors paid attention to the influence of a magnetic field to the angular and spectral characteristics of MOR in the middle-wave and long-wave IR, and the results of theoretical and experimental studies of thermal radiation of MOR are also presented and discussed.

On base of the obtained results it is described a number of possible applications of the effect. It is shown that this effect opens up the wide possibilities for developing both new controllable magneto-optical devices and methods of determination of the structure parameters.

2. Model and theory

Let us consider a magneto-optical Fabry-Perot resonator that consists of two non-absorbing mirrors with the reflection coefficients ρ_1, ρ_2 and a magneto-optical medium inside. The mirrors are spaced by a distance d. The medium is characterized by an isotropic at $H = 0$ complex refractive index $N = n + i\chi$ ($\chi << n$). An external magnetic field is perpendicular to the surface xy of the resonator.

The unpolarized light with wavelength λ and intensity I_0 falls on MOR at angle θ_1 (see Fig.2). Since the incident light is unpolarized, it contains equal quantities of the linearly polarized components with any planes of polarization. Propagating in the resonator the wave refracts and reflects back in a volume and splits into the series of coherent among themselves secondary waves j_ξ ($\xi = 0,1,2...$). Their coherence is determined by a coherence of their corresponding linearly polarized components. When the light is crossing the MOR, the planes of polarization of the linearly polarized components rotate. It is shown in the balloons of Fig.2.

For calculations of the light propagation with the Faraday rotation the matrix method of multi-beam summation (Morozhenko & Kollyukh, 2009) is used. A matrix-vector of electric field of an arbitrary linearly polarized component (see Fig.2) of the incident wave is (Yariv & Yen, 1984):

$$\mathbf{E}_0 = E_0 \cdot \begin{bmatrix} \cos(\beta) \\ \sin(\beta) \end{bmatrix}, \tag{1}$$

where β is an azimuth of the component. Time dependence of $\mathbf{E}_0$ is omitted. A propagation of the linearly polarized components in the MOR is described by following matrices:

$$\mathbf{R}_1 = \begin{bmatrix} r_1^{(s)} & 0 \\ 0 & -r_1^{(p)} \end{bmatrix}, \mathbf{R}_2 = \begin{bmatrix} r_2^{(s)} & 0 \\ 0 & -r_2^{(p)} \end{bmatrix}, \mathbf{M}_1 = \begin{bmatrix} t_1^{(s)} & 0 \\ 0 & t_1^{(p)} \end{bmatrix},$$

$$\mathbf{M}_2 = \begin{bmatrix} t_2^{(s)} & 0 \\ 0 & t_2^{(p)} \end{bmatrix}, \mathbf{F} = \begin{bmatrix} \cos(\varphi) & \sin(\varphi) \\ -\sin(\varphi) & \cos(\varphi) \end{bmatrix} e^{ik_z d}, \tag{2}$$

where $r_{1,2}^{(s),(p)}$ and $t_{1,2}^{(s),(p)}$ are the reflection and transmission amplitudes respectively for s- and p-polarizations, $k_z = (2\pi / \lambda)\sqrt{N^2 - \sin^2(\theta_1)}$, φ is a single-trip Faraday rotation angle.

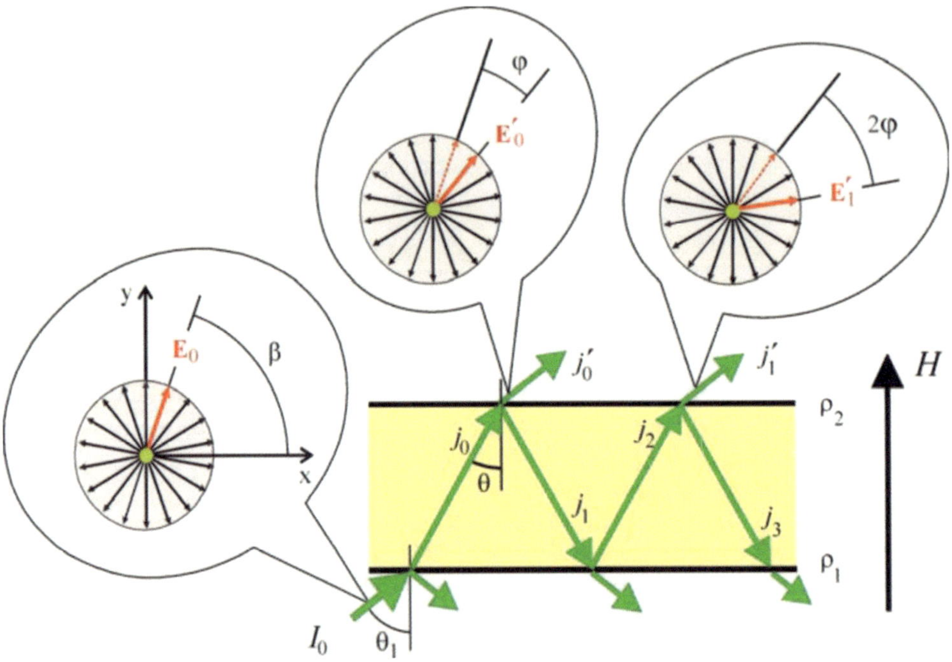

Fig. 2. Propagation of natural light in a magneto-optical Fabry-Perot resonator. Balloons show the azimuth distribution of the linearly polarized components. They are shown as at normal incidence for visualization and convenience.

The matrices $\mathbf{R}_{1,2}$ and $\mathbf{M}_{1,2}$ describe reflection and refraction respectively, $\mathbf{F}$ is a matrix of passage with the Faraday rotation.

The transmitted secondary waves are:

$$\mathbf{E}'_{\xi} = \mathbf{M}_2 \left(\mathbf{F}\mathbf{R}_1\mathbf{F}\mathbf{R}_2\right)^{\xi} \mathbf{F}\mathbf{M}_1\mathbf{E}_0 . \tag{3}$$

The sum of the waves $\mathbf{E}'_{\xi}$ is the sum of matrix series

$$\mathbf{E}' = \mathbf{M}_2 \left(\sum_{\xi=0}^{\infty} (\mathbf{F}\mathbf{R}_1\mathbf{F}\mathbf{R}_2)^{\xi} \right) \mathbf{F}\mathbf{M}_1\mathbf{E}_0 . \tag{4}$$

It is easy to make sure to certain that for eigenvalue L of the matrix $\mathbf{F}\mathbf{R}_1\mathbf{F}\mathbf{R}_2$ a condition $|L| < 1$ is always true. Hence the sum in Eq. (4) can be replaced with the expression (Lancaster & Tismenetsky, 1985):

$$\mathbf{E}' = \mathbf{M}_2 \left(\mathbf{I} - \mathbf{F}\mathbf{R}_1\mathbf{F}\mathbf{R}_2\right)^{-1} \mathbf{F}\mathbf{M}_1\mathbf{E}_0 , \tag{5}$$

where **I** is the unity matrix. This summarized wave is a component of a total summarized wave $j' = \sum j'_{\xi}$. Since the separate components are not coherent, to determine the total transmission T it is necessary to sum up the intensities of all transmitted components and divide it into I_0:

$$T = \frac{1}{2}\sum_{i,j} \left| u_{ij} \right|^2 , \tag{6}$$

where u_{ij} are the elements of a matrix $\mathbf{M_2}\left(\mathbf{I} - \mathbf{FR_1FR_2}\right)^{-1}\mathbf{FM_1}$.

By doing the foregoing operations, it is possible to obtain an equations for reflection R:

$$R = \frac{1}{2}\sum_{i,j} \left| w_{ij} \right|^2 . \tag{7}$$

Here w_{ij} are the elements of a matrix $\left(\mathbf{M_1}\left(\mathbf{I} - \mathbf{FR_2FR_1}\right)^{-1}\mathbf{FR_2FM_1} + \mathbf{R_1}\right)$.

In case of thermal radiation (TR), the light is not external and radiated by a volume. The intensity of a "primary" wave, radiated at angle θ in a small solid angle $d\Omega$ and reached a boundary is

$$J_0 = (1 - \eta) n^2 W \cos(\theta) d\Omega , \tag{8}$$

where W is the Planck function, $\eta = \exp(-\alpha d / \cos(\theta))$, $\alpha = 4\pi\chi / \lambda$ is an absorption coefficient. Since n and α are isotropic, this wave is unpolarized.

The intensity of TR that the MOR radiates from the mirror ρ_1 at angle θ_1 ($\sin(\theta_1) = n \cdot \sin(\theta)$) in a solid angle $d\Omega_1 = n^2 d\Omega \cos(\theta) / \cos(\theta_1)$ is

$$I_{TR} = \frac{1}{2}\sum_{i,j} \left(\left| q_{ij} \right|^2 + \left| g_{ij} \right|^2 \right) (1 - \eta) W \cos(\theta_1) d\Omega_1 , \tag{9}$$

where q_{ij} and g_{ij} are the elements of the matrix $\mathbf{M_1}\left(\mathbf{I} - \mathbf{FR_2FR_1}\right)^{-1}$ and $\mathbf{M_1}\left(\mathbf{I} - \mathbf{FR_2FR_1}\right)^{-1}\mathbf{FR_2}$ respectively.

In Equations (6), (7) and (9) the matrix elements "$i1$" describe the peculiarities of s-polarized part of light, and the elements "$i2$" describe the peculiarities of p-polarized one.

Equation (9) is the Kirchhoft's law for a magneto-optical Fabry-Perot resonator in a magnetic field. The factors to the left of the Planck function are an emissivity (A) of the resonator. The emissivity describes all TR features that are related to the dielectric and geometric properties of a heated object. For this reason to analyze the peculiarities of A sometimes is more convenient, than intensity of TR.

3. Results and discussions

3.1 Theoretical results

The calculations were carried out for the resonator with the reflection coefficients of the mirrors $\rho_1 = \rho_2 = 0.6$. Modeling magneto-optical medium has a complex refractive index $N = 3 + i \cdot 7.1 \cdot 10^{-3}$ and thickness $d = 50$ μm.

Fig.3 shows the theoretical dependence of the angular distribution of the MOR transmission (a) and reflection (b) on the single-trip Faraday rotation angle for unpolarized light. It is seen, at $H = 0$ the angular distributions of transmission and reflection have a lobe-like character and correspond to a number of the interference maxima (lobes) and minima with the high contrast.

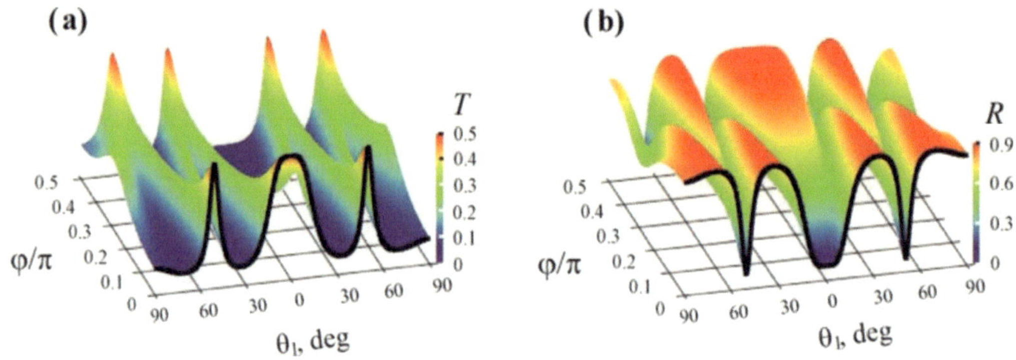

Fig .3. (a) Theoretical dependence of the angular distribution of the MOR transmission on magnetic field. Incident light is unpolarized. $\lambda = 10 \quad \mu$ m.
(b) Theoretical dependence of the angular distribution of the MOR reflection on magnetic field. Incident light is unpolarized. $\lambda = 10 \quad \mu$ m.

In the magnetic field the lobes of transmission split into two secondary lobes, which diverge and decrease in amplitude as the field is increasing. For $\varphi = \pi / 4$, the contrast of the angular distribution reaches a minimal value.

With a further increasing field the secondary lobes begin to merge pairwise with the neighboring one and at $\varphi = \pi / 2$ the angular distribution of the transmission takes a pronounced lobe-like character again. However, there is an inversion of the interference extrema: positions of the lobes correspond to the minima at $H = 0$.

Behavior of the angular distribution of reflection in the magnetic field differs from that discussed above for transmission distribution. In this case the interference minima split. Maxima lobes appear in their place with intensity increasing with increasing of the magnetic field.

When $\varphi = \pi / 4$, the amplitudes of the original lobes (at $H = 0$) are almost equal to the amplitude of the appearing ones and the angular distribution of R become practically homogeneous.

At $\varphi = \pi / 2$ the angular distribution of R has a lobe-like character again, but with the inversion of the resonance extrema.

The cause of the changes of the angular distributions of T and R is a change of the conditions of the multibeam interference in the magnetic field. The magnetic field redistributes the polarization planes of the waves of light inside the resonator. As a result, the interference of the transmitted and reflected waves can be suppressed or phase shifted by a phase difference of π .

In Fig.4 the directional diagrams of the resonator's TR are shown for $\lambda = 10 \ \mu$ m at different values of φ. With $\varphi = 0$ the directional diagram has a muti-peaked antenna pattern with a narrow central main lobe and an axially symmetric radially diverging secondary (conical) lobe. The intensities of the lobes and their ratio are dependent on the resonator parameters such as the coefficients of reflection $\rho_{1,2}$, absorption of the magneto-optical medium (Guga et al. 2004), MOR temperature and wavelength. In order to eliminate the last two parameters, the directional diagrams are shown in arbitrary units.

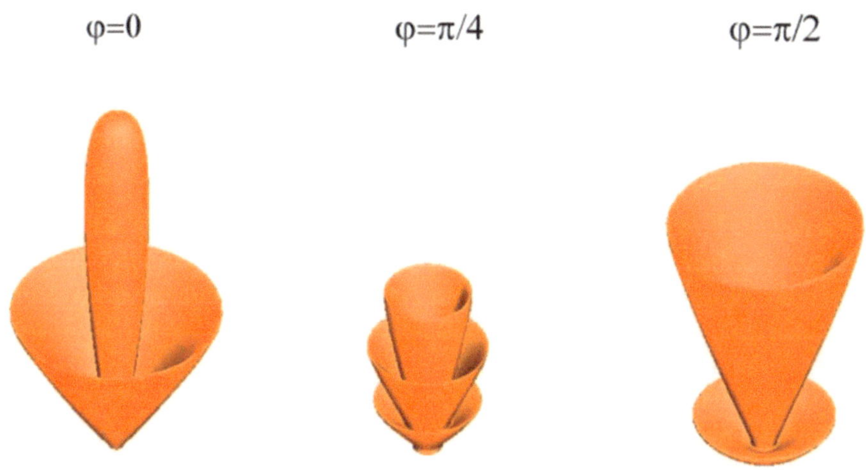

Fig. 4. Modification of TR directional diagram for $\lambda = 10 \ \mu$ m at different values of φ .

In the magnetic field, when $\varphi = \pi / 4$, the directional diagram supports a greater number of relatively weak minor lobes, making the radiation more uniform, making it more like TR from a nonresonant object.

When the Faraday rotation reaches a value $\varphi = \pi / 2$ the directional diagram takes on a clear multi-lobed structure again, but with no axially directed maximum radiation. The central lobe is missing and the axial radiation is minimum. Only flared minor lobes are present. Therefore, the magnetic field determines the directional diagram of TR and this property of resonator can be used to generate controlled sources of infrared radiation.

Fig. 5 presents the calculated fringes of constant inclination of the transmitted light when MOR is illuminated by an unpolarized polychromatic $\lambda = 9.668 - 10 \ \mu$ m source. The calculations are limited to the first interference order. As it is seen, at $\varphi = 0$ the fringes have a form of contrasting rings, painted in accordance with the wavelength of the interference lobe (the scale of wavelengths is shown on the right).

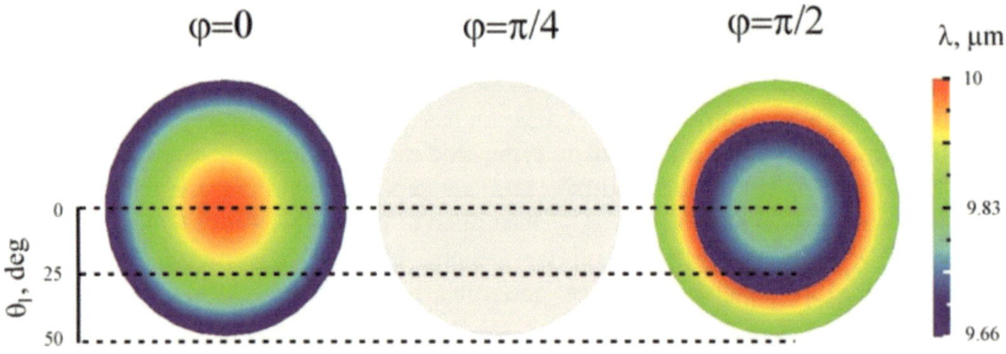

Fig. 5. Fringes of constant inclination of the MOR. The incident light is unpolarized. $\lambda = 9.668 - 10 \ \mu$ m. The wavelength scale is shown on the right.

When $\varphi = \pi / 4$ the angular dispersion of the resonator disappears. The transmitted light does not have the selected wavelengths (painted by grey) and the MOR transmission is close to the transmission of a non-resonator.

In the magnetic field when $\varphi = \pi / 2$ the fringes of constant inclination are clearly visible again. However, their color distribution has changed. In this case the two orders of interference are observed: the truncated blue end first order and the second one when $\theta_1 > 33^0$. Such changes of the angular dispersion are caused by the inversion of the interference extrema in polychromatic light.

3.2 Experimental results

For the measurements of spectra, the free plane-parallel plates of n-InAs were used as a simple Fabry-Perot resonator. The high value of the refraction factor $n \approx 3$ caused a value of the reflection coefficients of the faces - mirrors $\rho_1 = \rho_2 \approx 0.25$.

Semiconductor n-InAs is isotropic in the absence of a magnetic field. The high concentration of the free electrons $N_q = (1.3 \div 1.4) \cdot 10^{18}$ cm^{-3} made possible to carry out the measurements in a classical magnetic field: the Landau splitting energy is assumed to be small as compared to the thermal electron energy.

The plates were cut from a single crystal bar, then ground and subsequently polished on the broad faces. The 10×10 mm^2 samples had thickness $d = 80$ and 100 μ m; the deviation from plane-parallelism was no greater than a few seconds of arc.

The plate was placed between the magnet poles so that the magnetic field was directed normally to the broad faces of the sample. The measurements of spectra were carried out by Fourier-spectrometer (FTIR) with a resolution of 2 cm^{-1}, the aperture of the inlet of the optical equipment did not exceed 2.5^0. The experimental setups for investigating spectra of transmission and thermal radiation are shown in Fig. 6 and Fig. 7.

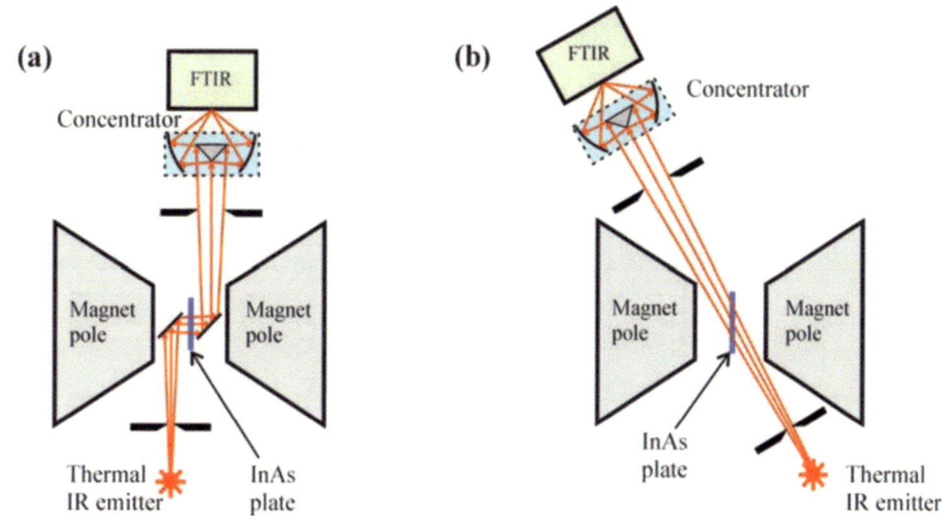

Fig. 6. Experimental setups for investigating transmission spectra of a plane-parallel n-InAs plate in a magnetic field at normal (a) and oblique incidence (b) of unpolarized light.

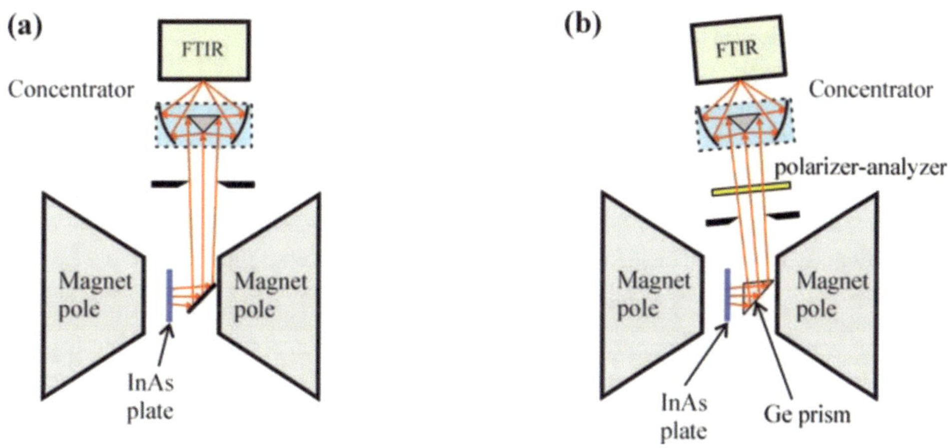

Fig. 7. Experimental setups for investigating spectra of thermal radiation of a plane-parallel n-InAs plate in a magnetic field: (a) - for investigating the total thermal radiation spectra; (b) - for analysis of the circular polarized modes of thermal radiation.

3.2.1 Transmission spectra

Fig. 8 shows the spectral dependencies of transmission of the plate at normal incidence in the absence of the magnetic field (blue line) and in the magnetic field (red line). As it is seen, the dependence $T(\lambda)$ at $H = 0$ has an oscillating form, that is typical for the Fabry-Perot resonator. Positions of the maxima and minima can be estimated from the interference conditions $\lambda_{\max} k = 2n(\lambda)d$ and $\lambda_{\min}(2k+1) = 4n(\lambda)d$ respectively ($k = 1,2,3...$). Unfortunately, the semiconductor n-InAs has a strong dispersion of the absorption coefficient: $\alpha \propto \lambda^3$ (Madelung, 2004). For this reason contrast of the interference pattern decreases when λ increases.

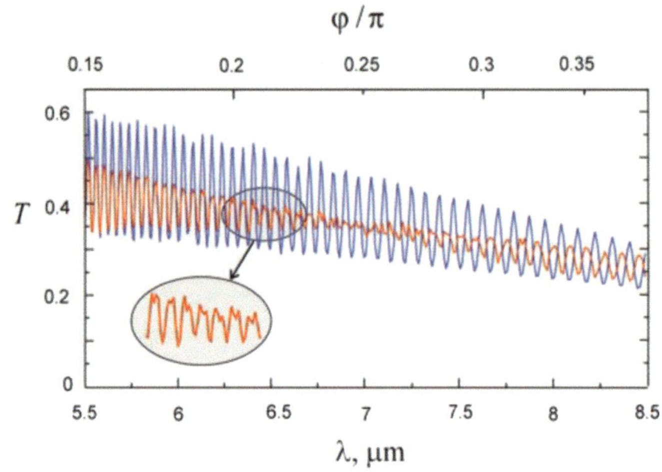

Fig. 8. Experimental transmission spectra of the free plane-parallel *n*-InAs plate at a normal incidence of unpolarized light. $d = 100$ μ m. Blue line: $H = 0$; red line: $H = 24$ kG

In the magnetic field the oscillating spectrum transformed into an oscillating spectrum with a link. A splitting of the maximums into two secondary ones is clearly visible in the spectral range $6.5 < \lambda < 7.3$ μ m. In the range $7.3 < \lambda < 7.6$ μ m (the link range) the secondary maxima are equidistant. Since their amplitudes are less than the zero-field ones, this is revealed as the doubled quantity of illegible interference maxima in the spectrum.

With a further increasing λ ($7.6 < \lambda$ μ m) the secondary maxima merge and the contrast grows. However, the phase of the interference extrema is inverted relative to the $H = 0$ case.

Let us regard this effect from the point of view of Faraday rotation. Since the value of φ depends on wavelength, the analysis of the spectral dependence makes possible to analyze the changes of the interference pattern in the dependence on the value of Faraday rotation at the constant magnetic field. The value of φ, that is shown on the upper scale, was calculated according to the expression:

$$\varphi = \pi \frac{d}{\lambda} \sqrt{\varepsilon_\infty} \left(\left(1 - \frac{\lambda^2 / \lambda_p^2}{(1 + \lambda / \lambda_c)} \right)^{1/2} - \left(1 - \frac{\lambda^2 / \lambda_p^2}{(1 - \lambda / \lambda_c)} \right)^{1/2} \right), \tag{10}$$

where $\lambda_c = 2\pi m^* c^2 / qH$ is a wavelength of the cyclotron oscillations, $\lambda_p = c\sqrt{\varepsilon_\infty m^*} \big/ q\sqrt{N_q}$ is a wavelength of the plasma oscillations, q is the electron charge, ε_∞ is the RF permittivity, *m** is an electron effective mass, *c* is the velocity of light.

The waves j'_ξ (see Fig.2) coming out at $\theta_1 = 0$ remain unpolarized. However the planes of polarization of their components are rotated at 2φ relatively to the corresponding components of the wave $j'_{\xi-1}$ (for example, the marked component E'_1 are rotated relatively to the E_0' on Fig.2). It leads to the violation of the conditions of interference between them. The result of their superposition is, in general, an elliptically polarized wave. And as a result, the contrast of an interference pattern decreases.

When $\varphi = \pi / 4$ the planes of polarization of the components are rotated at $2\varphi = \pi / 2$, i.e. are orthogonally related. The result of their interaction is not intensification or attenuation, but, in general, an elliptically polarized wave. In this area of wavelengths there is a link in the spectrum of T. The difference of the contrast from zero at $\lambda \approx 7.5$ μm is explained by interference of waves j'_ξ, $j'_{\xi+2}$, $j'_{\xi+4}$ ... . Their electrical fields are parallel but differ greatly in amplitudes.

At a further increasing λ the angle 2φ exceeds the value $\pi / 2$. It results in appearance of interference extrema. At $2\varphi = \pi$ the electrical fields of the coherent components become parallel again, however their phase difference is $2\delta + \pi$ where $\delta = 2\pi nd / \lambda$ is a phase difference when $H = 0$. It is a reason of the inversion of interference extrema.

With a normal incidence of light the transmission of resonator in a magnetic field can be described by the sufficiently compact analytical expression, which is obtained from Eq. (6):

$$T = \frac{1}{2} \cdot (1 - \rho_1)(1 - \rho_2)\eta \left(\frac{1}{Z(\delta_c)} + \frac{1}{Z(\delta_e)} \right) = T_c + T_e . \tag{11}$$

Here $Z(\delta_{c,e}) = 1 - 2\eta\sqrt{\rho_1 \rho_2} \cos(2\delta_{c,e}) + \eta^2 \rho_1 \rho_2$, $\delta_c = \delta + \varphi$, $\delta_e = \delta - \varphi$.

As it is seen, transmission spectrum is the sum of two independent terms, which describe "compressed" (T_c) and "extended" (T_e) spectrum with the increased (δ_c) and reduced (δ_e) on φ phase difference respectively. Since these spectra have different interference fringe spacing, depending on the relationship δ_c and δ_e they can be in phase, antiphase or in-between state. They can also be in the same state relative to the zero-field spectrum. The diverging of the maxima T_c and T_e is observed in the complete transmission spectrum as a splitting of the interference maxima. One secondary maximum corresponds to T_c and the other one corresponds to T_e.

When these spectra are in antiphase, they substantially compensate each other. In this region the link is observed. With a further increasing φ the maxima T_c and T_e begin to approach and merge, being, however, in antiphase to the original spectrum. The inversion of interference extrema is observed here.

Note that similar processes determine behavior of the discussed above angular dependences. However, for oblique incidence the reflection of real mirrors depend on the angle of incidence. In this case, they are also polarizing elements. This fact complicates the description of the process, though its essence remains the same.

Fig. 9 presents the experimental transmission spectra of the plate at angle of incidence $\theta_1 = (70 \pm 1)^0$. Note two features of this spectrum. First, in this case the link is located in the region $\varphi \approx 0.18\pi$. Second, there is a sagging of the transmission spectrum in the field relative to the zero-field spectrum.

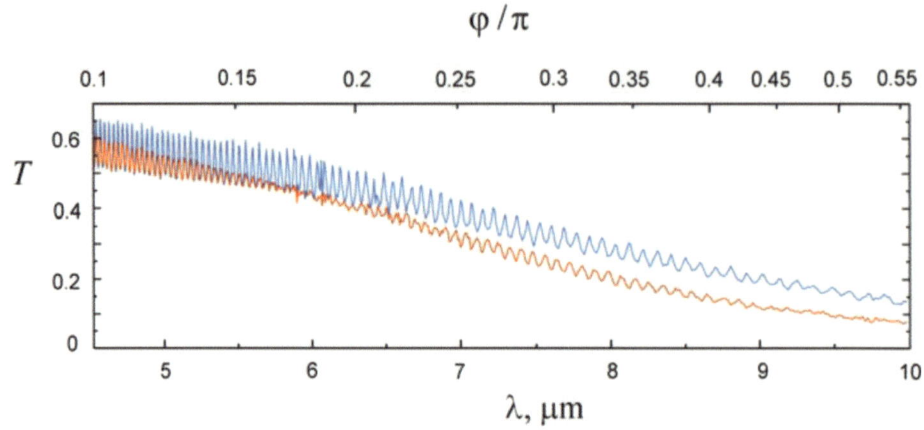

Fig. 9. Experimental transmission spectra of the free plane-parallel n-InAs plate at an angle of incidence $\theta_1 = (70 \pm 1)^0$. Light is unpolarized. Blue line: $H = 0$; red line: $H = 24$ kG

These features are caused by the polarizing action of the plate faces, whose reflection becomes anisotropic at oblique incidence: reflection of s- polarized component predominates above the reflection of the p- polarized one. For this reason the entered to the resonator light is partially polarized with the predominance of p- polarization. Faraday rotation leads to redistribution of the polarization of waves. With the approach the opposite face the light radically differs in degree and type of polarization from that entered. Interaction of light with the constantly changing polarization characteristics with the anisotropic mirrors leads to features of both the interference pattern and the absolute values of the resonator transmission.

Though the sagging of the spectral dependence T is not related to the interference effects, this fact is sufficiently interesting. Let us consider this fact in detail. For this, after excluding reflection for p-polarized light and interference, from Eq. (6) one can obtain the dependence of the transmission of a nonresonant magneto-optical sample on φ at the angle of incidence, equal to the Brewster angle ($\theta_1 = \theta_B$):

$$T^* = \eta \frac{2(1-\rho) + \rho^2 \cos^2(\varphi)\left(1 - \eta^2 \cos(2\varphi)\right)}{2\left(1 - \eta^2 \rho^2 \cos^4(\varphi)\right)}, \tag{12}$$

where $\rho = (n\cos(\theta) - \cos(\theta_1))^2 / (n\cos(\theta) + \cos(\theta_1))^2$.

Note that the angle of incidence 70^0, which was used in the experiment, is very close to the Brewster angle. As it is seen from Eq. (12), the value of the transmission of unpolarized light at $\theta_1 = \theta_B$ is a function of the single-trip Faraday rotation angle. This does not lead to violation of the energy conservation law, since the decrease of energy of transmitted light is compensated by an increase of the energy of reflected and absorbed light.

Thus, the Faraday rotation not only changes the conditions of interference, but also redistributes an energy between the transmitted and reflected light.

3.2.2 Spectra of thermal radiation

Fig. 10 presents the measured spectra of TR of the plate in the absence of the magnetic field (blue line) and in the magnetic field. The experimental setup for this investigation is shown in Fig. 7a. Behavior of the TR spectrum is qualitatively similar to the behavior of the transmission spectrum. Here it is also observed the splitting of maxima into two secondary ones, which pass into the link.

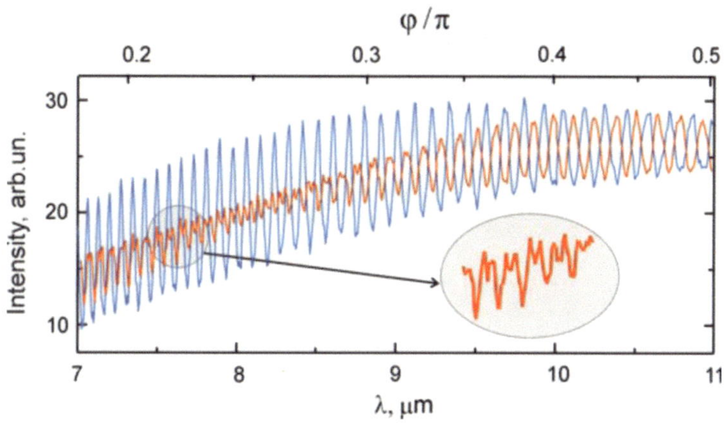

Fig. 10. Experimental spectra of thermal radiation of a free plane-parallel n-InAs plate at normal incidence, $d = 100$ μ m. Blue: $H = 0$; red: $H = 18$ kG; temperature is 355 K.

With a further increasing λ and, respectively φ, the secondary maxima begin to merge, forming clear interference pattern. In this case the inversion of interference extrema occurs too.

As for T, it is possible from Eq. (9) to obtain an analytical expression for the emissivity of MOR from a mirror ρ_1 at the normal incidence:

$$A = \frac{1}{2}\cdot(1-\rho_1)(1+\eta\rho_2)(1-\eta)\left(\frac{1}{Z(\delta_c)}+\frac{1}{Z(\delta_e)}\right). \tag{13}$$

It is seen that TR also consists of two independent modes with different spectral dependences. Taking into account that the value of the single-trip Faraday rotation angle is:

$$\varphi = \frac{\pi}{\lambda} d\left(n_+ - n_-\right), \tag{14}$$

where $n_\pm$ are refractive indexes of a right-hand and a left-hand circular polarized light, for the linear approximation $n = (n_+ + n_-)/2$ Eq. (13) can be converted to the form:

$$A = \frac{1}{2}\cdot(1-\rho_1)(1+\eta\rho_2)(1-\eta)\left(\frac{1}{Z(\delta_+)}+\frac{1}{Z(\delta_-)}\right). \tag{15}$$

Here $\delta_\pm = 2\pi n_\pm d / \lambda$. It becomes obvious that the TR modes are right-hand and a left-hand circular polarized. Using this fact, we can experimentally separate them and measure separately.

For separation and analysis of circular polarized light a germanium total internal reflection prism with base angles of 43° was used. Its principle of operation is analogous to that of Fresnel prism: it brings about a phase difference of $\pi/2$ between the perpendicular components of radiation. Since the polarization planes of right- and left-hand polarized radiation are mutually perpendicular, the required radiation modes was separated by applying a polarizer-analyzer and recorded by the FTIR. Besides, the prism served as a deflecting element to take the sample radiation out of the gap between the magnet poles (see Fig.7b).

Fig. 11 shows the spectra of the right- and left-hand circular polarized modes of TR, respectively. One can see easily that the oscillation phases of these spectral patterns do not coincide. The oscillation phase of the right-hand circular polarized mode takes the lead over that of the left-hand circular polarized mode, and at $\lambda \approx 9.2$ μm they are in antiphase.

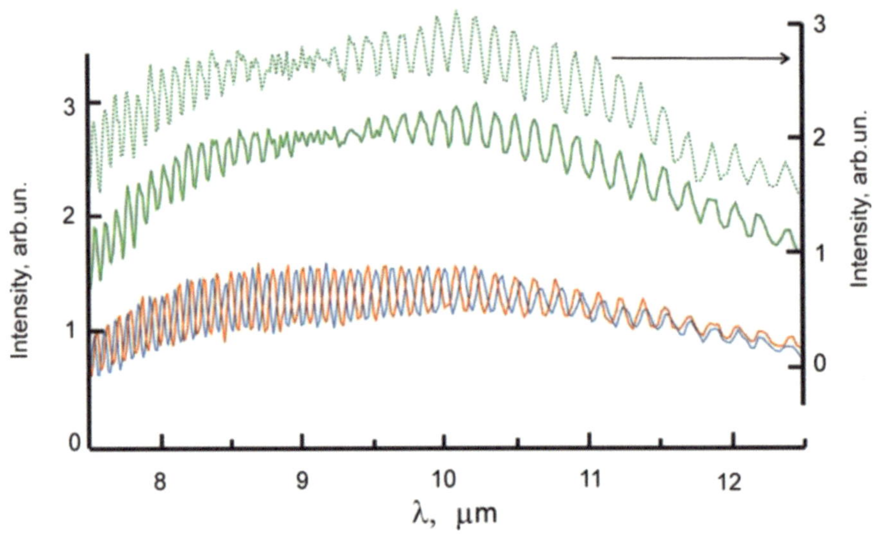

Fig. 11. Experimental spectra of thermal radiation of a free plane-parallel n-InAs plate in the magnetic field $H = 15$ kG. $d = 100$ μm. Red and blue lines are a right-hand and a left-hand circular polarized mode respectively; green line is a sum of the modes; green dot line is a total TR spectrum (a right axis). Temperature is 375K.

The solid green line is an arithmetical sum of the right-hand and left-hand circular polarized modes spectra; it agrees well with the total TR spectrum recorded without a polarizer-analyzer (the dot green line). Some insignificant distinction between amplitudes of this spectrum is due to the losses introduced by the polarizer-analyzer.

4. Applications

4.1 Determination of parameters of solids

The experimental investigations of the optical properties of solids make possible to obtain the extensive information about their physical properties and parameters. Among the magneto-optic methods of the investigations the Faraday effect is in highlight. It makes it possible to determine, for example, the effective mass of free carriers in the semiconductors, its temperature and concentration dependences and thus to conclude about the form of energy bands.

The applied methods of determination the magnitude of the Faraday effect consist in the measurement of intensity of linearly polarized light that is transmitted through the sample and polarizer-analyzer. The changes of intensity, which appear in a magnetic field, are compensated by the turning of polarizer-analyzer or are measured by a register.

Since the measurements have an absolute nature, an error in the measurement of rotation angle is determined by error in the determination of a light intensity change and an intensity of light itself. Besides the random errors of measurements the systematic errors can arise, that is caused by the imperfection of the used polarization devices, the inhomogeneity of the samples and their surfaces, the multiple reflection of light in thin crystals and others. The last forces researchers to guard, for example, to make the samples in the wedge form.

According to the results of investigations given above the multibeam interference is not the parasitic effect, which puzzles the researchers of the Faraday effect. Moreover, the presence of interference together with the Faraday rotation makes possible to determine the magnitude of the Faraday rotation angle rapidly and reliably using affordable equipment.

It is simple to obtain from Eq. (11) and Eq. (12) the envelope functions (that determine the dependence on λ of the oscillation maxima and minima, respectively) for the oscillating functions of transmission ($T^{(-)}$, $T^{(+)}$) and emissivity ($A^{(-)}$, $A^{(+)}$). They have the following form:

$$T^{(\pm)} = \frac{(1-\rho_1)(1-\rho_2)\eta}{1 \pm 2\eta\sqrt{\rho_1\rho_2}\cos(2\varphi) + \eta^2\rho_1\rho_2} \tag{16}$$

$$A^{(\pm)} = \frac{(1-\rho_1)(1+\eta\rho_2)(1-\eta)}{1 \pm 2\eta\sqrt{\rho_1\rho_2}\cos(2\varphi) + \eta^2\rho_1\rho_2} \tag{17}$$

Applying the same transformation to the Eq. (7) it can be obtained analogous envelope functions for the reflection:

$$R^{(\pm)} = \frac{\left(\rho_1 \pm 2\eta\sqrt{\rho_1\rho_2}\cos(2\varphi) + \eta^2\rho_2\right)}{1 \pm 2\eta\sqrt{\rho_1\rho_2}\cos(2\varphi) + \eta^2\rho_1\rho_2} \tag{18}$$

Solving the equations $T^{(-)} = T^{(+)}$, $A^{(-)} = A^{(+)}$ and $R^{(-)} = R^{(+)}$ it is easy to determine that the envelope functions cross at a point $\varphi = \pi / 4$. This result confirms above empirical conclusion that the link position corresponds to the Faraday rotation angle $\varphi = \pi / 4$. This is a convincing base for use a registration of the spectral dependence T, either R or TR for the determination of the value of Faraday rotation from the position of link.

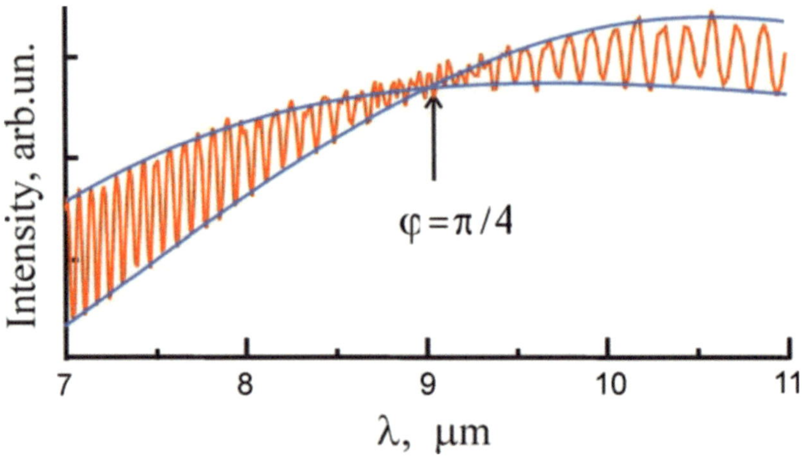

Fig. 12. Experimental spectra of thermal radiation of a free plane-parallel n-InAs plate and envelope lines (blue lines), $d = 80$ μm. $H = 18$ kG; temperature is 355K.

Fig.12 shows the experimental spectra of TR of the free plane-parallel n-InAs plate and envelope lines (blue lines). The link has a certain extent in the spectrum. Determination its position "by sight", causes a certain error. This error is insignificant. In this case the determined "by sight" spectral extent of the link is approximately $8.9-9.25$ μm, which introduces error into the determination of the Faraday rotation angle $\varphi = 45^0 \pm 1.7^0$ deg. The determined according to the point of intersection of the envelopes functions spectral position of the value $\varphi = \pi / 4$ is equal to 9.07 μm.

The accuracy of the determination of semiconductor parameters from the value φ depends also on the accuracy of the determination of the resonator thickness d and its refractive index n. The d value is determined by technological process and can be known with the high accuracy but n can greatly changes depending on the type and level of doping, as well as on the wavelength. And in this case the presence of interference becomes very useful factor again. As it is known, the interference fringe spacing is determined by the factor nd / λ. Knowing precise value d it is possible to determine n very accurately by analyzing the spectral characteristic without magnetic field. For the used in the experiment sample $n = 3.08$ in the range $\lambda = 8.9 - 9.2$ μm.

Using the experimental data, the effective mass of electrons in applied semiconductor n-InAs at a temperature of 355 K was determined by the described method. Its value $m^* / m_q = 0.04$ (m_q is the electron mass) matches the reference data (Madelung, 2004).

The advantages of the described method are obvious:

- the measurements can be carried out by using the unpolarized light. This makes it possible to exclude the polarizers and analyzers and thus to simplify both an optical scheme and registering equipment.
- the analysis of spectra without field and in the field makes possible to determine several parameters of the material at once: refractive index and its dispersion, the thickness of magneto-optical layer, the value of φ.

- caring out the measurements with several magnitudes of a magnetic field, it is easy to determine the spectral dependence of φ. This makes it possible to determine such important parameters of semiconductor as the frequency of plasma oscillation, the concentration of the free charge carriers and doping impurity.
- the measurements have not absolute, but relative nature. The accuracy of the method is determined exclusively by the accuracy of the determination of magnitude of H and by the quality of the record of interference pattern, and it does not depend on the position of base line. This fact is very important. This makes it possible to get rid of many systematic errors, added both by optical background and measuring optical and electronic equipment into the value of T or R or TR.
- the method makes possible to carry out the investigations successfully when a magneto-optical layer is located on an opaque substrate. In this case φ cannot be determined by the classical scheme of the Faraday effect investigation. However, as it was shown, TR contains all necessary information and can be used successfully. For this purpose it is necessary only to heat the investigated sample to temperature, higher than background.

4.2 Sources of IR radiation

Currently, infrared radiation sources are used extensively in researches, systems of gas analysis, spectroscopy, medicine etc. While in near-infrared ($\lambda < 2$ μm) the light-emitting diodes (LEDs) are used successfully, in medium-wave (MWIR) and long-wave IR (LWIR) an external quantum efficiency still abruptly reduces.

At present, in these ranges they use cheap and reliable thermal emitting elements (globar, Nernst pin et al.) with modulation of continuous emission by mechanical modulators. The disadvantages of these sources are nonselectivity of emission and the impossibility to create the pulse-periodic structure of output light flux. In addition, they require optical filters or monochromator. This leads to an increase of electric energy demand, growth of weight and dimensions of the device.

However, the use of medium-wave and long-wave ranges of IR spectrum significantly expands the scope of application of optical instruments. It is caused by the following:

- in these ranges there are the atmospheric transparency windows, and thus, the light waves can spread over long distances;
- many substances have characteristic features in these ranges which allows to detect and recognize them with great accuracy;
- the LWIR range encloses the maximum of thermal radiation of objects with temperature 10^0-100^0 C, i.e., of most of ambient objects. It is very important for systems of analysis, control and monitoring.

For creation of the modern systems of IR engineering and optoelectronics it is necessary that the source could work in the pulse or in the pulse-periodic regimes of generation of noncoherent radiation. Also there is a special interest in the sources with smooth tuning of spectral characteristic.

The researchers pin their hopes on the quantum well based LEDs, which construction and composition makes possible to expand the radiation range to $\lambda \approx 8-10$ μ m (Das et al, 2008; Yang et al, 1997) and even to create LED with two color spectral characteristic of emission (Das, 2010). However, the intensity of emission of such devices at room temperatures in LWIR does not exceed several μ W.

The promising concept of the problem solution is to use the non-luminescent (thermal) semiconductor sources as the IR emitters. The control over their TR intensity is exerted via variation of the absorption coefficient beyond the semiconductor fundamental absorption edge by varying the free charge carrier concentration. The physical principles of operation and constructions of some semiconductor TR sources are described, for example, in (Malyutenko, Bolgov et al., 2006). The advantage of the IR sources is that they do not require additional modulation facilities. However, they have a broad ($\lambda \approx 2-18$ μ m) spectrum of radiation. Therefore the additional filters are needed when narrow-band radiation is required.

New potentialities in realization of controllable narrow-band thermal sources of IR radiation appear when structures with coherent TR are used as a radiation elements. Recently, there are two different approaches to achieve the coherent thermal radiation, highly directional in a narrow spectral range. The first one is to use the materials in which there are the surface-phonon (plasmon) polariton waves: polar dielectrics, doped semiconductors or metals (Biener et al., 2008; Celanovic et al., 2005; Greffet et al., 2007; Lee et al., 2008). Since surface waves decay exponentially from the interface, the conversion from the evanescent mode to traveling mode is achieved by properly designing a periodical microstructure, for example, grating on an emitting material. Their TR is characterized by a strong peak at a certain frequency typical for the surface polariton excitations. The appropriate control can be provided by varying the material of the radiating structure (phonon mechanism) or the free carrier concentration in the same material (plasmon mechanism).

The second approach is to use semitransparent plane-parallel mono- or multilayer resonator structures (Drevillon et al., 2011; Jérémie & Ben-Abdallah; 2007, Kollyukh et al., 2003; Laroche et al., 2006; Lee & Zhang, 2007; Morozhenko & Kollyukh, 2009). Such resonators are very applicable for development of the controllable narrow-band IR sources. TR from the resonators occurs for both polarizations. It makes it possible to increase an intensity of radiation.

Application of the magneto-optical materials enables to change dynamically the optical characteristics of resonator by an external magnetic field and to control parameters of their TR.

Fig. 13 shows the theoretical spectral dependencies of emissivity of a magneto-optical resonator in the different magnetic fields, that is shown in φ units. Since the thickness of resonator is small in comparison with the wavelength, the zero-field spectrum (at $\varphi = 0$) is a number of the narrow widely distant behind each other lines. Their maxima are practically equal to 1, that corresponds to the TR intensity of the blackbody at the same temperature.

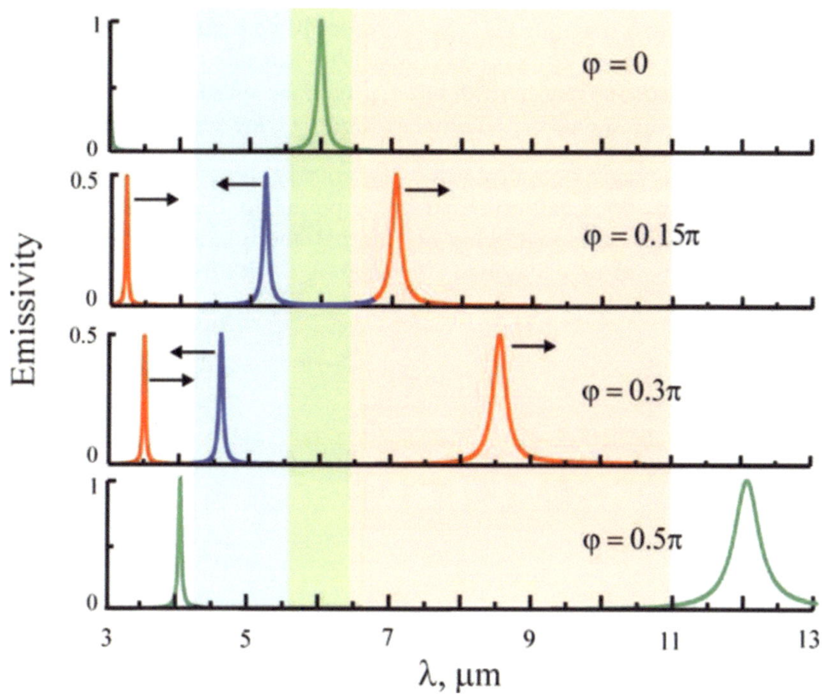

Fig. 13. Calculated emissivity spectra of a magneto-optical Fabry-Perot resonator at normal propagation of TR. $\rho_1 = 0.95$, $\rho_2 = 1$, $d = 1$ μm, $n = 3$, $\alpha = 300$ cm^{-1}.

The line width of the radiation ($\Delta\lambda$) is determined by the parameters of resonator. When the resonator Q factor is high and condition $\varsigma = \eta\sqrt{\rho_1\rho_2} \approx 1$ is satisfied, $\Delta\lambda$ on the half-height of emissivity are described by the expression:

$$\Delta\lambda \approx \frac{\lambda_{\max}(1-\varsigma)}{\pi\sqrt{\varsigma}} \tag{19}$$

Here it is assumed that the dispersion n is disdainfully small in the spectral range of line.

As it is seen on Fig.13, in the magnetic field these lines split into two narrow lines with maxima 1/2, which are diverge into the red and blue range of the spectrum. In the under consideration spectral range it is possible to mark out several characteristic areas: the area "of amplitude modulation" of the emission line, which corresponds the zero-field line (it is shown by green color), and two areas "of control of radiation spectrum" (blue and pink colors). These areas are named in accordance with their possible application for developing of the narrow-band IR source with the controllable characteristics. Let us consider each of them.

4.2.1 Source with the amplitude modulation of intensity

The development of IR sources with the internal modulation of intensity is very urgent for such equipments as, for example, optical IR gas analyzers. In the optical gas analyzers a gas concentration is measured by the magnitude of light absorption in the characteristic absorption band. The advantage of the gas analyzers in comparison with the other types (electrochemical, thermocatalytic, semiconductor) is caused by the following factors: proximity and nondestructive nature of the measurements; selectivity; quick-action and the ability to carry out measurements in real time; a uniquely wide range of measurement.

To realize all these advantages of optical gas analyzers it is necessary to have a narrow-band light source with radiation maximum that corresponds to the absorption band of the measured gas, and with the possibility of internal modulation of intensity. The last characteristic makes possible to exclude the modulating device from the construction of gas analyzer.

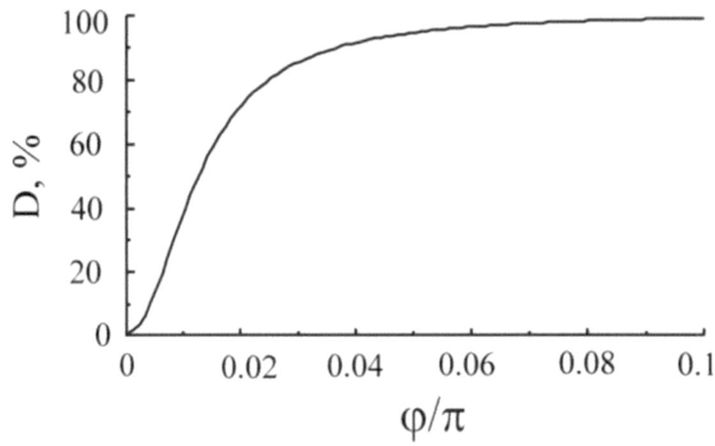

Fig. 14. Dependence of the amplitude modulation index of TR in the spectral range of the zero-field line $5.95 < \lambda < 6.05$ μ m on the single-trip Faraday rotation angle.

This source of MWIR and LWIR radiation can be realized, if the observation is carried out in a spectral range of the zero-field line. By application of a magnetic field the secondary lines of TR leave this range and the intensity of radiation becomes practically zero. The dependence of amplitude modulation index $D = (I_{TR}^0 - I_{TR}^H)/(I_{TR}^0 + I_{TR}^H)$ (here I_{TR}^0 and I_{TR}^H are the intensities of TR at $H = 0$ and $H \neq 0$ respectively) on the Faraday rotation angle in a spectral range of the zero-field line ($5.95 < \lambda < 6.05$ μ m) is shown on Fig. 14. As it is seen, unity modulation of the intensity of the radiation line is reached already at $\varphi \approx \pi / 10$.

In addition, as it can be seen from Fig.11, this source of IR radiation is multicolored, that is very important, for example, for testing multi wavelength IR sensors.

4.2.2 Sources with controllable spectral characteristic

Narrow-band sources with a tunable radiation spectrum are, actually, the integral spectroscopic device, which includes a source and monochromator in one device. Demand on such sources is obvious: the modern technologies make possible to create the compact chips of information processing, super-dense receiving matrices, fibre-optic paths. However, a presence in the spectroscopic devices of a dispersion element (prism, diffraction grating) with a necessary optical base nullifies all attempts of miniaturization and compactness of the devices.

The frequency-controlled lasers partially fill this niche. But their range of smooth tuning is insignificant. Since such lasers have large sizes, significant consumption of electric power and are expensive, they remain the special-purpose instruments and do not solve the problem of developing of the cheap compact spectroscopic device of general-purpose.

Application of the magneto-optical resonator as the radiating element is the promising way of creating of that sort of sources. As it was shown, blue ($4.3 < \lambda < 5.6$) and red ($6.4 < \lambda < 11$) ranges in Fig.13 are the ranges in which a change of the spectral position of a radiation line is realized. In the blue range $\lambda_{\max}$ of a secondary line is shifted to the short-wave side with an increasing magnetic field. For the order of interference k, one gets

$$\lambda_{\max}^{blue} = \frac{nd}{k + \varphi / 2\pi} \tag{20}$$

In the red range, the shift of $\lambda_{\max}$ occurs into the long-wave side:

$$\lambda_{\max}^{red} = \frac{nd}{k - \varphi / 2\pi} \tag{21}$$

Fig. 15 shows the dependences of the spectral position of the TR line on the single-trip Faraday rotation angle for these two regions. The ranges of spectral tuning of emitter are bounded by the edge of an zero-field line from one side and by the edge of line at $\varphi = \pi / 2$.

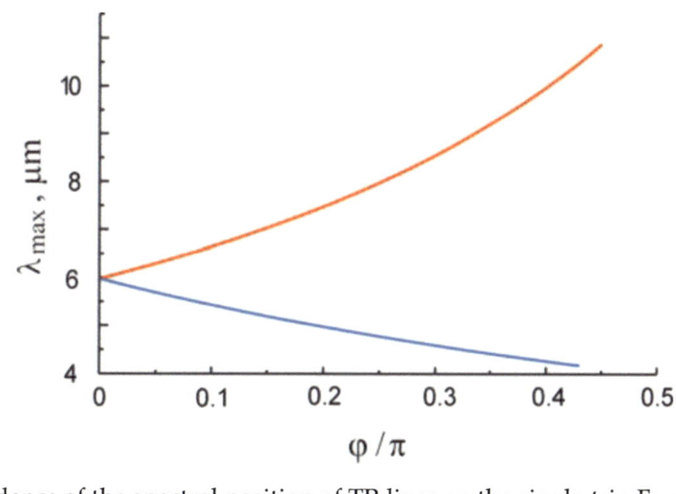

Fig. 15. Dependence of the spectral position of TR lines on the single-trip Faraday rotation angle in the blue (blue line) and red (red line) ranges.

Thus, the range of the radiation spectrum tuning (Λ) of this source is approximately equal to half of the interference fringe spacing at $H = 0$ and depends on the thickness of resonator and order of interference.

Dependences of the blue and red ranges on the thickness of resonator are shown on Fig.16 for the orders of interference 1, 2, and 3. It is seen that for realization of control of the source spectrum in a broad range it is necessary to have the strong Faraday rotation ($\varphi \le \pi / 4$) in the short optical base equal to several micrometers. In addition, that the resonator could work as a thermal IR source, it is necessary that its temperature exceeds the temperature of background (ambient temperature). This imposes the strict requirements on the magneto-optical medium of resonator.

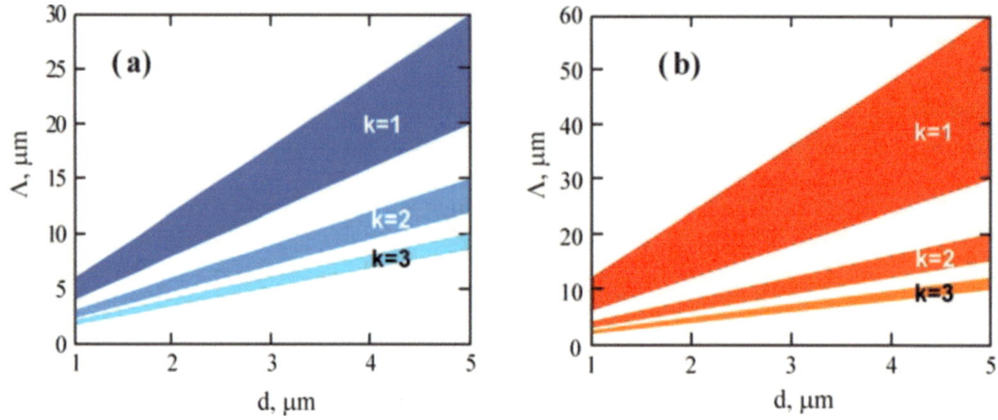

Fig. 16. Dependences of the blue (a) and red (b) ranges on the thickness of resonator for the orders of interference $k = 1,2$ and 3.

These devices can be applicable in many fields of engineering, science, medicine, technology etc.

4.2.3 Sources with control of spatial field of radiation

At present time the infrared optical-electronic systems, devices of thermal imaging and IR cameras are used in many areas of science and technology. Checking of their functionality, calibration and testing of their characteristics are an important task. Therefore the urgent problem is the development of methods of simulating and creation of the IR sources, spatial coordinates and intensity of radiation of which can dynamically change (imitators of the heated objects) (Williams, 1998).

Multielement radiation sources are used traditionally for this purpose. Now the two-dimensional (2D) devices based on electrically heated pixels have been developed and successfully applied (Pritchard et al., 1997; Robinson et al., 2000). Since the radiating element of these devices is the thermal resistor, they make it possible to create 2D field of radiation in a wide range of infrared spectrum. However, thermal control of radiation intensity of a pixel limits the system performance that does not exceed 100 Hz. In addition, in the thermal sources there is a problem of the thermal isolation between control circuit and radiating elements, and also between the adjacent radiating elements.

More promising are systems with the multielement luminescent radiating elements based on semiconductor laser diodes (Beasley et al., 1997; Cantey et al., 2008). These projectors are capable to generate dynamic infrared scenes in real-time. The ability to simulate high apparent temperatures is the result of luminescent infrared radiance emitted by the diode lasers. An operating spectral range of these projectors is $\lambda = 3-5$ μm, that corresponds to the first atmospheric transparency window.

In (Malyutenko et al., 2001) it was researched an array of IR sources based on a narrow-bandgap semiconductor and operating both on the principle of positive and negative luminescence excitation under conditions of the magnetoconcentration effect. The introduced device is capable of creating both positive and negative radiation contrasts relative to the background emission level.

A disadvantage of the all matrix approaches is the fact that adjacent radiating matrix elements are separated by the spaces, needed for the electrical and thermal isolation. This decreases the brightness of matrix. Furthermore, in the matrix sources there is a task of address control of separate elements.

A whole (not matrix) thermal source with large area (several cm^2) with a possibility of coordinate modulation of its emissivity can be released of these disadvantages. In (Malyutenko et al., 2003; Malyutenko, Bogatyrenko et al., 2006) it was proposed to use a translucent plate of wide-gap semiconductor Ge or Si (a screen) as 2d radiation source, and a coordinate modulation of the emissivity to achieve by the intrinsic photoeffect. This heated screen was locally illuminated by light of the visible or near IR ranges. As a result of increasing of free carriers concentration in the places of illumination the authors obtained a local increase of TR in MWIR and LWIR ranges.

In the present paper it is proposed to use as a screen a heated MOR with nonuniform thickness and to modulate its emissivity by an external magnetic field.

Let us consider a resonator with non-parallel mirrors (wedge MOR). Since the condition for interference maximum of TR contains a thickness, the interference fringes of TR appear on a surface, each of which is characterized by condition $d = const$ at fixed wavelength. They are called the fringes of constant thickness. Let the MOR thickness changes in the x direction by a simple linear law $d(x) = d_0 + ax$. In this case the position of the radiating area (x_{TR}) with a wavelength λ in a magnetic field is determined from (20) and (21) by expression:

$$x_{TR} = \left(\frac{k \pm \varphi / 2\pi}{n} \lambda - d_0 \right) \cdot a^{-1} . \qquad (22)$$

Sign "+" corresponds a secondary line of TR of the "blue" area and "-" corresponds the "red" area.

The spatial distribution of TR is one or more localized on the MOR surface radiative strips that are perpendicular to the axis x. Their position is determined by both the parameters of resonator and by value of Faraday rotation angle (by a magnetic field strength).

Fig.17. shows the calculated distribution of TR intensity on surface of the wedge MOR resonator at the different values of φ for $\lambda = 10$ μm. With an increasing magnetic field the radiating strip shifts smoothly in direction of larger thickness. The thickness of resonator is chosen here in such way that the rest of the field remains dark. Since this emitter is not multielement, but it is whole homogenous structure, the movement of the radiating strip is realized not discretely but smoothly.

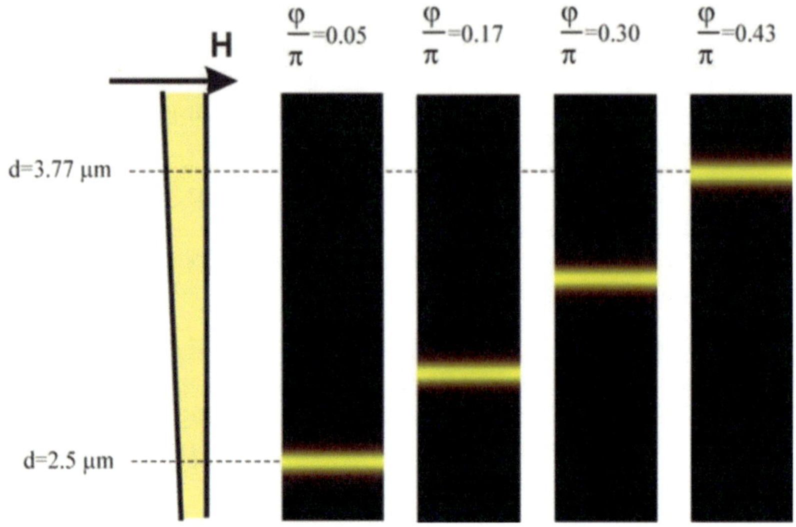

Fig. 17. Calculated fringes of constant thickness of TR of the wedge magneto-optical Fabry-Perot resonator (on the left). $\rho_1 = 0.95$, $\rho_2 = 1$, $n = 3$, $\eta = 0.97$, $\lambda = 10$ μm, $k = 1$.

Fig. 18. MOR with the discrete relief on the radiating surface (on the left) and calculated distribution of TR intensity on its surface at the different values φ. $\rho_1 = 0.95$, $\rho_2 = 1$, $n = 3$, $\lambda = 10$ μm, $k = 1$.

Devices of this type can be used as IR test patterns for calibrating optoelectronic devices with respect to spatial, temperature, and time resolution.

Fig.18 presents the field of TR of MOR with the discrete relief on the radiating surface. Applying the different value of a magnetic field, it is possible to create conditions for the maximum of interference for each of the elements of surface separately, leaving the rest of the surface not radiating. Thus this screen is capable of generating dynamic 2D infrared scenes in real-time by an external magnetic field.

5. Conclusion

In summary, the results of theoretical and experimental investigations of reflection, transmission and thermal radiation of a magneto-optical Fabry-Perot resonator in an external magnetic field are presented. Attention was paid to the investigation of both angular and spectral dependencies of the *T*, *R* and TR in the medium- and long-wave IR ranges for the unpolarized light.

It is established that under the conditions of multibeam interference a magnetic field substantially changes the characteristics of the transmission and reflection of unpolarized light as well as of the own thermal radiation of the resonator. It is shown that the changes also appear for the polychromatic light.

Authors produced a detailed explanation of this effect: the cause of the changes of MOR characteristics is a change of the conditions of the multibeam interference in a magnetic field. A magnetic field redistributes the polarization planes of the light waves inside the resonator. As a result, the interference of the transmitted, reflected and radiated waves can be suppressed or phase shifted by a phase difference of π. The produced in the paper theory is based on the matrix multi-beam summation considering the Faraday rotation effect. The calculation results have a good agreement with experimental data.

In the part "4. Applications" a number of the possible applications of MOR is described. It is shown that the presence of interference in the samples is a favorable factor for investigating the Faraday effect. The determination of the value of the Faraday rotation angle φ by registration the transmission or reflection spectra of unpolarized light or own TR is a convenient way, which has several advantages over the traditional methods. First, it makes it possible to simplify optical scheme and recording system considerably and, thus, to make these studies more available for researchers. Second, since the measurements have relative, but not absolute nature, an error in determining the value of φ is greatly reduced. Third, this way makes possible to determine several parameters of a plane-parallel magneto-optical sample at once. Fourth, analysis of a spectrum of TR makes possible to determine φ in a case, when the classical scheme of the Faraday effect is not applied. For example, when magneto-optical layer is located on the opaque substrate.

Considerable attention is paid to the possibility of creation the controlled IR sources of different purposes with the application of the magneto-optical structures, such as MORs or magnetophotonic crystals. The resonator properties of these objects cause the narrow-band spectrum and the narrow-beam directional diagram of their thermal emission. And the influence of an external magnetic field makes possible to change dynamically the intensity or spectral position of the radiation line, and also to relocate the local radiating regions on the emitter area.

These devices can be applicable in many fields of engineering, science, medicine, criminology, technology etc. The main applications of the IR sources with control of their emission spectrum is the IR spectroscopy and the gas analysis and monitoring of the environment. IR sources with control of spatial field of radiation are very important in order to check the functionality, calibration and testing of their characteristics of the different IR optical-electronic systems. They solve the urgent problem of development of the simulating methods and creation of the imitators of the heated objects, spatial coordinates and intensity of radiation of which can dynamically change: the dynamic IR scene projectors and scene simulating devices.

For realization of these sources it is necessary to realize the strong Faraday rotation in the short optical base. In addition, it is necessary that emitters' temperature exceeds the temperature of background. This imposes the strict requirements on the magneto-optical medium of the resonator structure. However, authors are assured that the contemporary high technologies are able to synthesize the material, which corresponds these requirements.

In conclusion we have to note, that a magneto-optical Fabry-Perot resonator is a simple case of a magnetophotonic crystal. Further theoretical and experimental investigations of the emitting properties of MPCs in a magnetic field will make possible to determine new peculiarity and effects, which also can be used for the creation of modern optical devices, which can work both with the polarized and unpolarized light in the IR spectral range.

6. References

Beasley, D.B.; Cooper, J.B. & Saylor, D.A. (1997) Calibration and nonuniformity correction of MICOM's diode-laser-based infrared scene projector. *Proc. SPIE.* Vol. 3084. pp. 91-101, ISBN: 9780819424990.

Biener, G.; Dahan, N.; Niv, A.; Kleiner, V. & Hasman, E. (2008). Highly coherent thermal emission obtained by plasmonic bandgap structures. *Appl. Phys. Lett.* Vol. 92, No 8, pp. 081913-1 - 081913-3, ISSN: 0003-6951.

Cantey, T. M.; Ballard, G. & Gregory, D. A. (2008) Application of type II W-quantum-well diode lasers for high-dynamic-temperature-range infrared scene projection. *Opt. Engin.* Vol. 47, Issue 8 pp. 086401, ISSN: 0091-3286.

Celanovic, I.; Perreault, D. & Kassakian, J. (2005). Resonant-cavity enhanced thermal emission *Phys. Rev. B,* Vol. 72, pp.075127-1-6, ISSN: 0163-1829.

Das, N. C. (2010). Infrared light emitting device with two color emission. *Solid State Electron.* Vol. 54, Issue 11, pp. 1381-1383, ISSN: 0038-1101.

Das, N. C.; Bradshaw, J.; Towner, F. & Leavitt, R. (2008) Long-wave (10 micron) infrared light emitting diode device performance. *Solid State Electron.*; Vol. 52, Issue 11, pp 1821-1824, ISSN: 0038-1101.

Dokukin, M. E.; Baryshev, A. V.; Khanikaev, A. B. & Inoue, M. (2009). Reverse and enhanced magneto-optics of opal-garnet heterostructures. *Opt. Express,* Vol. 17, No 11, pp. 9062-9070, SSN: 1094-4087.

Drevillon, J.; Joulain, K.; Ben-Abdallah, Ph.; & Nefzaoui, E. (2011). Far field coherent thermal emission from a bilayer structure, *J. Appl. Phys.* Vol. 109, pp. 034315-1 - 034315-7, ISSN: 0003-6951.

Fujikawa, R.; Baryshev, A.V.; Khanikaev, A.B.; Kim, J.; Uchida, H.; Inoue, M. & Mater. J. (2009). Enhancement of Faraday rotation in 3D/Bi:YIG/1D photonic heterostructures. *J. Mater. Sci. Mater. Electron.* Vol. 20, No. 1, pp. 493-497 , ISSN: 0022-2461.

Goto, T.; Baryshev, A. V.; Inoue, M.; Dorofeenko, A. V.; Merzlikin, A. M.; Vinogradov, A. P.; Lisyansky, A. A. & Granovsky, A. B. (2009a). Tailoring surfaces of one-dimensional magnetophotonic crystals: Optical Tamm state and Faraday rotation. *Phys. Rev. B,* Vol. 79, Issue 12, pp. 125103 - 125103-5, ISSN 1098-0121.

Goto, T.; Baryshev, A. V.; Inoue, M.; Dorofeenko, A. V.; Merzlikin, A. M.; Vinogradov, A. P.; Lisyansky, A. A. & Granovsky, A. B. (2009b). One-way electromagnetic Tamm states in magnetophotonic structures. *Appl. Phys. Lett.* Vol. 95, Issue 1, pp. 011101 011101-3, ISSN 0003-6951.

Greffet, J.J.; Laroche, M. & Marquier, F. (2007). *Microstructured Radiators Final Report.* Ecole Centrale, ASIN: B00383OLFQ, Paris.

Guga, K.Yu.; Kollyukh, O.G.; Liptuga, A. I.; Morozhenko, V. & Pipa, V.I. (2004) Features of thermal radiation of plane-parallel semiconductor wafers. *Semiconductors*, Vol. 38, No 5, pp. 507-511, ISSN: 1063-7826.

Inoue, M. & Fujii, T. (1997). A theoretical analysis of magneto-optical Faraday effect of YIG films with random multilayer structures. *J.Appl. Phys.*, vol.81, No. 8, pp. 5659–5661, ISSN: 0021-8979.

Jacob D., Vallet M., Bretenaker F., Le Floch A., & Le Naur R. (1995). Small Faraday rotation measurement with a Fabry–Perot cavity. *Appl. Phys. Lett.* Vol. 66, Issue 26, pp. 3546-3548, ISSN: 0003-6951.

Jérémie, D. & Ben-Abdallah, P.; (2007) Ab initio design of coherent thermal sources. *J. Appl. Phys.* Vol. 02, Issue: 11, pp.114305-12, ISSN: 0021-8979.

Kollyukh, O.G.; Liptuga, A.I.; Morozhenko, V. & Pipa, V.I. (2005). Magnetic-field modulation of the spectrum of coherent thermal radiation of semiconductor layers. *Phys. Rev.B.* Vol. 71, Issue 7, pp. 073306 - 073306-4, ISSN 1098-0121.

Kollyukh, O.G.; Liptuga, A. I.; Morozhenko, V. & Pipa, V.I. (2003) Thermal radiation of plane-parallel semitransparent layers. *Opt. Commun.* Vol. 225, pp.349-352, ISSN: 0030-4018.

Lancaster, P. & Tismenetsky, M. (1985) *The theory of matrices* (2nd ed.) Academic Press, ISBN: 0-12-435560-9, Orlando, FL.

Laroche, M.; Carminati, R. & Greffet, J.-J. (2006) Coherent Thermal Antenna Using a Photonic Crystal Slab. *Phys.Rev. Lett.*, Vol. 96, No 12, pp. 123903-1 - 123903-3, ISSN: 0031-9007.

Lee, B. J. & Zhang, Z. M. (2007). Coherent Thermal Emission From Modified Periodic Multilayer Structures, *J. of Heat Transfer*, Vol. 129, No 1, pp. 14-26, ISSN: 0022-1481.

Lee, B. J.; Wang, L. P. & Zhang, Z. M. (2008). Coherent thermal emission by excitation of magnetic polaritons between periodic strips and a metallic film, *Opt. Express*, Vol. 16, Issue 15, pp. 11328-11336, ISSN: 1094-4087.

Ling, H.Y. (1994). Theoretical investigation of transmission through a Faraday-active Fabry-Perot etalon. *J. Opt. Soc. Am. A*, Vol.11, No 2, pp. 754-758, ISSN 1084-7529.

Madelung, O. (2004). *Semiconductors: Data Handbook* (3rd ed.), Springer Verlag, ISBN: 978-3-540-40488-0, Berlin.

Malyutenko, V. K.; Bogatyrenko, V. V.; Malyutenko, O. Yu. & Chyrchyk, S. V. (2006). Cold background infrared scene simulation device. *Proc. SPIE.* Vol. 6208, pp. 240-248, ISBN: 9780819462640.

Malyutenko, V.K.; Bolgov, S. S. & Malyutenko, O.Yu. (2001). Multielement IR Sources with Alternating Contrast. *Technical Phys. Lett.*, Vol. 27, No. 8, pp. 644–646, ISSN: 1063-7850.

Malyutenko, V. K.; Bolgov, S. S. & Malyutenko, O. Yu. (2006). Above-room-temperature 3-12 μ m Si emitting arrays. *Appl. Phys. Let.*, vol. 88, pp. 211113-1 - 211113-3, ISSN 0003-6951.

Malyutenko, V.K.; Michailovskaya, K.V.; Malyutenko, O.Yu.; Bogatyrenko, V.V. & Snyder, D.R. (2003) Infrared dynamic scene simulating device based on light down-conversion. *IEE Proc. Optoelectron.*, Vol. 150, No. 4, pp. 391 - 394, ISSN: 1350-2433.

Morozhenko, V. and Kollyukh, O.G. (2009). Angular and spectral peculiarities of coherent thermal radiation of the magneto-optical Fabry-Perot resonator in magnetic field. *J. Opt. A*, Vol. 11, No. 8, pp. 085503-1 - 085503-6, ISSN 1464-4258.

Piller, H. (1966). Effect of internal reflection on optical Faraday rotation. *J. Appl. Phys.*, Vol. 27, No 2, pp. 763-768. ISSN 0003-6951

Pritchard, A.P., Lake, S.P., Balmond, M.D., Gough, D.W., Venables, M.A., Sturland, I.M., Crisp, G. & Watkin, S.C. (1997). Current status of the British Aerospace resistor array IR scene projector technology, *Proc. SPIE,* Vol. 3084, pp. 71-77, ISBN: 9780819424990.

Rheinhlander, B.; Neumann, H. & Gropp M. (1975). Electron effective masses in direct-gap $Al_xGe_{1-x}As$ epitaxial layers from Faraday rotation measurements. *Exp. Techn. Phys.*, Vol. 23, No 1, pp. 33-39, ISSN: 0948-2148

Robinson, R.; Oleson, J.; Rubin, L. & McHugh, S. (2000). MIRAGE: System overview and status. *Proc. SPIE.* Vol. 4027, pp. 387-398, ISBN: 9780819436535.

Rosenberg, R.; Rubinstein, C.B. & Herriott, D.R. (1964). Resonant optical Faraday rotation. *Appl. Optics*, Vol. 3, No 9, pp.1079-1083. ISSN: 0003-6935.

Srivastava, G. P.; Mathur, P. C.; Kataria, N. D. & Shyam, R. (1975). High-frequency effective mass of charge carriers in CdS. *J. Phys. D: Appl. Phys.* Vol. 8, No 5, pp. 523-529.

Stramska, H.; Bachan, Z.; Bysiemki, P. & Kołsodziejczak, J. (1968). Interband Faraday rotation and ellipticity observed at the absorption edge in silicon. *Phys. Stat. Sol. (b).* Vol. 27, No 1, pp. k25-k28, ISSN: 0370-1972.

Voigt, W. (1908) *Magneto- und Electro-Optic.* Teubner, Leipzig.

Vorobev, L.E.; Komissarov, V. S. & Stafeev, V. I. (1972). Faraday and Kerr effects of hot electrons in n-type InSb in the infrared (II). *Phys. Stat. Sol. (b),* Vol. 52, No 1, p. 25-37. ISSN: 0370-1972.

Wallenhorst, M.; Niemijller, M.; Dotsch, H.; Hertel, P; Gerhardt, R. & Gather, B. (1995) Enhancement of the nonreciprocal magneto-optic effect of TM modes using iron garnet double layers with opposite Faraday rotation. *J. Appt. Phys.* Vol. 77, No 7, pp. 2902-2905 ISSN: 0021-8979.

Williams, O. M. (1998). Dynamic infrared scene projection: a review. *Infrared Phys. Technol.* Vol. 39, Issue 7, pp. 473-486, ISSN: 1350-4495.

Yang, R.Q.; Lin, C-H; Murry, S.J.; Pei, S.S; Liu, H.C; Buchanan, M. & Dupont, E. (1997) Interband cascade light emitting diodes in the 5–8 μ m spectrum region. *Appl Phys Lett*; Vol. 70, Issue 15, pp. 2013-2015, ISSN: 0003-6951.

Yariv, A. & Yen, P. (1984). *Optical waves in crystals.* John Wiley & Sons, ISBN 0471091421, New York.

Zhu, R; Fu, S. & Peng, H. (2011). Far infrared Faraday rotation effect in one-dimensional microcavity type magnetic photonic crystals. *J. Magn. Magn. Mater.* Vol. 323, Issue 1, pp. 145-149. ISSN: 0304-8853.

5

Infrared Radiation Photodetectors

Ashraf S. A. Nasr[1,2]
[1]*Radiation Engineering Dept., NCRRT, Atomic Energy Authority, Cairo,*
[2]*College of Computer Science, Qassim University, Buryadah,*
[1]*Egypt*
[2]*Kingdom of Saudi Arabia*

1. Introduction

Nanotechnology is regarded world-wide as one of the key technologies of the *21st* century. Nanotechnological products and processes hold an enormous economic potential for the markets of the future. The production of ever smaller, faster and more efficient products with acceptable price-to-performance ratio has become for many industrial branches an increasingly important success factor in the international competition. The technological competence in nanotechnology will be a compelling condition to compete successfully with better procedures and products on high technology markets in the future. Due to its interdisciplinary cross-section character, nanotechnology will affect broad application fields within the ranges of chemistry/materials, medicine/life sciences, electronics/ information technology, environmental and energy engineering, automotive manufacturing as well as optics/analytics and precision engineering in various ways (Aboshosha et al., 2009a; Aboshosha et al., 2009b; Abou El-Fadl et al., 1998; Eladle et al., 2009; Nasr, El_mashade, 2006; El_mashade et al., 2003; El_mashade et al., 2005; Future Technologies Division, 2003; Nasr, 2011; Nasr, 2011b; Nasr, 2009; Nasr et al., 2009; Nasr, 2008; Nasr,2006; Nasr & Ashry; 2000).

One of the considered devices which had witness the development is IR photodetectors. A photodetector is an electronic device that converts incident photons into electric current. Photodetectors have a wide range of scientific and consumer applications and have played a central role in the development of modern physics during the last century (Ryzhii, 2003).

Photodetectors are classified according to their optical range which can be covered. As we can see in Fig. 1, the electromagnetic spectrum in the optical start regarding wavelength from UV, visible, and IR ranges (Buckner, 2008). But it is evident to notice that the visible range is small. So, it will be impeded some time into UV or/ and IR detectors. The main two types then are UV and IR photodetectors. The two types of semiconductor-based photodetectors will be reviewed here. The main problem is appearing at long wavelength optical range as the required energy gap should be small. As a result of the previous note, the required band gap materials which cover this region have very high cost. And, it is faces a complexity into the fabrication processes. The trend of the world was directed to nanotechnology (quantum) photodetectors to overcome the previous drawbacks. As a result, the review is directed to the quantum IR photodetectors (QIRPs) like quantum well, dot, and wire infrared photodetectors (QWIPs, QDIPs and QRIPs) respectively. They open the way to overcome main problems in commercial utilized IR photodetectors such as

HgCdTe photodetectors. Although QIRPs solve many problems but they suffer from other disadvantages that are illustrated into the following subsections.

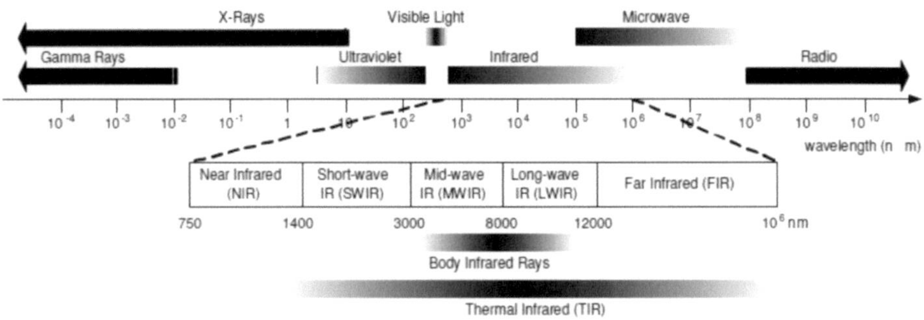

Fig. 1. The electromagnetic spectrum and the IR region (Buckner, 2008).

2. History of nanoscience & nanotechnology

Ever since man started taking control of the environment and shaping things to meet human needs, he has endeavored to understand matter at its fundamental level (Nai-Chang, 2008). Since the dawn of 21stcentury, it has become possible to study, design, and synthesize structures with the precision of one billionth of a meter (nanometer). Nanoscience is the study of the fundamentals principle of matter at scale of ~1-100 nm, while nanotechnology is the applications of such knowledge to making materials and devices (National Nanotechnology Initiative Workshop, 2004).

In December 1959, Feynman gave a visionary speech at Caltech, entitled "There is Plenty of Room at the Bottom" (URL http://www.its.caltech.edu/~feynman/). This speech was unique and held a defining place in the field now known as nanoscience and nanotechnology. In approximately 7,000 words, Feynman projected a vision that is only beginning to be realized today: *"What I want to talk about is the problem of manipulating and controlling things at a small scale… What I have demonstrated is that there is room – that you can decrease the size of things in a practical way. I now want to show that there is plenty of room. I will not now discuss how we are going to do it, but only what is possible in principle – in other words, what is possible according to the laws of physics."* Since the historical speech, there has been significant progress to date in this highly interdisciplinary research area now known as nanoscience and nanotechnology. Not only that Feynman's vision of manipulating and controlling things down to nano- and even atomic scale becomes realizable, but novel nano-scale devices and characterization tools have also revolutionized means to exploring new science and developing new technology in a wide variety of research fields. While steady progress towards miniaturizing physical structures and tools has been made since Feynman's visionary speech and collective efforts have reached a new pinnacle since former President Bill Clinton gave his special scientific and technological strategic speech at Caltech in 1999, the strong interdisciplinary nature of nanoscience and nanotechnology is still being constant redefined and is full of both excitement and uncertainties.

3. Photodetectors fabrication techniques

It is evident to explain the advanced fabrications processes before exposing to types of photodetectors as it will help us to recognize the development which has happened in the last decades.

3.1 General nano-fabrication classifications

In general, the nano-fabrication techniques may be categorized into two types: the "top-down" and "bottom-up" approaches. The top-down approach is primarily based on lithographic techniques, including the traditional optical and electron-beam lithography for processing inorganic materials such as semiconductors, metals and dielectric/ferroelectric materials. The bottom-up approach includes the nano-imprint lithography for cross-bar architecture (Nai-Chang, 2008; Jung et al., 2006), chemical lithography for self-assembled circuits, stamping techniques for processing organic materials, multilayer soft lithography (MSL) for handling fluidic samples, and scanning probe-based lithography (SPL) that offers novel means to manipulate matter down to molecular and even atomic scale. The nano-imprint lithography may involve the use of e-beam lithography in the processing of the mold for pattern transfer (Jung et al., 2006). Recently nano-scale pitched layers of GaAs/GaAlAs superlattices have been employed for the mold, which enable efficient fabrication of high-density metallic and semiconducting nanowires and crossbar arrays (Jung et al., 2006). Here, this chapter is concentrated into the top down approach for fabrication a semiconductor-based photodetectors.

3.2 Epitaxy method

The word epitaxy refers to the ordered growth of a crystalline material on top of a crystalline substrate, such that the crystal lattice of the new layer is in registration with that of the substrate, see Fig. 2 (Muhammad et al., 2007). Epitaxial semiconductor layers are used in most semiconductor device fabrication processes because of the ability to accurately control crystal composition and doping and to form atomically abrupt hetero-interfaces. The techniques for epitaxial growth have typically been divided into processes that deliver the molecules for crystal growth in the liquid phase (LPE), deliver in the gas phase, or processes that deliver the molecules in a vacuum chamber in molecular beams. Depending on the technique, the atoms for crystal growth arrive as elemental species or within precursor molecules that thermally or chemically decompose on the surface to leave the desired species for epitaxy (Donkor, 2004).

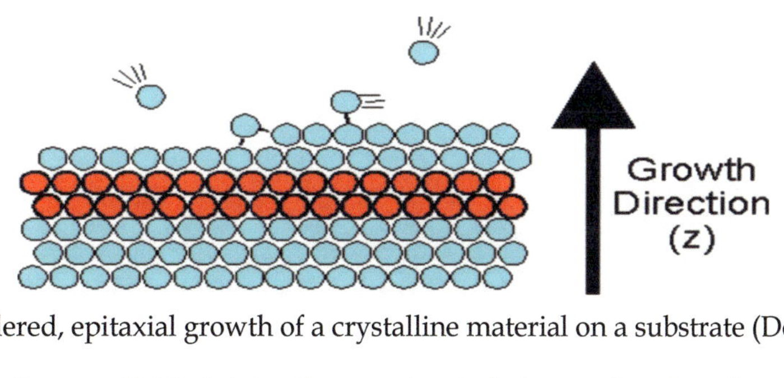

Fig. 2. The ordered, epitaxial growth of a crystalline material on a substrate (Donkor, 2004).

Epitaxial growth was divided into three major techniques. Frank and van der Merwe (Donkor, 2004) used elasticity theory to derive the concept of a critical misfit below which monolayer-by-monolayer growth appears. Volmer and Weber (Donkor, 2004), applying nucleation theory, assumed that crystalline films grew from 3D nuclei on the substrate and that their relative number and growth rate were determined by interfacial and surface free energies. The third model by Stranski and Krastanov (Brune, 2001) was based on atomistic calculations and assumed that initially a few pseudomorphic 2D layers are formed, on top of which 3D crystal with their natural lattice constant will grow. Each of the three scenarios is observed, they rise to the following labeling of the three growth modes of epitaxy:

i. Frank-van der Merwe (FW) growth mode (2D morphology, layer-by-layer or step- flow growth).
ii. Volmer-Weber (VW) growth mode (3D morphology, island growth).
iii. Stranski-Krastanov (SK) growth mode (initially 2D, after critical thickness, 3D morphology, layer-plus-island growth).

Various technique are used for growing epitaxial layer using these three modes, but the new crystal growth techniques been used are Molecular Beam Epitaxy, MBE (Liang et al., 2006; Liu, 1993; Krost et al., 1999; Coleman, 1997) and Metal Organic Chemical Vapor Deposition, MOCVD. MOCVD is also referred to, and used interchangeably with, MOVPE/OMVPE (Metallorganic /Organometallic Vapor Phase Epitaxy). MOCVD is a broader term that is applicable to the deposition of crystal, polycrystalline and amorphous materials. Both MBE and MOCVD have produce a wide range of very high-purity semiconductor materials with excellent optical and electrical properties.

3.3 Molecular Beam Epitaxy (MBE)

MBE is an Ultra-High-Vacuum (UHV)-based technique for producing high quality epitaxial structures with monolayer (ML) control. Since its introduction in the 1970s as a tool for growing high-purity semiconductor films, MBE has evolved into one of the most widely used techniques for producing epitaxial layers of metals, insulators and superconductors as well, both at the research and the industrial production level (Muhammad et al., 2007). The principle underlying MBE growth is relatively simple: it consists essentially of atoms or clusters of atoms, which are produced by heating up a solid source. They then migrate in an UHV environment and impinge on a hot substrate surface, where they can diffuse and eventually incorporate into the growing film as in Fig. 3.

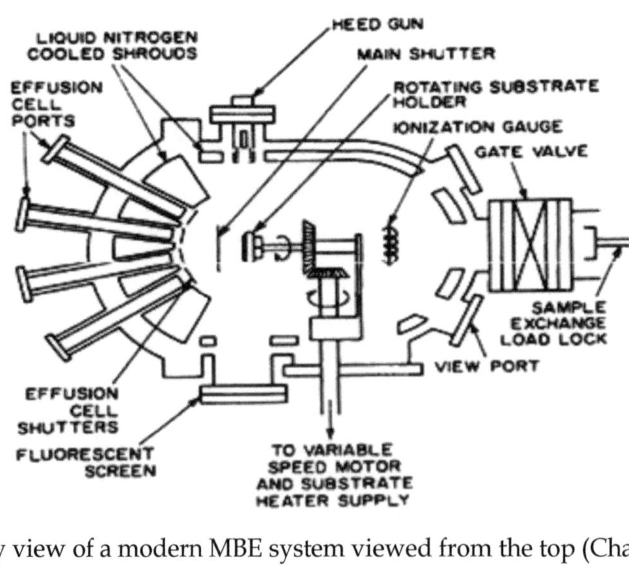

Fig. 3. Cut-away view of a modern MBE system viewed from the top (Chang & Kai, 1994).

The growth of uniform epitaxial films from multiple effusion cells requires special effusion cell geometry and continuous rotation of the substrate around an axis normal to the substrate surface. The substrate holder can feature rotation speeds up to 125 rpm. The control unit remotely orients the sample holder into any of hour positions: growth, transfer, electron beam (EB), auxiliary. The controller also allows remote continual adjustment of rotation speed. The modern commercial MBE growth chamber (Fig. 4) is often equipped with a rotary substrate manipulator capable of turning the substrate azimuthally during growth. With this feature the substrate can be heated more uniformly, resulting in epilayers of very good thickness and doping uniformity (Chang & Kai, 1994; Lee, 2005).

Fig. 4. MBE System at MIT, Cambridge (Lee, 2005).

3.4 Metal Organic Chemical Vapor Deposition (MOCVD)

The growth process in MOCVD is similar to MBE, but the atoms are carried in gaseous form to the substrate as shown in Fig. 5. For example, to grow epitaxial GaAs, the gasses can be trim ethyl gallium and arsine. These precursor molecules thermally decompose on the hot substrate, the resulting organic molecules evaporate, leaving the Ga and As atoms for incorporation into the crystal. Typical growth temperatures are 600°C, and the growth rates are on-the-order-of 1 molecular layer of GaAs per second. This growth rate allows the precursor gasses to be switched quickly enough to give very abrupt material interfaces. The growth process given by $(CH_3)_3Ga + AsH_3 \rightarrow GaAs + 3CH_4$

The most important organometallic compounds that have been studied are trim ethyl gallium (TMGa), trim ethyl aluminum (TMAl) and trim ethyl indium (TMIn). These sources should be easily synthesized and easily purified related to practicalities of using several different sources together (Coleman, 1997). In general, the organometallic constituents are transported to a heated substrate by passing a carrier gas, usually hydrogen or nitrogen or a mixture of the two, over or through the compound contained in a constant-temperature bubbler vessel. Most MOCVD growth of III-V compound semiconductors and alloys involves the use of hydrides, such as arsine or phosphine, for the column V species because they are already gaseous and supplied from simple cylinder-based delivery systems.

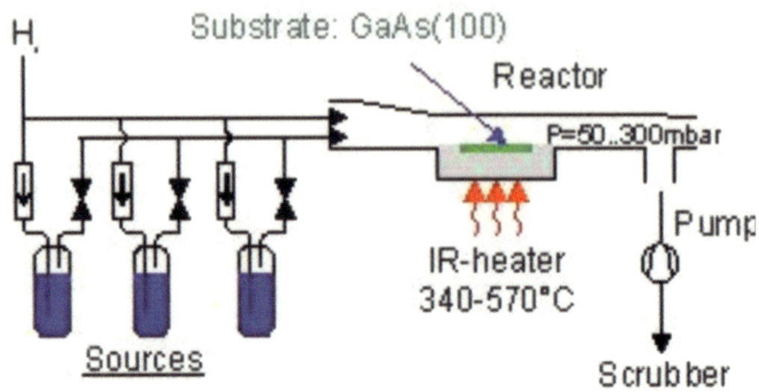

Fig. 5. Schematic view of a MOCVD reactor (Coleman, 1997).

MOCVD reactors consist of three major components: the reactor gas delivery system, the reaction chamber and the reactor safety infrastructure. The reactor delivery system or gas panel is a very clean, leak-free network of stainless-steel tubing, automatic valves and electronic mass flow controllers as shown in Fig. 6. Hydride delivery modules generally require a few valves and an electronic mass flow controller, since these sources are already provided as dilute, high pressure gases in gas cylinders. An important part of the main gas panel is the supply of carrier gases within a vent-run configuration.

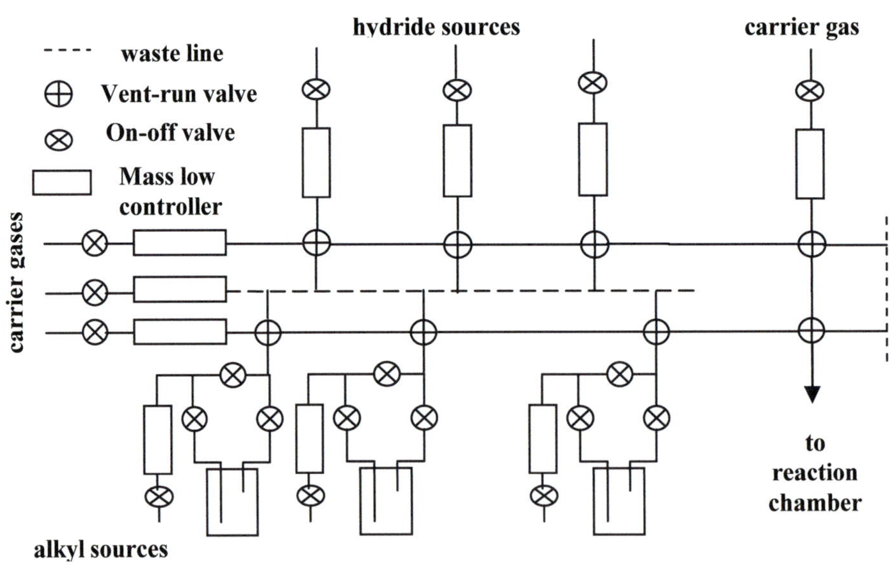

Fig. 6. Schematic diagram of MOCVD reactor delivery system gas panel, illustrating hydride delivery modules, alkyl delivery modules and the vent-run configuration (Coleman, 1997).

The reaction chamber is the vessel in which the source gases are mixed, introduced into a heated zone where an appropriate substrate is located, and the basic pyrolysis reactions take place. There are two basic reaction-chamber geometries (Manasevit, 1968) commonly used for the MOCVD growth of optoelectronic materials, see Fig. 7.

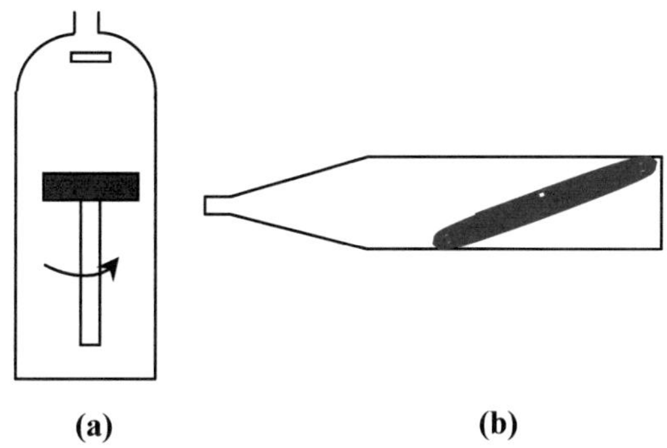

Fig. 7. Two common chamber geometry designs for MOCVD. (a) vertical (b) horizontal (Manasevit, 1968).

Both designs are cold-wall systems that reflect the basic pyrolysis nature of the process and make use of an indirectly heated (radio-frequency induction heated or infrared radiant heated) silicon carbide-coated graphited susceptor. The chamber itself can be quartz, stainless steel or quartz-lined stainless steel.

Safety is of paramount importance in the design and operation of MOCVD growth apparatus whether the sources used are hydride or not (Johnson, 1984). Hydrides pose the biggest risk because they are high-pressure toxic gases. The alkyls pose the next highest risk because, although they are toxic and pyrophoric, they are liquids and generally easier to handle. Ancillary risks include quartz reaction chambers (which are breakable), large volumes of explosive hydrogen gas, high temperatures, and the acids and solvents used for preparing for, and cleaning up after, a growth run. Handling these risks, falls into three categories applies equally well to all epitaxial growth process indeed to all semiconductor processing activities. The first is limited access, requires a higher level of authority and a correspondingly higher level of training. The second is training which required emergency-response situations such as hazardous-materials response and the third category is a hardware safety infrastructure. Automatic shutdown of source gases and a switch to inert purge gases should take place in the event of power failure when inadequate backup power is available.

The hydrodynamics of the reactor geometry play a key role in the nature of the process for MOCVD growth. In practical reactors, a large mismatch often exists between the inlet tube, which typically is a standard tube size of less than 10mm diameter and the characteristic dimensions of the reaction chamber. Thus, even in a relatively simple horizontal reaction chamber, the gas may have to travel well into the chamber before a simple parabolic profile stabilizes.

A heated zone is necessary to drive the pyrolysis reaction and provide for the desired materials deposition. The temperature gradient between the susceptor and the chamber ambient can be very large, often several hundreds of degrees Celsius. For example, the optimum MOCVD growth temperature for many III-V compounds and alloys falls in the range of 600-800 °C. Air or water cooling is often used to maintain the chamber walls at temperature close to room temperature.

The atomic force microscope (AFM) results of QW and QD InGaAs structure clearly show that the MOCVD reactor works so good, as shown in Fig. 8. Further more studies and more parameters should be taken for better characterization of the highly strained epilayer which is useful for quantum devices applications (Muhammad et al., 2007).

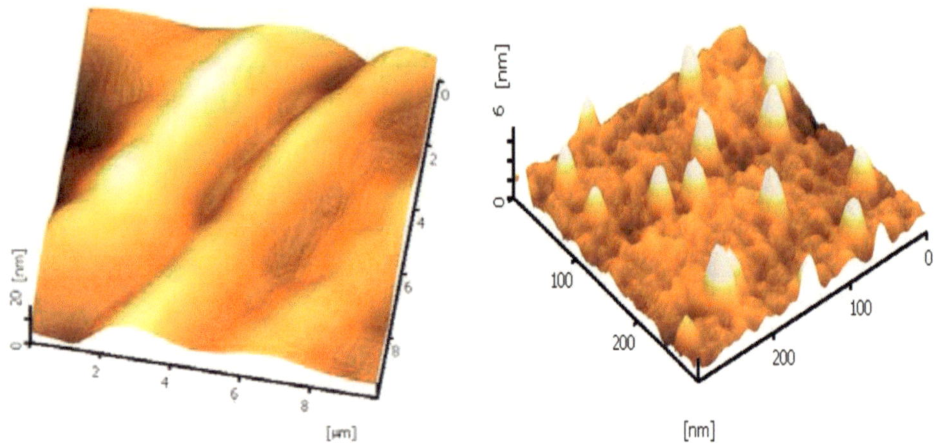

Fig. 8. AFM results of (a) Quantum well and (b) Quantum dot InGaAs on GaAs wafer (Muhammad et al., 2007).

By applying preceding conventional epitaxial crystal growth techniques, it is possible to gain precise orientation control during nanowire growth. The technique, vapor-liquid-solid epitaxy (VLSE), is particularly powerful in the controlled synthesis of high-quality nanowire arrays and single-wire devices. For example, ZnO prefers to grow along the [001] direction and readily forms highly oriented arrays when epitaxially grown on an a-plane sapphire substrate as shown in Fig. 9 (Pauzauskie & Yang; 2006).

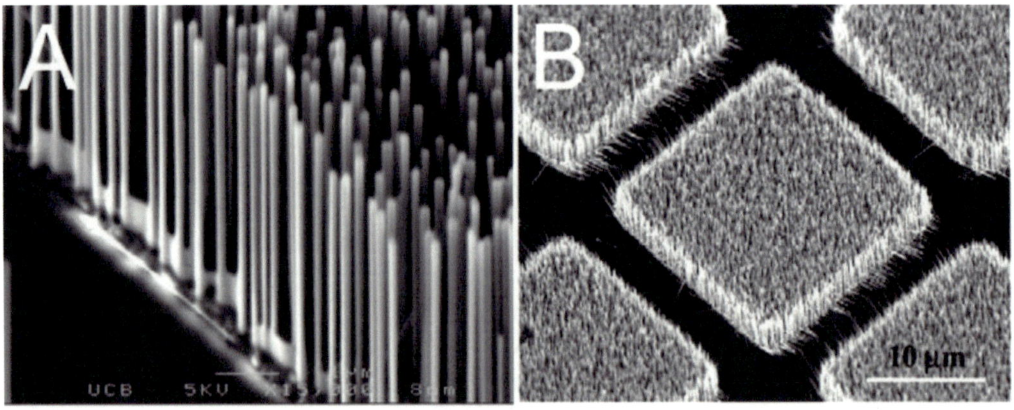

Fig. 9. Scanning Electron Microscope (SEM) images of (a) a [001] ZnO nanowire array on an a-plane sapphire wafer, and (b) [110] GaN nanowire arrays on (100) plane γ-LiAlO2 (Pauzauskie & Yang; 2006).

4. General principles of electroluminescent devices

To understand the physical meaning of the detection process, we must mention how light emitting and how we choose the layer structure of any device (Detector or source of light). Fig. 10 shows the layer structure and circuit diagram for a typical electroluminescent device. The device consists of several epitaxial layers grown on top of a thick crystal substrate. The epitaxial layers consist of a *p-n* diode with a thin active region at the junction. The diode is operated in forward bias with a current flowing from the *p-layer* through to the *n-layer* underneath. The luminescence is generated in the active region by the recombination of electrons that flow in from the *n-type* layer with holes that flow in from the *p-type* side (El_Mashade et al., 2003).

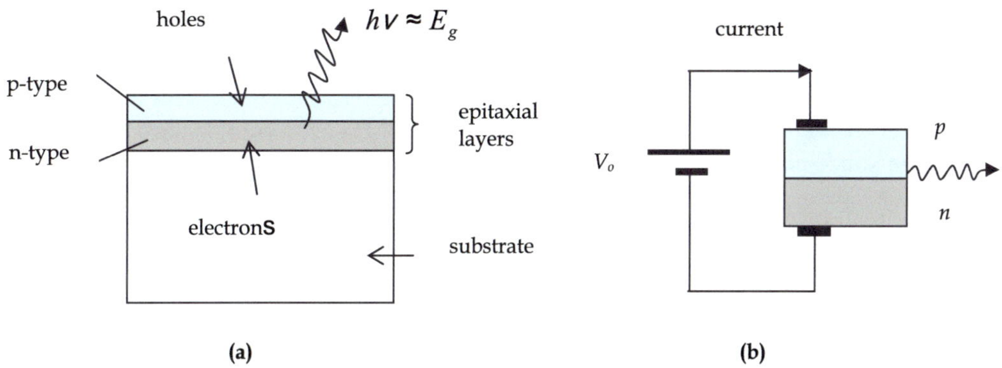

Fig. 10. (a) Layer structure and (b) circuit diagram for a typical electroluminescent device. The thin active region at the junction of the p-type and n-type is not shown, and the dimensions are not drawn to scale. The thickness of epitaxial layers will be only ~1µm, whereas the substrate might be ~500µm thick. The lateral dimensions of the device might be several millimetres (El_Mashade et al., 2003).

Any direct gap semiconductor can, in principle, be used for the active region, but in practice only a few materials are commonly employed. The main factors that determine the choice of the material are:

- The size of the band gap;
- Constraints related to lattice matching;
- The ease of p-type doping.

The first point is obvious: the band gap determines the emission wavelength. The second and third points are practical ones relating by which the way the devices are made. These are discussed further below.

These techniques are crucial to the successful growth of the high quality QIRPs structures described in the last section. The crystal growth conditions constrain the epitaxial layers to form with the same unit cell size as the substrate crystal. This means that the epitaxial layers will be highly strained unless they have the same lattice constant as the substrate, which meaning; we have 'lattice matching' between the epitaxial layers and the substrate. If this condition is not satisfied, crystal dislocations are likely to form in the epitaxial layers, which would severely degrade the optical quality. Fig. 11 plots the band gap of a number of III-V materials used in electroluminescent devices against their lattice constant. The lattice constants of the commonly used substrate crystal are indicated at the top of the figure. The materials separate into two groups. On the right we have the arsenic and phosphorous compounds which crystallize with the cubic zinc blende structure, while on the left we have the nitride compounds which have the hexagonal wurtzite structure (El_Mashade et al., 2003).

x	y	E_g(eV)	λ_g(μm)
0	0	1.35	0.92
0.27	0.58	0.95	1.30
0.40	0.85	0.80	1.55
0.47	1	0.75	1.65

Table 1. Band gap energy E_g (eV) and emission wavelength λ_g(μm) for several compositions of the direct band gap quaternary III-V alloy $Ga_xIn_{1-x}As_yP_{1-y}$. The composition indicated all satisfy the lattice-matching condition for InP substrates, namely x~0.74y.

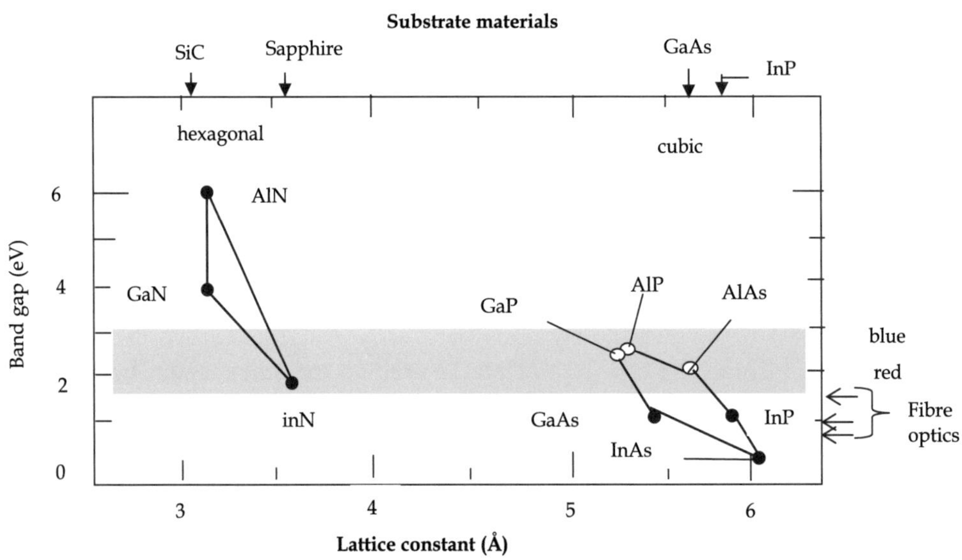

Fig. 11. Band gap of selected III-V semiconductors as a function of their lattice constant. The materials included in the diagram are the ones commonly used. The lattice constants of readily available substrate crystals are indicated along the top axis. The nitride materials on the left grow with the hexagonal wurtzite structure, whereas the phosphides and arsenides on the right have the cubic zinc blende structure (El_Mashade et al., 2003).

For many years, the optoelectronic industry has been mainly based on GaAs. GaAs emits in the infrared at 870 nm, and by mixing it with AlAs to form the alloy $Al_xGa_{1-x}As$, light emitters for the range 630-870 nm can be produced. Lattice-matched AlGaAs can easily be grown on GaAs substrates because of the convenient coincidence that the lattice constants of

GaAs and AlAs are almost identical. AlGaAs emitters are widely used in local area fiber optic networks operating around 850nm, and also for red LEDs. AlGaAs is an example of a 'ternary' alloy which contains three elements. 'quaternary' alloys such as $(Al_yGa_{1-y})_xIn_{1-x}P$ can also be formed. All of these arsenic and phosphorous alloys suffer from the problem that they become indirect as the band gap gets larger. This limits their usefulness to the red and near-infrared spectral range. Applications in the fiber optics industry require light emitting devices that operate around *1.3* µm and *1.55* µm. These are the wavelengths where silica fibers have the lowest dispersion and loss, respectively. Emitters for these wavelengths tend to be made from the quaternary alloy $Ga_xIn_{1-x}As_yP_{1-y}$. Lattice matching to InP substrates can be achieved if $x{\sim}0.74y$. This allows a whole range of direct gap compounds to be made with emission wavelengths varying for 0.92 µm to 1.65 µm (Table 1) (El_Mashade et al., 2003).

P-type doping is difficult in wide band gap semiconductor because they have very deep acceptor levels. The high value of effective mass of hole, m_h^* and the relatively small value of relative dielectric constant, $\varepsilon_{r,}$ increases the acceptor energies, and hence reduces the number of holes which are thermally excited into the valence band at room temperature. The low hole density gives the layer a high resistivity, which causes ohmic heating when the current flows and hence device failure. Nakamura's breakthrough came after discovering new techniques to activate the holes in p-type GaN by annealing the layers in nitrogen at 700C°. In the last sections we will describe how the use of quantum well layers has led to further developments in the field of elecctroluminescent materials (Sze, 1981; Fox, 2001).

5. Photodetectors performance parameters

The fundamental mechanism behind the photodetcetion process is optical absorption. Consider a semiconductor slab, shown schematically in Fig. 12. If the energy $h\nu$ of incident photons exceeds the bandgap energy, an electron-hole pair is generated each time a photon is absorbed by the semiconductor. Under the influence of an electric field set up by an applied voltage, electrons and holes are swept across the semiconductor, resulting in a flow of electric current (Levine, 1993). The photocurrent I_p is directly proportional to the incident optical power P_{in}, that is,

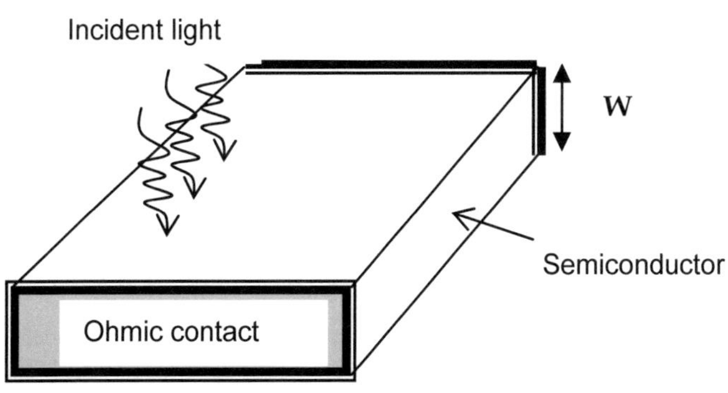

Fig. 12. A photoconductive detector

$$I_p = RP_{in} \tag{1}$$

where R is the responsivity of the photodetector (in unit of A/W). The responsivity R can be expressed in terms of a fundamental quantity η, called the quantum efficiency and defined as

$$\eta = \frac{electron - generation \quad rate}{photo - incidence \quad rate} = \frac{I_p / q}{P_{in} / h\upsilon} = \frac{h\upsilon}{q} R \tag{2}$$

where, $\hbar$ is the Plank's constant, q is the electron charge, υ is the light frequency.

substituting the value of I_p from Eq. (1) into Eq. (2)

i. The **Responsivity** R is given by:

$$R = \frac{\eta q}{h\upsilon} = \frac{\eta \lambda}{1.24} \tag{3}$$

where $\lambda = c / \upsilon$, c is the speed of light, and λ is expressed in micrometers. This relation shows that the responsivity of a photodetectors increases with the wavelength λ simply because the same current can be generated with photons of reduced energy. Such linear dependence on λ is not expected to continue forever, since eventually the photon energy becomes too small to generate electrons. In semiconductors this happens when the photon energy becomes less than the bandgap energy. The quantum efficiency η then drops to zero.

The dependence of η on λ enters through the absorption coefficient a. If the facets of the semiconductor slab in Fig. 12 are assumed to have an antireflection coating, the power transmitted through the slab of width W is $P_{tr} = \exp(-\alpha W) P_{in}$. The absorbed power is thus given by:

$$P_{abs} = P_{in} - P_{tr} = [1 - \exp(-\alpha W)] P_{in} \tag{4}$$

Since each absorbed photon creates an electron-hole pair,

ii. **The quantum efficiency** η is given by:-

$$\eta = P_{abs} / P_{in} = 1 - \exp(-\alpha W) \tag{5}$$

As expected, η becomes zero when $a=0$. On the other hand, η approaches 1 if $a\,W>>1$.

Photodetectors can be broadly classified into two categories: photoconductive and photovoltaic. A homogeneous semiconductor slab with ohmic contacts (see Fig. 12) acts as a simple kind of photoconductive detector. Little current flows when no light is incident because of low conductivity of semiconductor. Incident light increases conductivity through electron-hole generation, and allows current to flow in proportion to the optical power. Photovoltaic detectors operate through a built in electric field that opposes current flow in the absence of light. Electron-hole pairs generated through light absorption are swept across the device by the built-in electric field, resulting in a flow of electric current. Reverse biased p-n junctions fall in the category of photovoltaic detectors and are commonly used for light wave systems because of their high sensitivity and fast response.

Since infrared absorption due to intersubband transitions was first observed in multiple-quantum-well (MQW) structures, QWIPs and arrays based on this principle have become a competitive infrared technology. The figure of merit used to evaluate the performance of most QWIPs is the specific *detectivity (D*)*, which is a measure of the signal-to-noise ratio (SNR). In calculating the detectivity at a certain electrical frequency, the output noise current at this frequency must be known.

iii. The estimated dark current limited peak detectivity values, is defined as (Shen, 2000; Liu, 1992):

$$D^* = \frac{R\sqrt{A\Delta f}}{i_n} \quad cmHz^{1/2}/w \tag{6}$$

Where R is the responsivity, A is the area detector, i_n is the noise current can be estimated by $i_n^2 = 4eI_d g\Delta f$ Substituting in Eq. (6), the detectivity can be defined as:

$$D^* = \frac{R\sqrt{A\Delta f}}{\sqrt{4eI_d g}} \tag{7}$$

Theoretically, the dark current limited detectivity is proportional to $E_F \exp(-E_F / 2k_B T)$, which predicts a large change detectivity for change in the temperature.

6. Types of photodetectors

6.1 Photodiodes

A reverse-biased p-n junction consists of a region, known as the depletion region, that is completely devoid of free charge carriers and where a large built-in electric field opposes flow of electron from *n*- to *p*-side (and of holes from *p*- to *n*-side). When such a p-n junction is illuminated with light on one side, say the p-side (see Fig. 13), electron-hole pairs are created through absorption. Because of the large built-in electric field, electrons and holes generated inside the depletion region accelerate in opposite directions and drift to the *n*- and *p*-sides, respectively. The resulting flow of current is proportional to the incident optical power. Thus, a reverse-biased p-n junction acts as a photodetector and is referred to as the p-n photodiode. Considerable improvement in the performance of such photodiodes is possible through suitable modifications. Two such devices are known as the p-i-n photodiode and the avalanche photodiode (APD). This section starts with a discussion of simple p-n photodiodes and then considers p-i-n and APD devices (Zhao, 2006).

6.2 p-n photodiodes

Fig. (13- a) shows the structure of a p-n photodiode. The semiconductor structure device was based on a junctions formed between the same material with different doping, for example; p-type or n-type on GaAs, is known as homojunction. As shown in Fig. 13- b, incident light is absorbed not only inside the depletion region but also outside it. Photons absorbed inside the depletion region generate electron-hole pairs which experience a large electric field and drift rapidly toward the p- or n-side depending on the electric charge (Fig. 13- c). The resulting current flow constitutes the photodiode response to the incident optical power with Eq. (1).

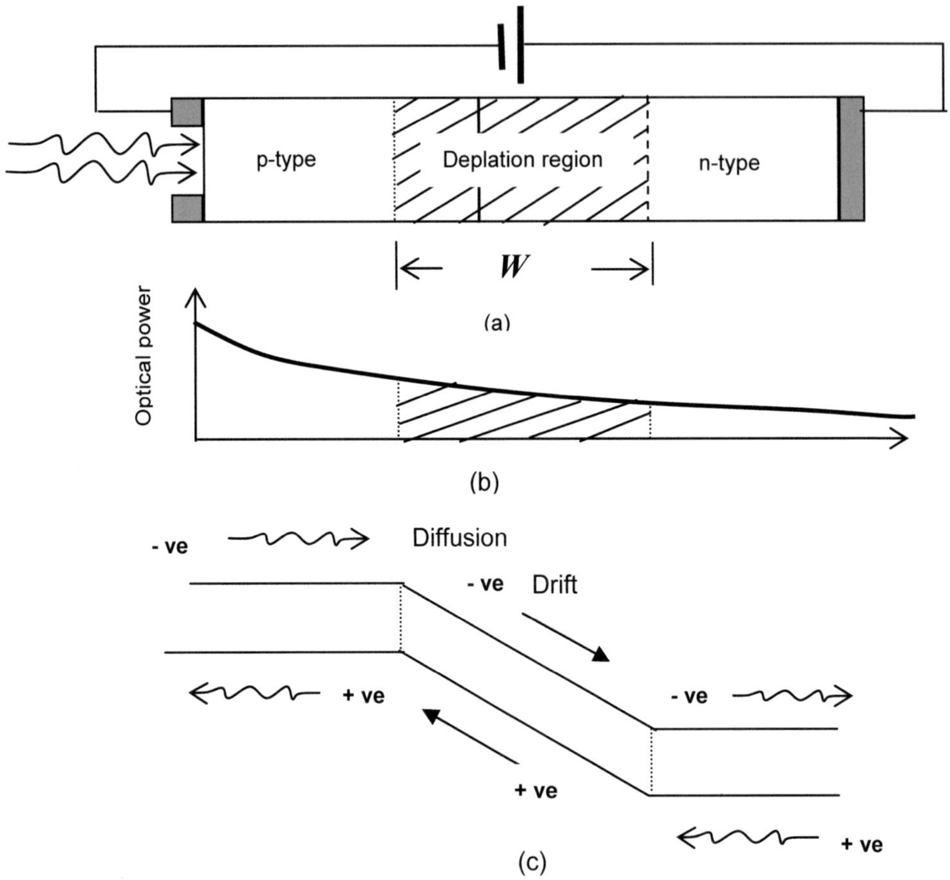

Fig. 13. (a) p-n photodiode and the associated depletion region under reverse bias. (b) Variation of optical power inside the photodiode. (c) Energy-band diagram showing carrier movement through drift inside the depletion region and through diffusion outside it (Zhao, 2006).

The response time is governed by the transit time. If W is the width of the depletion region and v_s is the average drift velocity, the transit time is given by (Fox, 2001):

$$\tau_{tr} = W / v_s \tag{8}$$

Typically, $W \sim 10\mu m$, $v_s = 10^7$ cm/s, and $\tau_{tr} \sim 100ps$. This value is small enough that p-n photodiodes are capable of operating up to bit rates of about 1Gb/s. Both W and v_s can be optimized to minimize τ_{tr}. The depletion-layer width depends on the acceptor and donor concentrations N_A and N_D, used to make the p-n junction and is given by (Sze, 1981):

$$W = [\frac{2\varepsilon}{q}(V_{bi} + V_o)(\frac{1}{N_A} + \frac{1}{N_D})]^{1/2} \tag{9}$$

where ε is the dielectric constant, V_{bi} is the built-in voltage, and V_o is the applied voltage. The value of V_{bi} depends on the semiconductor material and is $1.1V$ for GaAs. The velocity

v_s depends on the applied voltage but attains a maximum value (called the saturation velocity) in the range 5-10×10^{-6} cm/s, depending on the material used for the photodiode.

The limiting factor for p-n photodiodes is the presence of a diffusive component in the photogenerated current. The physical origin of the diffusive component is related to the absorption of incident light outside the depletion region. Electrons generated in the p-region have to diffuse to the depletion-region boundary before they can drift to the n-region; similarly, holes generated in the n-region must diffuse to the depletion-region boundary. Diffusion is an inherently slow process; carriers take a nanosecond or larger to diffuse over a distance of about $1 \mu m$. Fig. 15 shows how the presence of diffusive component can distort the temporal response of a photodiode. In practice, the diffusion contribution depends on the bit rate and becomes negligible if the optical pulse is much shorter than the diffusion time. It can also be reduced by decreasing the widths of p- and n-regions and increasing the depletion region width so that most of incident optical power is absorbed inside it. This approach adopted for p-i-n photodiodes will be discussed next.

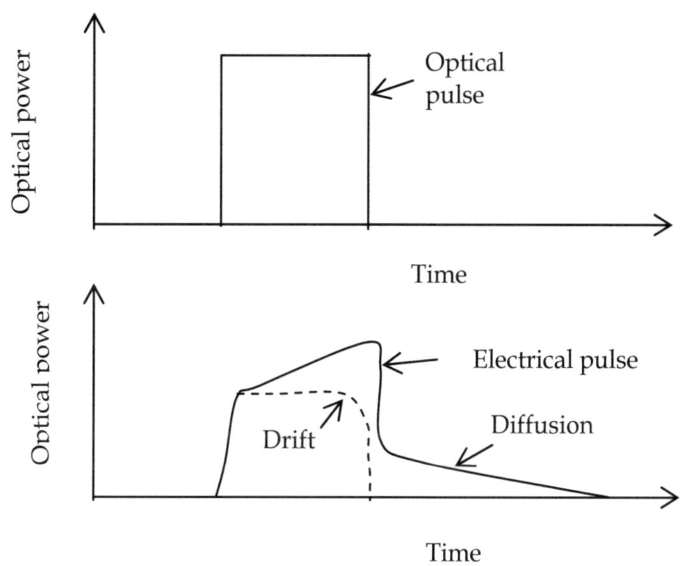

Fig. 14. Response of a p-n photodiode to a rectangular optical pulse when both drift and diffusion contribute to the detector current (Zhao, 2006).

6.3 p-i-n photodiodes

A simple way to increase the depletion-region width is to insert a layer of undoped (or lightly doped) semiconductor material between the p-n junctions. Since the middle layer consists of nearly intrinsic material, such a structure is referred to as the p-i-n photodiode (Zhao, 2006). Fig. (15-a) shows the device structure together with the electric field distribution inside it under reverse-bias operation. Because of the intrinsic nature of the middle layer, this layer offers a high resistance, and most of the voltage drop occurs across it. As a result, a large electric field exists in the middle of i-region. In essence, the depletion region extends throughout the i-region, and its width W can be controlled by changing the middle layer thickness during fabrication. The main difference from the p-*n* photodiode is that the drift component of the detector current dominates over the diffusion component simply because most of the incident power is absorbed inside the i-region of the p-i-n photodiode. Since the depletion width W can be tailored in p-i-n photodiodes, a natural question is how large W should be. The optimum value of W depends on a compromise between responsivity and the response time. The responsivity can be increased by increasing W so that the quantum efficiency η approaches 100%. However, the response time also increases, as it takes longer for carriers to drift across the depletion region. For indirect-bandgap semiconductor such as *Si* and *Ge*, typically W must be in the range *20-50* μm to ensure reasonable quantum efficiency. The bandwidth of such photodiodes is then limited by slow response associated with a relatively long transit time (*> 200ps*). By contrast, W can be as small as *3-5μm* for photodiodes that use direct bandgap semiconductors such as InGaAs. The transit time for such photodiodes is in range τ_{tr}=*30-50ps* if we use $v_s = 1\times10^7 cm/s$ for the saturation velocity. Such values of τ_{tr} correspond to a detector bandwidth Δf=*3-5GHz* if we define the bandwidth as $\Delta f_t = (2\pi\tau_{tr})^{-1}$. Bandwidth as high as *20 GHz* have been achieved for optimized p-i-n photodiodes [6]. Even higher values are possible (*up to 70 GHz*) with a narrow i-region *(<1μm),* but only at the expense of a lower quantum efficiency and responsivity (Sze, 1981).

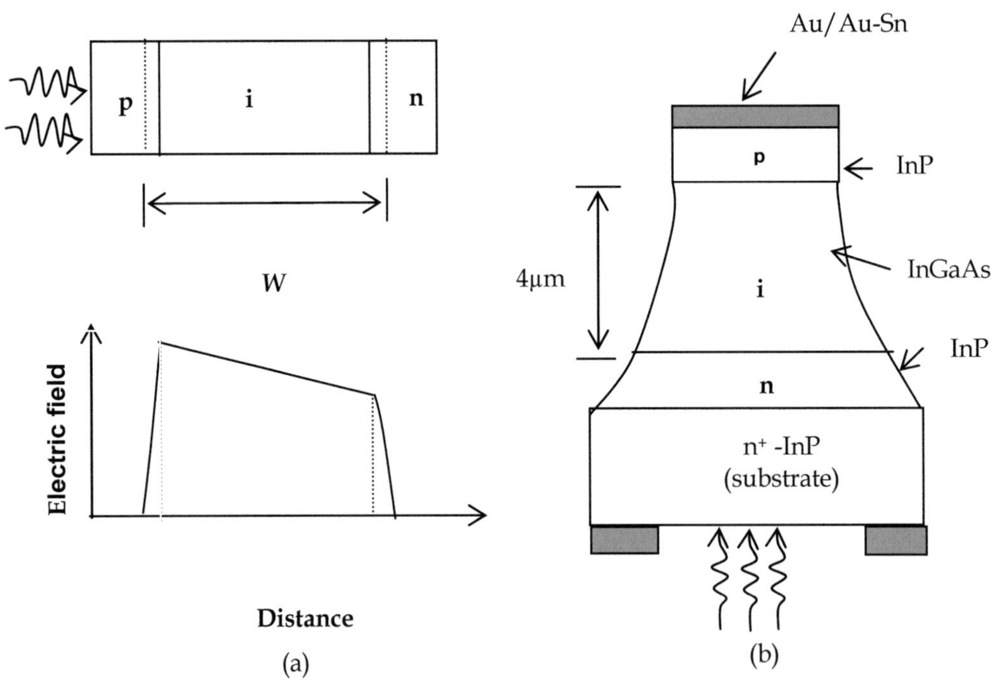

Fig. 15. (a) A p-i-n photodiode together with the electric field distribution inside various layers under reverse bias. (b) Design of an InGaAs p-i-n photodiode (Tucker et al., 1986).

The performance of p-i-n photodiodes can be considerably improved by using a double-heterostructure design. Similar to the case of semiconductor lasers, the middle i-type layer is sandwithched between the p-type and n-type layers of a different semiconductor whose bandgap is chosen such that light is absorbed only in the middle i-layer. A p-i-n photodiode commonly used for lightwave applications uses InGaAs for the middle layer and InP for the surrounding p-type and n-type layers (Sze, 1981; Tucker et al., 1986). Fig. 15 -b shows such an InGaAs p-i-n photodiode. Since the band gap of InP is $1.35eV$, InP is transport for light whose wavelength exceeds $0.92\ \mu m$. By contrast, the bandgap of lattice-matched InGaAs material is about $0.75\ eV$, a value that corresponds to a cutoff wavelength of $1.65\mu m$. The middle InGaAs layer thus absorbs strongly in the wavelength region 1.3-$1.6\mu m$. The diffusive component of the detector current is completely eliminated in such a heterostructure photodiode simply because photons are absorbed only inside the depletion region. The quantum efficiency η can be made almost 100% by using an InGaAs layer of several micrometers thick. The front facet is often antireflection-coated to minimize reflections. InGaAs photodiodes are quite useful for lightwave systems operating near 1.3- and $1.55\mu m$ wavelengths and are often used in optical receivers.

6.4 Avalanche photodiodes

All detectors require a certain amount of minimum current in order to operate reliably. The current requirement translates into a minimum power requirement through $P_{in}=I_p/R$ (see Eq. (1)). Detectors with a large responsivity R are preferred, since they require less optical power. The responsivity of p-i-n photodiodes is limited by Eq. (3) and takes its maximum value $R = q / h\nu$ for $\eta=1$. APDs can have much larger values of R, as they are designed to provide an internal current gain in a way similar to photomultiplier tubes. They are generally used in optical communication systems for which the amount of optical power that can be spared for the receiver is limited. The physical phenomenon behind the internal current gain is known as impact ionization (Stillman & Wolfe; 1977; Melchior, 1977).

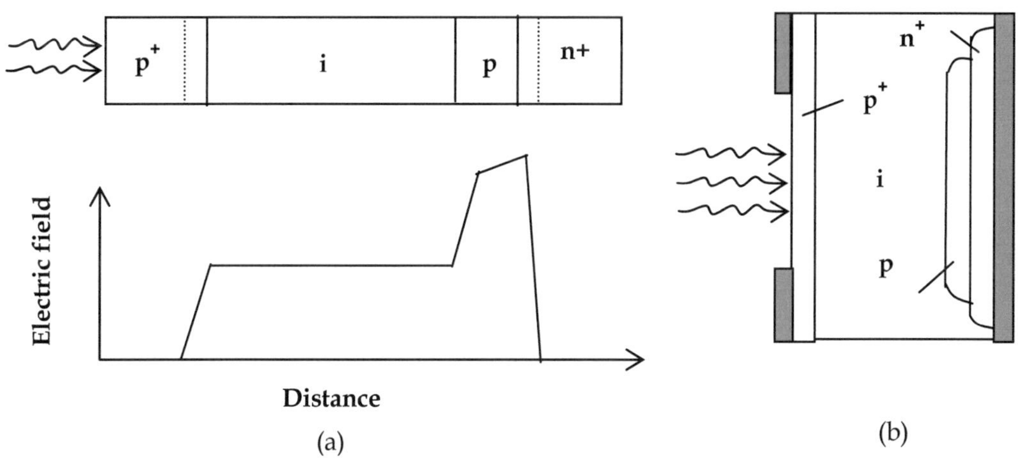

Fig. 16. (a) an APD together with the electric-field distribution inside various layer under reverse bias. (b) Design of a silicon reach-through APD [42, 43]

Under certain conditions, an accelerating electron can acquire sufficient energy to generate a new electron–hole pair. The energetic electron gives a part of its kinetic energy to another electron in the valance band that ends up in the conduction band, leaving behind a hole. The net result of impact ionization is that a single primary electron, generated through absorption of photon, creates many secondary electrons and holes, all of which contribute to the photodiode current. Of course, the primary hole can also generate secondary electron-hole pairs that contribute to the current. The generation rate is governed by two parameters, a_e and a_h, known as the impact-ionization coefficients of electrons and holes, respectively. The numerical value of these coefficients depends on the semiconductor material and on the electric field that accelerates electrons and holes. Values on the order of $1\times10^4\ cm^{-1}$ are obtained for an electric field in the range $2\text{-}4\times10^5\ V/cm$. Such large fields can be realized by applying a high voltage (>*100V*) to the APD.

APDs differ in their design from that of p-i-n photodiodes mainly in one respect. An additional layer is added through which secondary electron-hole pairs are generated by impact ionization. Fig. 16-a shows the APD structure together with the variation of electric field in various layers. Under reverse bias, a high electric field exists in the p-type layer sandwiched between i-type and n^+-type layers. This layer is referred to as the gain region or the multiplication region, since secondary electron-hole pairs are generated here through impact ionization. The i-layer still acts as a depletion region in which most of the incident photons are absorbed and primary electron-hole pairs are generated. Electrons generated in the i-region cross the gain region and generate secondary electron-hole pairs responsible for the current gain. The current gain or the multiplication factor M is given by (Stillman & Wolfe; 1977; Melchior, 1977):

$$M = \frac{1-k_A}{\exp[-(1-k_A)\alpha_e d]-k_A} \tag{10}$$

where d is the thickness of the gain region, $k_A = \alpha_h / \alpha_e$. The APD gain is quite sensitive to the ratio of the impact-ionization coefficients. Where $\alpha_h = 0$ so that only electrons participate in the avalanche process, $M = \exp(\alpha_e d)$, and the APD gain increases exponentially with d. On the other hand, when $\alpha_h = \alpha_e$ so that $k_A = 1$ in Eq. (10), $M = (1-\alpha_e d)^{-1}$. The APD gain then becomes infinite for $\alpha_e d = 1$, a condition known as the avalanche breakdown. Although higher APD gain can be realized with a smaller gain region when $\alpha_e and \alpha_h$ are comparable, the performance is better in practice for APDs in which either $\alpha_e >> \alpha_h$ or $\alpha_h >> \alpha_e$, so that the avalanche process is dominated by only one type of charge carrier. Because of the current gain, the responsivity of APD is enhanced by the multiplication factor M and is given by

$$R_{APD} = MR = M(\eta q / h\nu) \tag{11}$$

It should be mentioned that the avalanche process in APDs is intrinsically noisy and results in a gain factor that fluctuates around the average value. The intrinsic bandwidth of an APD depends on the multiplication factor M. This is easily understood by noting that the transit time τ_{tr} for an APD is no longer given by Eq. (8) but increases considerably simply because generation and collection of secondary electron-hole pairs take additional time. A particularly useful design is known as reach-through APD (see Fig. 16-b) because the depletion layer reaches to the contact layer through the absorption and multiplication regions. It can provide high gain ($M \approx 100$) with low noise and relatively large bandwidth.

6.5 Advanced APD structures

It is difficult to make high-quality APDs that operate in the wavelength range *1.3-1.6μm* that is of interest for most lightwave systems. A commonly used material is InGaAs (lattice-matched to InP) with a cutoff wavelength of about *1.65μm*. However, the impact-ionization coefficients α_e *and* α_h for InGaAs are comparable in magnitude (Campbell, 1983). As a result, the bandwidth is considerably reduced, and the noise is also high. Furthermore, because of a relatively narrow bandgap, InGaAs undergoes tunneling breakdown at electric fields of about 1×10^5 *V/cm*, a value that is below the threshold for avalanche multiplication. This problem can be solved in heterostructure APDs by using an InP layer for the gain region because quite high electric fields ($>5\times10^5$ *V/cm*) can exist in InP without tunneling breakdown. Since the absorption region (i-type InGaAs) and the multiplication region (n-type InP layer) are separate in such a device, this structure is known as separate absorption and multiplication *(SAM)* APD. Since $\alpha_h > \alpha_e$ for InP, the APD is designed in such away that holes initiate the avalanche process in an *n-type* InP layer, and K_A is defined as $K_A = \alpha_e / \alpha_h$. Fig. (17- a) shows a mesa-type SAM APD structure. One problem with a SAM APD is related to the large bandgap difference between InP (E_g=*1.35 eV*) and InGaAs (E_g =0.75 *eV*). Because of a valence-band step about *0.4 eV*, holes generated in the InGaAs layer are trapped at the heterojunction interface and are considerably slowed before they reach to the multiplication region (InP layer). Such an APD has an extremely slow response and relatively small bandwidth. This problem can be solved by using another layer between the absorption and multiplication regions whose bandgap is intermediate to those of InP and InGaAs. The quaternary material InGaAsP, the one used for semiconductor lasers, can be tailored to have a bandgap anywhere in the range *0.75-1.35 eV* and is ideal for this response. It is even possible to grade the composition of InGaAsP over the distances of roughly *10-100 nm*. Such APDs are called SAGM APDs, where SAGM indicates separate absorption, grading, and multiplication regions.

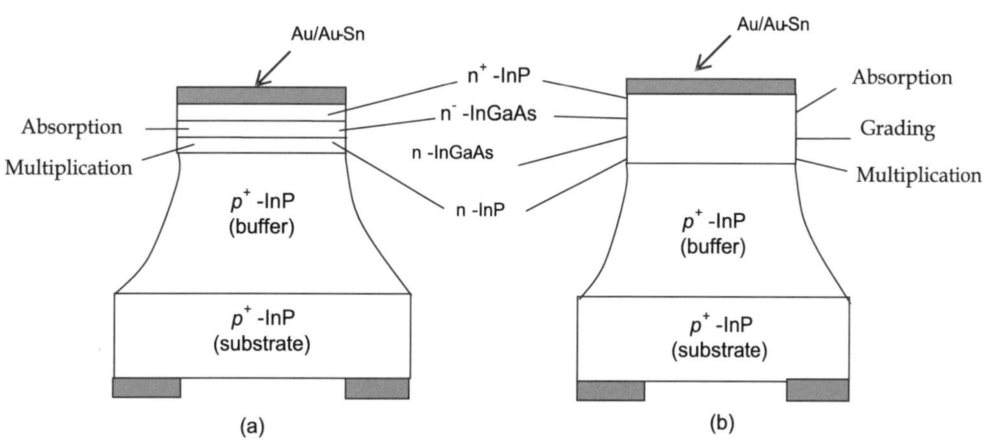

Fig. 17. Design (a) a SAM APD and (b) a SAGM APD (Kasper & Campbell, 1987; Tarif, 1991).

Fig. 17- b shows the design of an InGaAs APD with the SAGM structure. The use of InGaAsP grading layer improves the bandwidth considerably. This device exhibited a gain-bandwidth product of about 70 GHz for *M>12*. Values up to *100GHz* have been demonstrated by using a "charge region" between the grading and multiplication regions (Kasper & Campbell, 1987; Tarif, 1991).

A different approach to the design of high-performance APDs is provided by multiquantum-well (MQW) or superlattice structure (Tarif, 1991) [46]. The major limitation of SAM APDs results from nearly equal values of α_e and α_h resulting in K_A=0.7. A MQW design offers the possibility of reducing the value of K_A from the bulk value of nearly unity. The absorption and multiplication regions in MQW APDs alternate and consist of thin layers (~10nm) of different bandgap semiconductor materials. In this sense, the APD design is quite similar to MQW semiconductor lasers except that the device parameters are optimized to result in a lower value of K_A. This approach was first demonstrated for GaAs/AlGaAs MQW APDs and resulted in a considerable enhancement of the electron impact-ionization coefficient (Capasso, 1985). It is less successful for InGaAs/InP material system. Progress has been made through the so-called staircase APDs, in which the InGaAsP layer is compositionally graded to form a sawtooth kind structure in the energy-band diagram that looks like a staircase under reverse bias (Capasso, 1985). Another technique for making high-speed APDs uses alternate layers of InP and InGaAs layers for the grading region (Capasso, 1985). However, the ratio of widths of the InP to InGaAs layers varies from zero near the absorbing region to almost near the multiplication region. Since the effective bandgap of a quantum well depends on the quantum-well width (InGaAs layer thickness), a graded "pseudo-quaternary" compound is formed as a result of variation in the layer thickness. The performance of InGaAs APDs has improved considerably as a result of many such advances.

6.6 UV photodetectors

With their direct bandgap tunable from 200 to 365 nm, $Al_xGa_{1-x}N$ alloys are among the best candidates for ultraviolet (UV) photodetectors. Applications for such devices include solar UV detection, visible blind detection, missile warning, space communication and flame sensing, see Fig. 18 (Norbert & Jurgen,2005; NASA, 1999; Hsueh et al.; 2007). Actually, photomultiplier tube light detectors are the most efficient devices for flame sensing, however, they suffer from their high price, a limited lifetime of 1 to 2 years, a large size and the need of high voltage power supply. For utilization under room light or solar light, the UV flame detector should be able to selectively sense the flame wavelength below 280 nm, meaning that it should be able to detect light intensities below 1 nW/cm^2 with a rejection ratio between UV and visible light of six orders of magnitude. Also, the flame detector should be able to operate in a high temperature field and a response speed on the order of few *ms* is required for safety switch applications. Thanks to the use of a low-temperature-deposited buffer layer and the realization of p-type GaN (Norbert & Jurgen, 2005), the nitride research encounters a vast development. Achievement of photoconductors, photodiodes and phototransistors has been reported. However, for low-intensity detection a high photocurrent on dark current ratio (PC/DC) is needed in order to avoid misdetection, especially for a device operating in a wide range of temperatures. Photocurrent values close to the expected level have been reported, however, for dark current level, the reported results are still far from theoretical limit. For Schottky, *p-n* or *p-i-n* photodiodes, one of the origins of high leakage current was attributed to the high level of threading dislocations (10^9 to 10^{10} cm^{-2}) present in the current GaN epilayer. Recently, it was reported that the insertion of low-temperature-deposited GaN or AlN inter-layers between high-temperature-grown GaN layers reduced the threading dislocations. The insertion of a high quality intrinsic layer in devices such as PIN photodiodes or phototransistors is believed to be one of the key points to improve the performance of those UV photodetectors (Pernot et al., 1999). Comparison of different types of UV photodetectors is denoted in table 2 (Waleker et al., 2000).

Photodetector type	Advantages	Disadvantages	Challenges
photoconductors	-Easy to fabricate -Internal photoelectric gain	-Low speed -Large dark current -Large Johnson noise	-Interdigitated patterns require enhanced resolution
p-n or p-i-n photodiodes	-Low or zero dark current -High speed -High impedance (good for FPA readout circuitry) -Compatible with planar processing technology (for FPA) -For p-i-n photodiodes, easy optimization of quantum efficiency and speed with I layer	-Speed is limited by minority carried diffusion time and storage time -Speed and quantum efficiency trade-off	-Etching is necessary to expose various layers -Ohmic contacts necessary to both n- and p-type material
MSM or Schottky photodiodes	-High efficiency -High speed -Easy to fabricate	-Require high absorption coefficient -No-sharp cut-off (below bandgap response) -Front side illumination needed	-Schottky contact is needed

Table 2. Comparison of different types of UV photodetectors (Waleker et al., 2000).

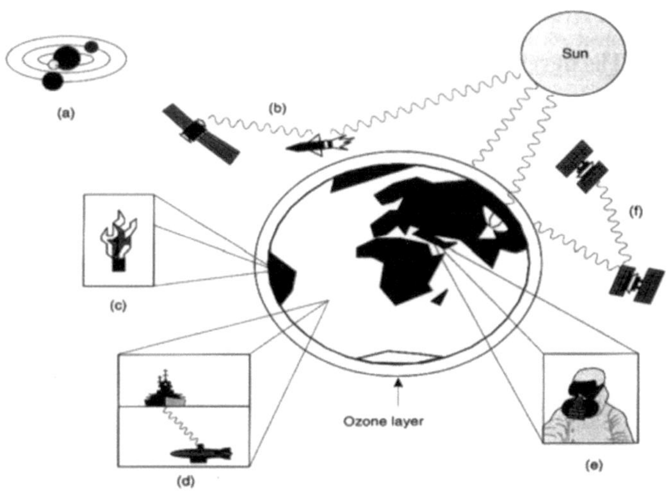

Fig. 18. Applications for UV photodetectors: (a) UV astronomy, (b) solar blind detectors, (c) flame detection and engine monitoring (d) underwater UV communications, (e) chemical/ biological battlefield reagent detectors, (f) space communications secure from Earth (Pernot et al., 1999).

6.6.1 Future and recent advances of UV photodetectors

Many astronomical objects produce orders of magnitudes more photon fluxes at optical wavelengths than they do in the vacuum UV. In order to eliminate this huge background contribution and substantial source of noise solar-blind detector and imaging systems are required. A comprehensive study of UV imaging detectors is given by (Norbert & Jurgen, 2005). Several specific detector types can be used for vacuum UV astronomy in the future, multidimensional detectors, semi-conductive array (e.g. CCDs) and microchannel plate detector (MCP). Recent research teams are studied the UV nanowires. They found that the conductivity of the ZnO nanowires was very sensitive to UV light. Very recently, researchers reported the growth of vertical, well aligned ZnO nanowires on ZnO :Ga and TiN buffer layers by reactive evaporation Hsueh et al.; 2007).

The further development of technologies for this detector is the basis to enhance the performance for UV applications. The current technology can be classified in two categories: Solid-state devices based on silicon or wide bandgap semiconductors and photoemissive devices, coupled with a gain component and an electron detector. The detective quantum efficiencies (DQE) for various UV detectors as MCPs, CCDs, Electron-Bombarded CCDs and for the expected DQEs for future A1GaN solid-state sensors are shown in Fig.19.

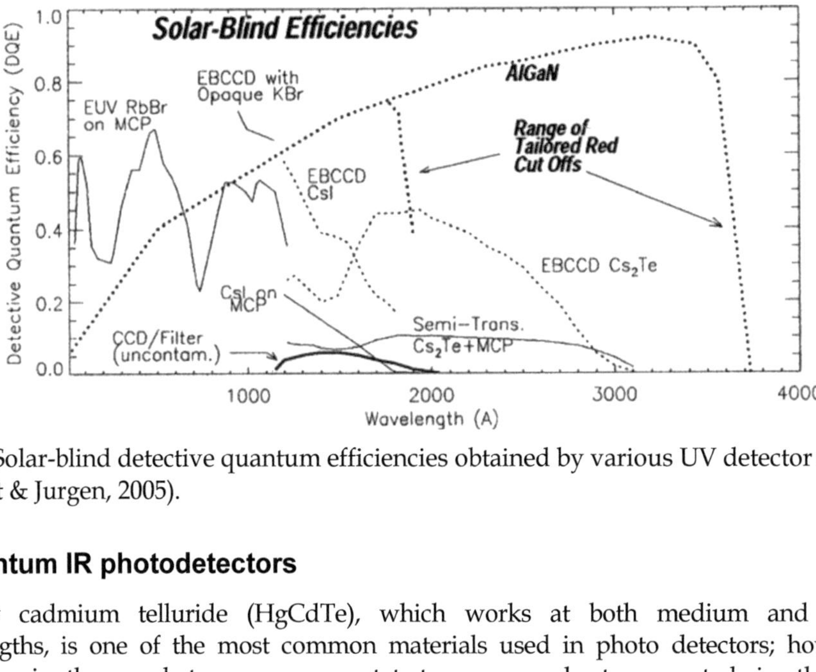

Fig. 19. Solar-blind detective quantum efficiencies obtained by various UV detector adopted (Norbert & Jurgen, 2005).

7. Quantum IR photodetectors

Mercury cadmium telluride (HgCdTe), which works at both medium and long IR wavelengths, is one of the most common materials used in photo detectors; however its limitation is the need to use a cryostat to manage heat generated in the device (www.nanoforum.org, 2007). Here, the focusing will be directed to the quantum IR detectors.

7.1 Idea of nanotechnology devices - Quantum confined structures

In this section, it will be given an overview of the optical properties of quantum confined semiconductor structures. These are artificial structures in which the electrons and holes are confined in one or more directions (Fox, 2001; www.nanoforum.org, 2007).

The optical properties of solids do not usually depend on their size. Ruby crystal, for example, have the same red colour irrespective of how big they are. This statement is only true as long as the dimensions of the crystal are large. If we make very small crystals, then the optical properties do in fact depend on the size. A striking example of this is semiconductor doped glasses. These contain very small semiconductor microcrystals within a colourless glass host, and the colour of the filter can be altered just by changing the size of the crystals.

The size dependence of the optical properties in very small crystals is a consequence of the quantum confinement effect. The Heisenberg uncertainty principle tells us that if we confine a particle to a region of the x axis of length Δx, then we introduce an uncertainly in its momentum given by:

$$\Delta p_x \sim \frac{\hbar}{\Delta x} \tag{12}$$

where $\hbar$ is the Plank's constant.

If the particle is otherwise free, and has a mass m, the confinement in the x direction gives it an additional kinetic energy of magnitude

$$E_{confinement} = \frac{(\Delta p_x)^2}{2m} \sim \frac{\hbar^2}{2m(\Delta x)^2} \tag{13}$$

This confinement energy will be significant if it is comparable to or greater than the kinetic energy of the particle due to its thermal motion in the x direction. This condition may be written:

$$E_{confinement} \sim \frac{\hbar^2}{2m(\Delta x)^2} > \frac{1}{2}K_B T \tag{14}$$

And tells us that quantum size effects will be important if

$$\Delta x \sim \sqrt{\frac{\hbar^2}{mK_B T}} \tag{15}$$

This is equivalent to saying that Δx must be of the same order of magnitude as the de Broglie wavelength $\lambda_{deB} \equiv p_x / \hbar$ for the thermal motion. The criterion given in Eq. (15) gives us an idea of how small the structure must be if we are to observe quantum confinement effects. For an electron in a typical semiconductor with $m^*_e = 0.1m_o$ at room temperature, we find that we must have $\Delta x \sim 5nm$. Thus, a 'thin' semiconductor layer of thickness 1μm is no thin by the standards *of the electrons*. It is in fact a bulk crystal which would not exhibit any quantum size effects. To observe quantum size effects we need thinner layers. Table 3 summarizes the three basic types of quantum confined structures that can be produced. The structures are classified as to whether the electrons are confined in one, two or three dimensions. These structures are respectively called:

- quantum wells (1-D confinement);
- quantum wires (2-D confinement);
- quantum dots (3-D confinement).

Table 3 also lists the number of degrees of freedom associated with the type of quantum confinement. The electrons and holes in bulk semiconductors are free to move within their respective bands in all three directions, and hence they have three degrees of freedom. The electrons and holes in a quantum well, by contrast, are confined in one direction, and therefore only have two degrees of freedom. This means that they effectively behave as two-dimensional (2-D) materials. Similarly, quantum wire structures have 1-D physics, while quantum dots have '0-D' physics. This last point means that the motion of the electrons and holes is quantized in all three dimensions, so that they are completely localized to the quantum dot.

structure	Quantum confinement	Number of free dimensions
Bulk	none	3
Quantum well/superlattice	1-D	2
Quantum wire	2-D	1
Quantum dot/box	3-D	0

Table 3. Number of degree of freedom tabulated against the dimensionality of the quantum confinement.

The very small crystal dimensions required to observe quantum confinemt effect have to be produced by special techniques.

- Quantum well structures are made by techniques of advanced epitaxial crystal growth.
- Quantum wire structures are made by lithographic patterning of quantum well structures, or by epitaxial growth on patterned substrates.
- Quantum dots structures can be made by lithographic patterning of quantum well or by spontaneous growth technique.

Epitaxial techniques are very versatile, and they allow the growth of a great variety of quantum well structures as described in the first subsection of techniques of fabrications. Fig. (20-b) shows one such variant derived from the single well structure shown in Fig. (20-a). The crystal consists of a series of repeated GaAs quantum wells of width *d* separated from each other by AlGaAs layers of thickness *b*. This type of structure is either called a multiple quantum well (MQW) or a superlattice, depending on the parameters of the system. The distinction depend mainly on the value of *b*. MQWs have *b* values, so that the individual quantum wells are isolated from each other, and the properties of the system are essentially the same as those of single quantum wells. But in the case of a superlattice the values of *b* and *d* are equal. They are often used in optical applications to give a usable optical density. It would be very difficult to measure the optical absorption of a single *10nm* thick quantum well, simply because there is so little to absorb the light. By growing many identical quantum wells, the absorption will increase to a measurable value (Levine, 1993).

7.2 Comparison between long wavelength photodetectors

An important application of detectors is in the area of the detection of long wavelength radiation (λ ranging from *5-20μm*). If a direct band to band transition is to be used for such detectors, the bandgap of the material has to be very small. An important material system in which the bandgap can be tailored from *0* to *1.5 eV* is the *HgCdTe* alloy. The system is widely used for thermal imaging, night vision application, etc. However, the small bandgap *HgCdTe* is a very "soft' material which is very difficult to process. Thus the device yield is rather poor. The quantum well intersubband detector offers the advantages of long wavelength detection using established technologies such as the GaAs technology (Piotrowski, 2004).

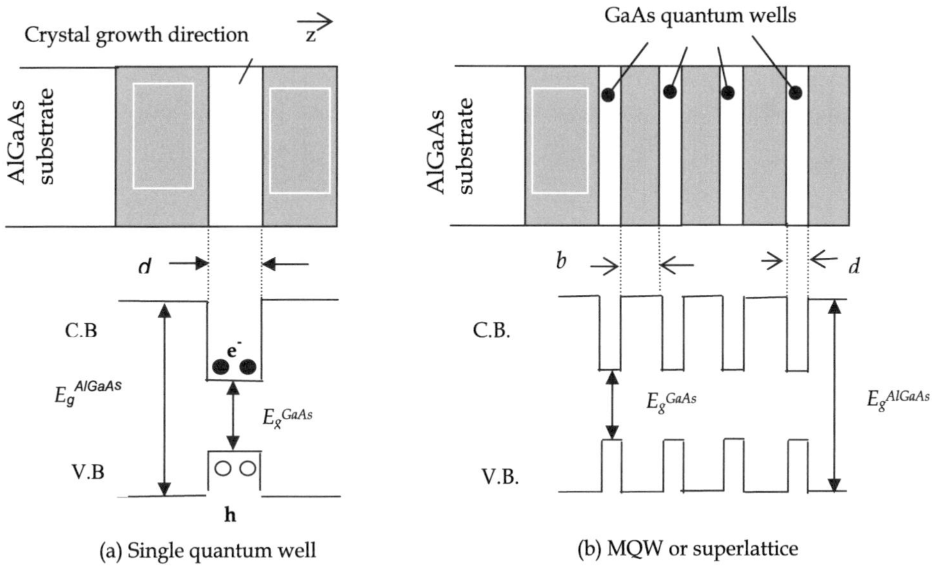

Fig. 20. (a) Schematic diagram of a single GaAs/AlGaAs quantum well. The quantum well is formed in the thin GaAs layer sandwiched between AlGaAs layers which have a larger bandgap. The lower half of the figure shows the spatial variation of the conduction band (C.B.) and the valence band (V.B.). (b) Schematic diagram of GaAs/AlGaAs multiple quantum well (MQW) or superlattice structure. The distinction between an MQW and a superlattice depends on the thickness b of the barrier separating the quantum wells (Levine, 1993).

In Fig. 21, we show a quantum well which is doped so that the ground states has a certain electron density and the excited state is unoccupied. When a photon with energy equal to the intersubband separation impinges upon the quantum well, the light is absorbed and the ground state electron is scattered into the excited state. For the absorption process to produce an electrical signal, one must have the following conditions satisfied:

1. The ground state electrons should not produce a current. If this is not satisfied, there will be a high dark current in the detector. The electrons in the ground state carry current by thermionic emission over the band discontinuity. At low temperatures this process can be suppressed. Now, there are many methods to obtain a high-operating-temperature infrared Photodetectors as in (Piotrowski & Rogalski, 2007).
2. It should be possible to extract the excited state electrons from the quantum well so that a signal can be produced. The excited electron state should, therefore, be designed to be near the top of the quantum well barrier, so that the excited electrons can be extracted with ease by an applied electric field.

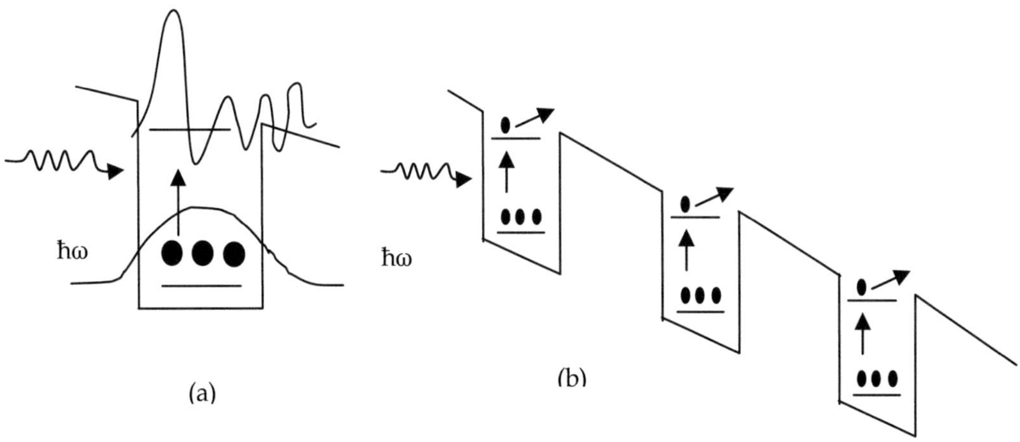

Fig. 21. (a) A schematic of the electron wave function in the ground state and the excited state. The photon causes transitions to the excited state from where the electrons are collected are shown in (b)

Band to band absorption detectors	Quantum well intersubband detectors
Narrow bandgap materials are "soft" and difficult to process. Therefore, device yield is poor	Established GaAs or InP based technologies can be used to produce high yield devices.
Dark current can be low at low temperatures, since an undoped material is used in the active region.	Dark current is usually high, since there are free carries in the quantum well regions.
	In the conduction band, quantum wells, vertical incident light is not absorbed.
Vertical imaging can be done	Mirrors have to be used, corrugated QW, or valence band quantum wells have to be used.
It is difficult to fabricate tunable detectors	Tunability can be achieved by altering the quantum well parameters.

Table 4. Comparison of the band to band and intersubband detectors for long wavelength detection

In the case of quantum well, the light is absorbed only if it is incident z-polarized. Here z is the quantum well growth direction. Thus, for vertical incidence, there is no absorption. This is an important drawback for such detectors. One way to overcome this problem is to use etched mirrors on the surface to reflect vertically incident light so that it has a z-polarization. Another way to avoid this is to use the intersubband transitions in the valence band where, due to the mixed nature of the HH and LH states, the z-polarization rule is not valid. However, the poor hole transport properties and the difficulty in reducing the dark current reduce the detector performance.

Quantum wire and quantum dot IR photodetectors overcome this problem as they are response to the normal incident IR rays (Nasr, 2009; Nasr et al., 2009; Nasr, 2008). In Table 4, a comparison of the band to band and intersubband detectors for long wavelength detection is shown (El_Mashade, 2003).

Fig. 22 shows the transition of electron in conduction band of QWIPs and each of photocurrent and dark currents. The inset shows a cross-section transmission electron micrograph of a QWIP sample (Gunpala et al., 2001).

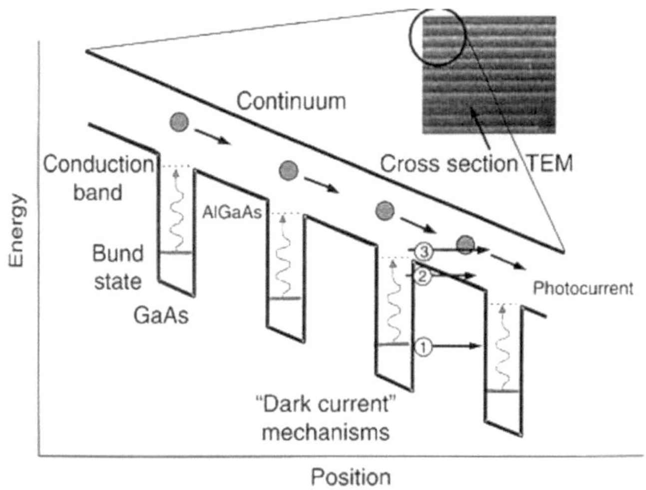

Fig. 22. Schematic diagram of the conduction band in a bound-to-quasibound QWIP in an externally applied electric field. Absorption of IR photons can photoexcite electrons from the ground state of the quantum well in the continuum, causing a photocurrent. Three dark current mechanisms are also shown: ground state tunneling (1); thermally assisted tunneling (2); and thermionic emission (3). The inset shows a cross-section transmission electron micrograph of a QWIP sample (Gunpala et al., 2001).

7.3 Quantum dot and wire infrared photodetectors

To eliminate the limitations of QWIP detectors, there has been significant interest in IR detectors with lower dimensions such as quantum wires (QR) (2-D) and quantum dots (QD) (3-D) (Das & Singaraju, 2005). The use of quantum wires and quantum dots for IR detection provides many advantages. From these advantages, first one, the sensitivity to normal incident IR. Second one, higher optical gain provided by QDIPs and QRIPs. Hence, the life time of the excited states in the QDs is expected to be longer arisen reduced capture probability due to the phonon bottleneck and formation of repulsive potential barriers by charged QRs or QDs. Third, The discreteness of the QR and QD energy spectrum which lead to expected IR wide band covering (long, medium, short IR region) (Nasr & El_Mashade, 2006; Phillips et al., 1999). Although of these strengths, there are drawbacks. The major one is a high value of dark current obtained as illustrated in theoretical and experimental results (Nasr et al., 2006; Ryzhii et al., 2000). In a previous work, we declared that QRIPs gives a smaller and acceptable values of dark current compared with QDIPs (Nasr & El_Mashade, 2006). This opens the way to overcome mentioned problem (Nasr, 2007).

The main features of operation of (QDIP) in conduction band are illustrated into Fig. (23-a) and can be summarized to the following:

1. Thermo- and photoexcition of electrons from bound states in QDs into continuum states,
2. Capture of mobile electrons into QDs (transition from continuum into bound states),
3. Transport of mobile electrons between the potential "hills" formed by charged QDs,
4. Injection of extra electrons from the emitter contact because of the redistribution of the potential in the QDIP active region caused by the change in the charges accumulated by QDs,
5. Collection of the excited and injection electrons by the pertinent contact.

The main features that differentiate QDIPs from QWIPs are associated with the discreteness of the QD energy spectrum leading to a dependence of the thermoactivation energy on the following (Ryzhii & Khmyrova, 2001):

The QD sizes, Strong sensitivity to normally incident radiation, phonon bottleneck effect (or similar effects) in the electron capture, and limitations imposed on the QD filling by the Pauli principle, the existence of the passes for mobile electrons between QDs, the activation character of the electron capture due to the repulsive potential of charge QDs and thermionic nature of the injection (Nasr & El_Mashade, 2006).

In the case of quantum wire IR photodetectors, The QRIPs under illumination by infrared radiations causes the photoexcitation of QRs because of the electron transitions from the bound states in QRs into the continuum states above the inter-QR barriers, the principle of QRIPs operation is shown in Fig. (23-b) (Nasr, 2007; Ryzhii & Khmyrova, 2001; Maksimović, 2003). The device structure is formed from quantum wires active layer separated by barrier layer. The overall structure is terminated by the ohmic contacts, emitter and collector. The device operation can be declared from the energy band in the current flow direction (z-axis) and carrier confinement directions (x- and y- axis) which contains in our case one energy state as seen in figure (23- b). When IR radiation is incident near the junction, the mobile charges (electrons) into the quantum wire active layer is moved to a higher energy state above or equal the barrier energy. These higher energy electrons can now cross into the barrier layer. Hence, photocurrent can be obtained by applying biasing voltage across the ohmic contacts to collect these generated photocarriers.

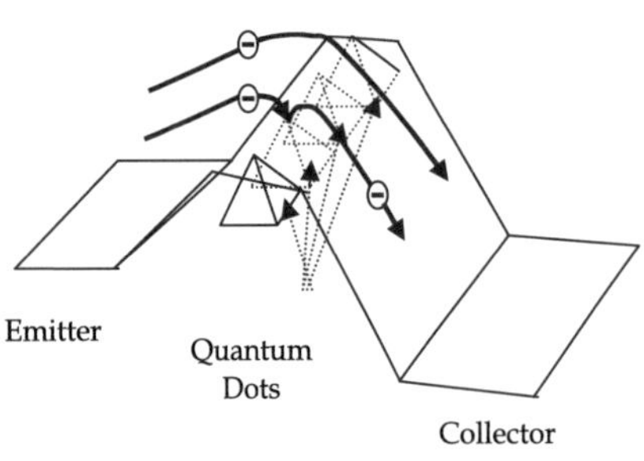

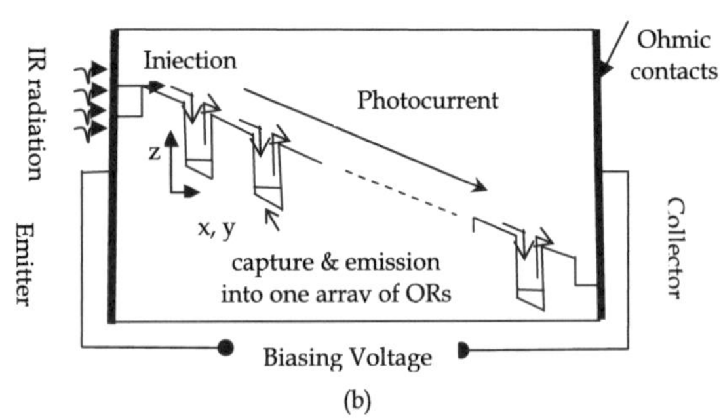

Fig. 23. (a) Conduction band profile in the QDIP and (b) QRIPs principle operation scheme (Nasr, 2007; Ryzhii & Khmyrova, 2001; Maksimović, 2003).

8. Modern advances and applications of IR photodetectors

8.1 Commercially available 320×256 LWIR focal plane arrays

Infrared imaging systems that work in 8-12 μm (LWIR) band have many applications, including night vision, navigation, flight control, early warning systems, etc. Several research groups have demonstrated the excellent performance of QWIP arrays. For example, Faska et al. (Gunpala et al., 2001) have obtained very good images using a 156×256 bound-to-miniband MQW FPA. The first 256×256 LWIR hand-held imaging camera was demonstrated by (Guapala et al., 1997). The device structure of this commercially a available FPA consisted of a bound-to-quasibound QWIP containing 50 periods of a 45 Å well of GaAs (doped $n=4\times10^{17}$ cm^{-3}) and a 500A barrier of $Al_{0.3}Ga_{0.7}As$. Ground state electrons are provided in the detector by doping the GaAs well layers with Si. This Photosensitive MQW structure is sandwiched between 0.5 μm GaAs top and bottom contact layers doped $n=5\times10^{17}$ cm^{-3}, grown on a semi-insulating GaAs substrate by MBE. Then a 0.7-μm thick GaAS cap layer on top of a 300Å $Al_{0.3}Ga_{0.7}As$ stop-etch layer was grown in stu on top of the device structure to fabricate the light coupling optical cavity.

The detectors were back illuminated through a 45° polished facet and a responsivity spectrum is shown in Fig. 24. The responsivity of detector peaks at 8.5 μm and the peak responsivity (Rp) of the detector is 300mA/W at bias V_B=-3V. The spectral width and the cutoff wavelength are $\Delta\lambda/\lambda$=10% and λ_c=8.9 μm, respectively. The measured absolute peak responsivity of the detector is small, up to about V_B=-0.5V. Beyond that it increase nearly linearly with bias reaching R_P=380mA/W at V_B=-0.5V. The peak quantum efficiency was 6.9% at bias V_B=-1V for a 45° double pass. The lower quantum efficiency is due to the lower well doping density (5×10^{17} cm^{-3}) as it is necessary to suppress the dark current at the highest possible operating temperature (Guapala et al., 1997).

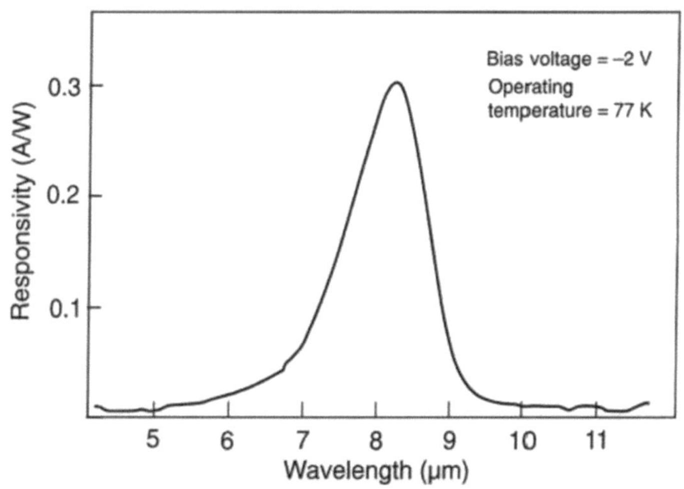

Fig. 24. Responsivity spectrum of a bound-to-quasibound LWIR QWIP test structure at temperature *T=77K*. The spectral response peak is at *8.3 μm* and the long wavelength cutoff is at *8.8 μm* (Guapala et al., 1997).

8.1.1 A 640×486 long-wavelength camera

By the same previous procedure, the camera can be established. A 640×486 QWIP FPA hybrid was mounted onto an 84-pin lead-less chip carrier and installed into a laboratory dewar which is cooled by liquid nitrogen to demonstrate a LWIR imaging camera (Guapala et al., 2001). Fig. 25 shows a frame of video image taken with this long-wavelength camera. This image demonstrates the high sensitivity of the camera. This high yield is due to the excellent GaAs growth uniformity and the mature of GaAs processing technology.

Fig. 25. This picture was taken at the night (around midnight) and it clearly shows where automobiles were parked during the daytime. This image demonstrates the high sensitivity of the 640×486 long-wavelength QWIP staring array(Guapala et al., 1998).

8.2 640×486 long-wavelength dualband imaging camera

The LWIR and very long-wavelength infrared (VLWIR) dualband QWIP device structure described in this section processed into interlace simultaneously readable dualband FPAs (i.e., odd rows for colour and the even rows for the other colour) (Guapala et al., 1999).. The device structure consists of a 30 periods stack, of VLWIR QWIP structure and a second 18 periods stack of LWIR QWIP separated by a heavily doped 0.5 μm thick intermediate GaAS contact layer. The first stack (VLWIR) consists of 30 periods of a 500Å $Al_xGa_{1-x}As$ barrier and a 60 Å GaAs well. Since the dark current of this device structure is dominated by the longer wavelength portion of the device structure, the VLWIR QWIP structure has been designed to have a bound-to-quasibound intersubband absorption peak at 14.5 μm. The second stack (LWIR) consists of 18 periods of a 500 Å $Al_xGa_{1-x}As$ barrier and a narrow 40Å GaAs well. This LWIR QWIP structure has been designed to have a bound-to-continuum intersubband absorption peak at 8.5 μm, since photo current and dark current of the LWIR device structure is relatively small compared to the VLWIR portion of the device structure, this whole dualband QWIP structure is then sandwiched between 0.5 μm GaAs top and bottom contact layers doped with $n=5\times10^{17}cm^{-3}$, and has been grown on a semi-insulating GaAs substrate by MBE. Then a 300Å $Al_{0.3}Ga_{0.7}As$ stop-etch layer and a 1.0 μm thick GaAs cap layer were grown *in situ* on top of the device structure.

GaAs wells of the LWIR and VLWIR stacks were doped with $n=6\times10^{17}$ and 2.5×10^{17} cm^{-3} respectively. All contact layers were doped to $n=5\times10^{17}$ cm^{-3}. The GaAs well doping density of the LWIR stack was intentionally increase by a factor of two to compensate for the reduced number of quantum wells in the LWIR stack (Guapala et al., 1999). It is worth noting that, the total (dark + photo) current of each stack can be independently controlled by carefully designing the position of the upper state, well doping densities, and the number of periods in each MQW stack. All of these features were utilized to obtain approximately equal total currents from each MQW stack. Two different 2D periodic grating structures were designed to independently couple the 8-9 and 14-15μm radiation into detector pixels in even and odd rows of the FPAs, the FPA fabrication process is described in detail into(Guapala et al., 1999). Video images were taken at a frame rate of 30 Hz at temperatures as high as T=74K, using a capacitor having a charge capacity of 9×10^{6} electrons (the maximum number of photoelectrons and dark electrons that can be counted in the time taken to read each detector pixel). Fig. 26 shows simultaneously acquired 8-9 and 14-15 μm images using this two-colour camera (Guapala et al., 1999).

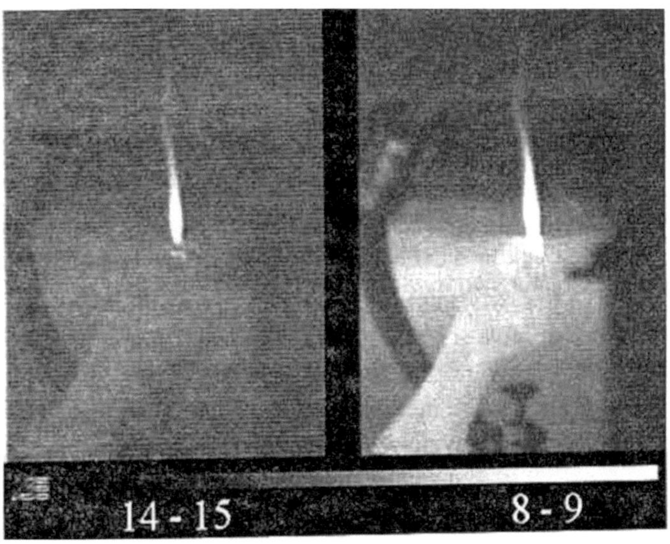

Fig. 26. Both pictures show (flame-simultaneously acquired) two-colour images with the 640×486 two-color QWIP camera. Image on the left is from 14-15μm IR and the image on the right is from 8-9 μm IR. Pixel pitch of the FPA is 25μm. The 14-15 μm image is less sharp due to the diffractions limited spot size being larger than the pixel pitch of the FPA (Guapala et al., 1999).

8.3 Broad-band QWIPs

A broad-band MQW structure can be designed by repeating a unit of several quantum wells with slightly different parameters such as quantum well width and barrier height. The first device structure (shown in Fig. 27 demonstrated by (Bandara et al., 1998a) has 33 repeated layers of GaAs three-quantum-well units separated by L_B ~575 Å thick $Al_xGa_{1-x}As$ barriers (Bandara et al., 1998b).

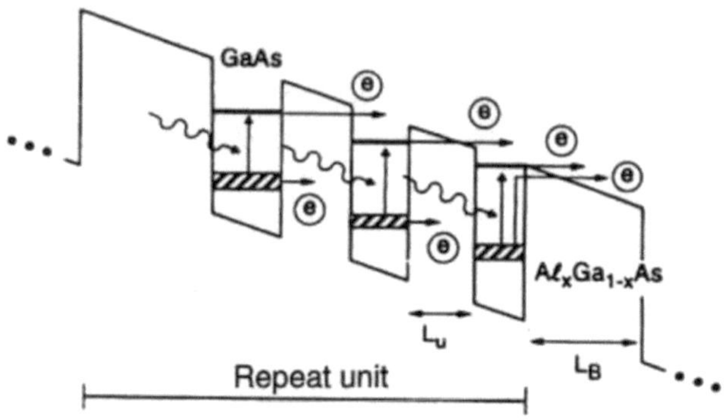

Fig. 27. Schematic diagram of the conduction band in broadband QWIP in an externally applied electric field. The device structure consists of 33 repeated layers of three-quantum-well units separated by thick $Al_xGa_{1-x}As$ barriers. Also shown are the possible paths of dark current electrons and photocurrent electrons of the device under a bias (Bandara et al., 1998a).

The well thickness of the quantum wells of three-quantum-well units are designed to respond at peak wavelengths around 13, 14, and 15μm respectively. These quantum wells are separated by 75Å thick $Al_xGa_{1-x}As$ barriers. The Al mole fraction (x) of barriers throughout the structure was chosen such that the λ=13μm quantum well operates under bound-to-quasibound conditions. The excited state energy level broadening has further enhanced due to overlap of the wave functions associated with excited states of quantum wells separated by thin barriers. Energy band calculations based on a two band model shows excited state energy levels spreading about *28 meV*. The unique characteristic of this instrument (besides being small and efficient) is that it has one instrument line shape for all spectral colours and spatial field positions. By using broad-band QWIP arrays with wavelength out to 16μm, the next version of this instrument could become the first compact, high resolution thermal infrared, hyper-spectral imager with a single spectral line shape and zero spectral smile. Such an instrument is in strong demand by scientists studying earth and planetary science.

8.4 Applications

8.4.1 Fire fighting

Video images were taken at a frame rate of 60Hz at temperature as high as *T=70K*, using a ROC (over current resistor) capacitor having a charge capacity of 9×10^6 electrons. This infrared camera helped a Los Angles TV news crew get a unique perspective on fires that raced through the Southern California seaside community of Malibu in October, 1996. The camera was used on the station's news helicopter. This portable camera features IR detectors which cover longer wavelengths than previous portable cameras could. This allows the camera to see through smoke and pinpoint lingering hotspots which are not normally visible (Guapala et al., 2001). This enabled the TV station to transmit live images of hotspots in areas which appeared innocuous to the naked eye. These hotspots were a source of concern and difficult for firefighters, because they could flare up even after the fire appeared to have subsided. Fig. 28 shows the comparison of visible and IR images of a just burned area as seen by the news crew in nighttime. It works effectively in both daylight and nighttime conditions. The event marked the QWIP camera's debut as a fire observing device.

(a) Visible (b) QWIP

Fig. 28. Comparison of visible and IR images of a just burned area as seen by a highly sensitive visible CCD camera and the long wavelength IR camera at nighttime. (a) Visible image from a CCD camera. (b) Image from the 256×256 portable QWIP camera. This portable camera features IR detectors covers longer wavelengths than previous portable cameras could (Guapala et al., 2001).

8.4.2 Volcanology

Recently, the camera has been used to observe volcanoes, mineral formations, weather and atmospherics conditions. This QWIP camera was taken to the Kilauea Volcano in Hawaii. The objectives of this trip were to map geothermal feature. The wide dynamic range enabled us to image volcanic features at temperatures much higher (300-1000K) than can be imaged with conventional thermal imaging systems in the 3-5μm or in visible. Fig. 29 shows the comparison of visible and IR images of the Mount Kilauea Volcano in, Hawaii. The IR image of the volcano clearly shows a hot lava tube running underground which is not visible to the naked eye (Guapala et al., 2001).

(a) Visible (b) QWIP

Fig. 29. Comparison of visible and IR images of the mount Kilauea Volcano in, Hawaii. (a) Visible image from a highly sensitive CCD camera. (b) Image from the 256×256 portable QWIP camera. This demonstrates the advantages of long wavelength IR in geothermal mapping (Guapala et al., 2001).

8.4.3 Medicine

Studies have determined that cancer cells exude nitric oxide. This causes changes in blood flow in tissue surrounding cancer that can be detected by a sensitive thermal sensor. Recently, OmniCorder Technologies, Inc., Stony Brook, N.Y. has developed an instrument called the BioScan System TM based on developed QWIP FPA technology. In this instrument a mid format QWIP FPA is used for Dynamic Area Telethermometry (DAT)(Bandara et al., 1998b). DAT has been used to study the physiology and patho-phy-siolgy of cutaneou perfusion, which has many clinical applications. DAT involves accumulation of hundreds of consecutive IR images and fast Fourier Transform (FFT) analysis of the biomodulation of skin temperature, and of the microhomogeneity of skin temperature (HST, which, measures the perfusion of the skin's capillaries). The FFT analysis yields the thermoregulatory frequencies and amplitudes of temperature and HST modulation.

To obtain reliable DAT data, one needs an IR camera in the larger 8μm range (to avoid artifacts of reflections of modulated emitters in the environment) a repetition rate of 30 Hz allowing accumulation of a maximal number of images during the observation period (to maximize the resolution of the FFT). According to these researches on the longer wavelength operation, higher spatial resolution, higher sensitivity and greater stability of the QWIP FPAs made it the best choice of all IR FPAs.

The two technologies work together to image the target area and to provide the physician with immediate diagnostic information. It causes no discomfort to the patient and uses no ionizing radiation. Basically, the digital sensor detects the IR energy emitted from body, thus "seeing" the minute differences associated with blood flow changes.

This camera has also been used by a group of researchers at the University of Southern California in brain surgery, skin cancer detection, and leprosy patients. In brain tumour removal a sensitive thermal imager can help surgeons to find small capillaries that grow towards the tumour due to angiogenesis (see Fig. 30). In general cancerous cells have high metabolic rate and these cancerous cells recruit new blood supply; one of the characteristic of a malignant lesion. Therefore, cancerous tissues are slightly warmer than the neighboring healthy tissues. Thus, a sensitive thermal imager can easily detect malignant skin cancers (see Fig. (31- a, and b)).

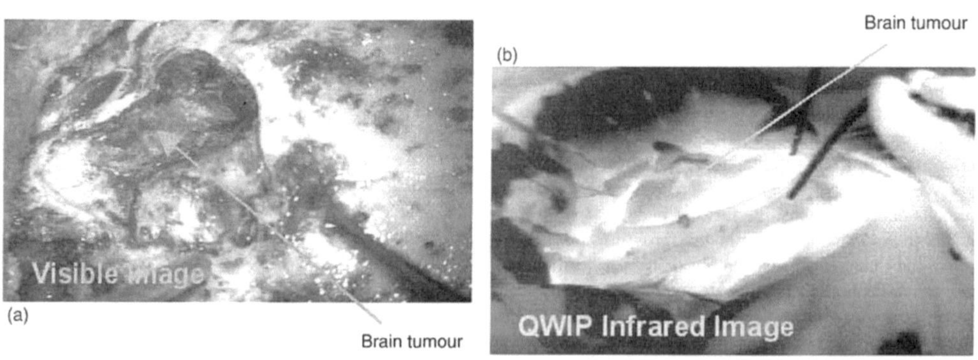

Fig. 30. (a) Visible image of a brain tumour (most of the cancerous cells are dead due to cancer sensitive drugs). (b) The thermal IR image clearly discriminate the healthy tissues from dead tissues (Bandara et al., 1998b).

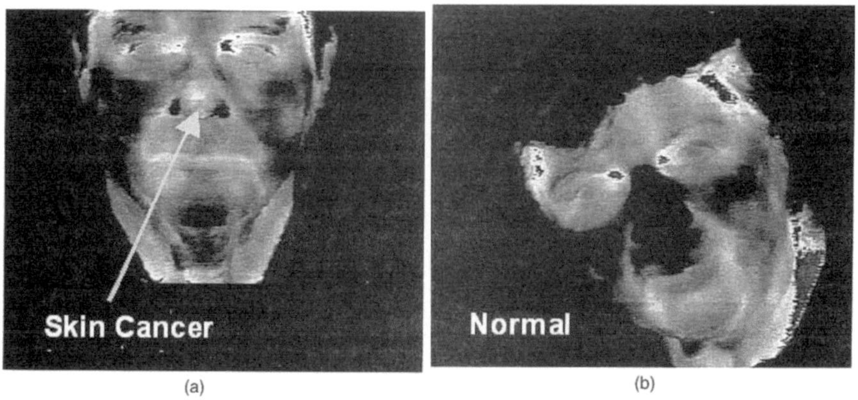

Fig. 31. (a) This clearly shows the tip of the nose is warmer than its surrounding tissues due to the enhanced metabolic activity (angiogenesis) of a skin cancer. (b) Shows a face with no skin cancer on the nose. Usually, nose and ears are colder relative to other parts of the face, because those are extending out the body (Bandara et al., 1998b).

8.4.4 Defence

It is not necessary to explain how real time IR is important in surveillance, reconnaissance and military operations. The QWIP RADIANCE was used by the researchers at the Ballistic Missile Defence Organization's innovative science and technology experimental facility in a unique experiment to discriminate and clearly identify the cold launch vehicle from its hot plume emanating from rocket engines.

Usually, the temperature of cold launch vehicles is a bout 250K, whereas the temperatures of the hot plume emanating from launch vehicle can reach 950 K. According to the plank's blackbody emission theory, the photon flux ratio of 250K and 950K blackbodies at 4μm is about 25,000, whereas the same photon difference at 8.5micm is about 115. Therefore, it is very clear that one must explore longer wavelengths for better cold-body versus hot plume discrimination, because the highest instantaneous dynamic range of IR cameras is usually 12-bits (i.e., 4096) or less. Fig. 32 shows an image of delta-II launch taken with QWIP RADIANCE camera. This clearly indicates the advantage of long-wavelength QWIP cameras in the discrimination and identification of cold launch vehicles in the presence of hot plume during early stage of lunch (Guapala et al., 2001).

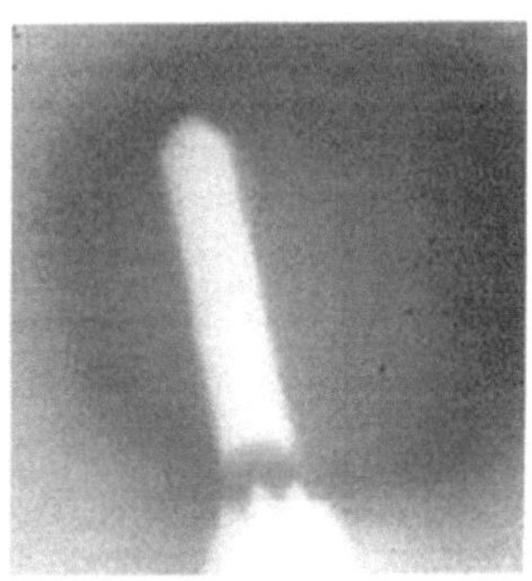

Fig. 32. Image of a Delta-II launch vehicle taken with the long-wavelength QWIP RADIANCE during the launch (Guapala et al., 2001).

8.4.5 Astronomy

In this section, astronomical observation with a QWIP FPA is studied (Guapala et al., 2001). A QWIP wide-field imaging multi-colour prime focus IR camera (QWICPIC) is utilized. Observations were conducted at the five meter Hala telescope at MT. Palomar with QWICPIC based on 8-9 μm 256×256 QWIPs FPA operating at $T=35K$. The ability of QWIPs to operate under high photon background without excess noise enables the instrument to observe from the prime focused with a wide 2′×2′ field of view, making this camera unique among the suite of IR instruments available for astronomy. The excellent 1/f noise performance of QWIP FPAs allows QWICPIC to observe in a slow scan strategy often required in IR observations from space. Fig. 33 compares an image of a composite near-IR image (a) obtained by Two Micron All Sky Survey (2MASS), with an 8.5 μm long-wavelength IR image (b) obtained with a QWIPFPA at primary focus of the Palomar 200-inch hale telescope. The S106 region displays vigorous star-formation obscured behind dense molecular gas and dust heated by star light. Thermal-IR imaging can be used to assess the prevalence of warm (~300K) dusty disks surrounding stars in such region. Formations of these disks are an evolutionary step in the development of planetary systems. These images demonstrate the advantage of large format, stable (low 1/f noise) LWIR QWIP FPAs for surveying obscured regions in search for embedded or reddened objects such as young forming stars.

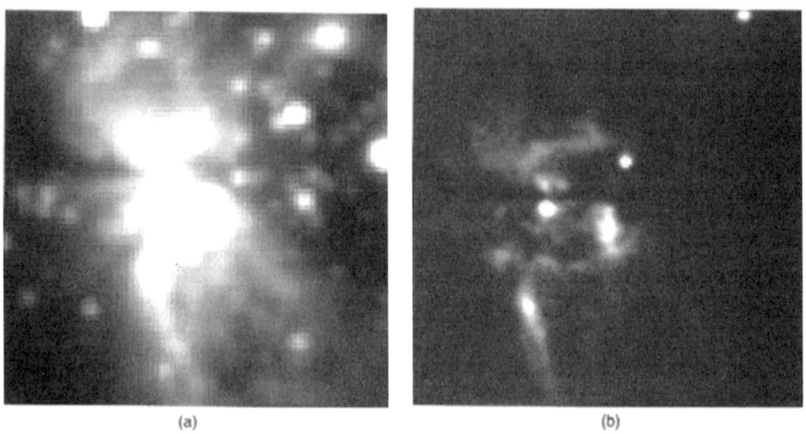

Fig. 33. Compares an image of a composite near-IR image (a) obtained by 2MASS, with an 8.5 μm long-wavelength IR image (b) obtained with a QWIPFPA at primary focus of the Palomar 200-inch hale telescope (Guapala et al., 2001).

9. Summery and conclusions

After previous discussions, one can interpret the importance of the photodetectors as a first part of any system-detection. This system-detection can be altered according to the covered range of photodetectors. By another meaning these applications are varied corresponding to their spectrum wavelengths. There are mainly two types, one of them is the infrared photodetectors which covered middle, long, ultra or far long wavelengths, (750- ~10^5nm). Another one is ultra violet photodetectors which covered shortest and short wavelengths, (~100 to 365 nm).

The study is concentrated to the semiconductor-based photodetectors. In the case of IR photodetectors, its interest is based on different fields of its applications such as night vision, medicine, space communications, missiles, and so on. The neediest of qualified IR detectors directed the researchers toward developing new constructions that denote a facility of mass production. The main advantage of the traditional commercial photodetectors is that they can be operated at room temperature. However the main problem is the complexity of fabrication. Moreover, the soften materials which it can be utilized. Other drawbacks can be recognized from traditionally p-i-n structure series is how one can balance between the response time; depending on the width of layer, and the required gain or amplification. So, the thinking is intended for overcome the previous problems. The ideas are how to obtain the crystal layers which offer a small energy gap between the valance and conduction bands. Also, how one can obtain both of fast detectors response time and gain. Moreover, the problem of fabrications blocked the growth in this direction of the bulk IR detectors. The solution is to transfer from inter-band translation (bulk) to intersubband translation (quantum). That means that electron or holes should be quantized. The number of freedom of mobile charge determines the type of IR photodetector. There are three types of quantum IR photodetectors (QIRPDs). The quantum dot IR photodetectors (QDIPs), the degree of freedom is zero. The quantum wire IR photodetectors (QRIPDs), the degree of freedom is one. The last type is quantum well IR photodetectors (QWIPs) the degree of freedom is two. Although the quantum detectors indicate an easier way of fabrication and mass production, it suffers from many drawbacks which depend on the nature of freedom into quantum detectors. For example, QWIPs suffer from coupling and lower gain. The coupling problem and gain are processed by much research groups and they are trying to overcome by utilizing corrugated quantum well and focal plane arrays (FPAs) as discussed before. In QDIPs, the problem of coupling is eliminated as it is response to the normal radiations. Meanwhile it suffers from high dark current. The last one is QRIDs, it denotes smaller dark current comparing with quantum dot detectors. It opens the way to enhance the performance of IR detectors. The performance parameters can be summarized to responsivity, quantum absorption efficiency, and detectivity. The theoretical details of the dependence of these photodetectors performance parameters into quantum detectors parameters are studied before in previous articles (Aboshosha et al., 2009; Aboshosha et al., 2009b; Abou El-Fadl et al., 1998; Eladle et al., 2009; Nasr, El_mashade, 2006; El_mashade et al., 2003; El_mashade et al., 2005; Future Technologies Division, 2003; Nasr, 2011; Nasr, 2011b; Nasr, 2009; Nasr et al., 2009; Nasr, 2008; Nasr,2006; Nasr & Ashry; 2000; Nasr, 2007). From the obtained results, the QWIPs give high responsivity which mean smaller values of dark current in corresponding to other types.

Another type of semiconductor-based detectors is UV ranged photodetectors. It is evident to consider the fact that, here, the energy gap is not small in contrast to IR detectors. So, the modern advances in UV photodetectors will be based on enhancing the p-n detector performance. Such as, the wide range which can be covered or higher temperature which can be operated or recent mixture composition which can be obtained.

On the other hand, One-dimensional UV semiconductor nanowires and nanorods have also attracted great attention because of their potential applications in both nanoscale electronic and optoelectronic devices. Compared with bulk or thin film photodetectors, 1-D photodetectors should be able to provide a large response due to their larger aspect ratio of length to diameter and high surface to volume ratio.

The importance of UV photodetector is related to its application as described before for solar UV detection, visible blind detection, missile warning, space communication and flame sensing.

The modern advanced IR phodetectros are built on utilizing QWIPDs FPAs which denote many features. For example, they give higher resolution and performance. Also, they become commercially available now.

The development in the case of QWIPs FPAs will be done by multi techniques. Such as, by solving quantum well by dot or/ and wire layer FPAs. Another technique can be recognized is to utilize a hybrid composition from well and dot layer which is known as quantum-dot-in–a-well (DWELL FPAs) detectors as given in the next section. This DWELL FPAs can combine the advantage of both of well and dot quantum types. That it can achieve a response to the normal incident IR rays in the case of QD and high responsivity in the case of QW.

10. Future prospects

Firstly, all applications considered before can utilize quantum dot and quantum wire infrared photodetectors in place of quantum well. That can be done by replacing each one well by one array of QDs or QRs. Practical applications of QDIPs are beginning to widespread in different fields. Later one, QRIPs is still under investigation phase but from my research background, it will give improved performance in comparison with QDIPs. Because of it denotes less than dark current and can cover wide range from IR spectrum (Nasr, 2007, Nasr, 2011). Moreover as discussed in (Nasr, 2009), the future of IR detectors is directed to exploit the advantages of FPAs in the reminder of IR quantum devices such as QDIPs and QRIPs. Additionally, recent articles are exposed to fabricate Long-Wave Infrared Quantum-Dots-in-a-Well detector as in (Pucicki et al., 2007; Andrews et al., 2009) and held comparison with FPAs as shown in Fig. 34. The results encourage developing a devices by the same way to combine the advantages of each quantum devices. Design and fabricate a broadband 8-14 μm infrared *1K×1K*, GaAs Quantum Well Infrared Photodetector (QWIP) imaging array is important point for enhance the performance as in (Buckner, 2008).

On the other hand, related to photodetectors that acquired our interest, there are quantum laser sources and modulators that move in the same direction and still need more studies now and in the next projects of nanotechnology devices.

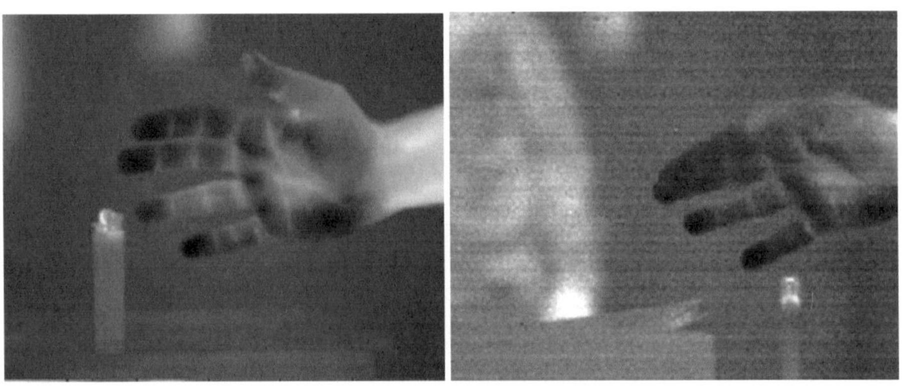

Fig. 34. Nonuniformity-corrected images obtained with the (left) QWIP and (right) DWELL FPA cameras at 60 K of a cigarette lighter on or near a book. The right image also has a person walking in the background (Andrews et al., 2009).

11. References

Aboshosha et al. (2009a). Precise Autonomous Rescuing relying on Integrated Visual and IR Detection Technique. *J. of Engineering & Computer Sciences; QUJECS*, Vol. 2, No.3

Aboshosha et al. (2009b). Improved Thermographic Object Detection Relying-on Image Processing and Sensorial Development of Quantum IR Photodetectors. *Menofia J. of Electronic Engineering Research (MJEER)*, Vol. 19, No.1&2, pp. 63-80

Abou El-Fadl et al. (1998). Computer Aided Analysis of Multimode Planar Graded-Index Optical Waveguides. the 2nd IMACS , International Multiconference CESA'98 Nabeul-Hammamet, Tunisia. April 1-4

Anbar et al. (1999). Babalola, "Clinical Applications of DAT Using a QWIP FPA Camera. *Proc. SPIE 3698*, p. 93-102

Andrews et al. (2009). Comparison of Long-Wave Infrared Quantum-Dots-in-a-Well and Quantum-Well Focal Plane Arrays. *IEEE Trans. on Electron Devices*, Vol.56, No. 3, 2009.

Bandara et al. (1998a). 10-16 μm Broad-Band Quantum Well Infrared Photodetectors. *Appl. Phys. Lett.*, Vol.72, pp. 2427-2429

Bandara et al. (1998b). 10-16 μm Broad-Band Quantum Well Infrared Photodetectors," Proc. SPIE, 3379, 396-401

Brune, H. (2001). Encyclopedia of Materials: Science and Technology. *Elsevier S*, pp. 3683-3693

Buckner, J. L. (2008). HyspIRI Technology Investment Overview. *Earth Science Technology Office*, NASA

Campbell et al. (1983). High-Performance Avalanche Photodiode with Separate Absorption,. 'Grading' and Multiplication Regions. *Electron Lett.* Vol.19, pp. 818-820

Capasso, F. (1985). Physics of Avalanche Photodiodes. *in Semiconductor and Semimetals*, part D, W. T. Tsang, Ed., Academic Press, Orlando, FL, Vol. 22, pp. 1-172

Chang, C. Y. & Francis Kai, (1994). GaAs High Speed Devices. *John Wiley & Sons*

Coleman, J. J. (1997). Metalorganic Chemical Vapor Deposition for Optoelectronic Devices" proceedings of IEEE, Vol.85

Das, B. & Singaraju, P. (2005). Novel Quantum Wire Infrared Photodetectors. *Infrared Physics & Technology*, Vol. 46, pp. 209-218

Donkor, E. (2004). GaAs-Based Nanodevices. *Encyclopedia of Nanoscience and Nanotechnology*, Vol. 3, pp. 703-717

El_Mashade et al. (2003). Characteristics Determination of Quantum Well Infrared Photodetectors. *IEEE Mediterranean Microwave Symposium*, pp. 29-40

El_Mashade et al. (2003).Theoretical Analysis of Quantum Dot Infrared Photodetectors. *J. Semicond. Sci. Technol.* Vol. 18, p. 891-900, UK

El_Mashade et al. (2005). Performance Analysis of Quantum Well Infrared Photodetectors. *J. of Optical Communications*, Vol. 3, pp. 98-110, Germany, Mar. 2005.

Eladl et al. (2009). Dynamic Characteristics of QWIP-HBT-LED Optoelectronic Integrated Devices" J. of Semiconductor Phys., Quantum Electronics & Optoelectronics, Vol. 12, No. 3. pp. 260-263

Fox, M., (2001). Optical Properties of Solids. *Oxford University*

Future Technologies Division, 2003. Applications of Nanotechnology in Space Developments and Systems. Published by *VDI Technology Center*, No. 47, Düsseldorf, Germany

Guapala et al. (1997). 9 μm Cutoff 256×265 *GaAs/* $Al_xGa_{1-x}As$ Quantum Well Infrared Photodetectors Hand-Held Camera. *IEEE trans. Electron devices*, Vol. 44, pp. 51-57
Guapala et al. (1998). Long-Wavelength 640×486 *GaAs/* $Al_xGa_{1-x}As$ Quantum Well Infrared Photodetectors Snap Shot Camera. *IEEE Trans. Electron Dev.*, Vol. 45, pp. 1890-1895
Guapala et al. (1999). 8-9 and 14-15 μm Two Color 640×486 *GaAs/* $Al_xGa_{1-x}As$ Quantum Well Infrared Photodetectors (QWIP) Focal Plane Array Camera. *Proc. SPIE*, 3698, pp. 687-697
Gunpala et al. (2001). Recent Development and Applications of Quantum Well Infrared Photodetectors Focal Plane Arrays. *J. Optoelect. Review*, Vol.9, No.2, pp.150-163
Hsueh et al. (2007). Crabwise ZnO Nanowire UV Photodetector Prepared on ZnO" : Ga/Glass Template. *IEEE Trans. on Nanotech.*, Vol. 6, No. 6, pp. 595-600
Johnson et al. (1984. A MOCVD Reactor Safety System for a Production Environment. *Journal Crystal Growth*, Vol. 68, pp. 497-501
Jung et al. (2006) . Circuit Fabrication at 17 nm Half-Pitch by Nanoimprint Lithography. *Nano Lett.* Vol. 6, No. 351
Kasper, B. L. & Campbell, J. C. (1987). Multigigabit-per-Second Avalanche Photodiode Lightwave Receivers. *J. Lightw. Technol.* LT-5, pp. 1351-1364
Krost et al. (1999). X-Ray analysis of Self-Organized InAs/InGaAs Quantum Dot Structure. *Cryst. Res. Technology*, Vol. 34, pp. 89-102
Lee et al. (2005. Epitaxial Growth of a Nanoscale, Vertical-facet, High Aspect Ratio, One-Dimensional Grating in III-V Materials for Integrated Photonics. *Appl. Phys. Lett.*, Vol. 87, 071110-1-3
Levine, B. F. (1993). Quantum Well Infrared Photodetectors. *J. Appl. Phys.*, Vol. 74, pp. 7-121
Liang et al. (2006). InGaAs Quantum Dots Grown on B-type High Index GaAs Substrates: Surface Morphologies and Optical Properties", Nanotechnology, 17, pp. 2736-2740.
Liu et al. (1993). Fabrication of Quantum Dots Using Insitu Etching and Regrowth During MBE Growth. *IEEE,* 1993.
Liu, H. C. (1992). Noise Gain and Operating Temperature of Quantum Well Infrared Photodetectors. *Appl. Phys. Lett.*, Vol. 61, pp. 2703-2705
Maksimović, M. (2003). Quantum Dot Infrared Photodetector as an Element for Free-Space Optical Communication Systems. Xi Telekomunikacioni Forum TEL For, Beograd, Sava centar, Vol.11, pp.25-27
Manasevit, H. M. (1968). Single-Crystal Gallium Arsenide on Insulating Substrates. *A Physics Letter*, Vol. 12, No 4, pp. 156-159
Melchior, H. (1977). Demodulation and. Photodetection Techniques. *Laser Handbook*, Vol.1, F. T. Arecchi and E. O. Schulz-Dubois, Eds., North-Holland, Amsterdam, 1972, pp. 725-835; Phys. Today, Vol. 30, No.12, 32
Muhammad et al. (2007). Epitaxial Method of Quantum Devices Growth. *Ibnu Sina Institute for Fundamental Science Studies*, Malaysia
Nai-Chang, Y. (2008). A Perspective of Frontiers in Modern Condensed Matter Physics" AAPPS Bulletin, Vol. 18, No. 2, pp. 1-19
NASA (1999). Report of UV/O Working Group
Nasr et al. (2006). "Dark Current Characteristics of Quantum Wire Infrared Photodetectors," IEE Proc. Optoelectronic J., Vol.1, No. 3, pp. 140–145
Nasr et al. (2009). Theoretical Characteristics of Quantum cascaded lasers. *J. of Engineering & Computer Sciences; QUJECS*, Vol. 2, No.2

Nasr, A. & Ashry, M. (2000). Computer Aided Analysis of TM- Multimode Planar Graded-Index Optical Waveguides and the Field Distribution. *Seventh Conference of Nuclear & Applications*, Cairo, Egypt 6-10

Nasr, A. & El_Mashade, M. (2006). Theoretical Comparison between Quantum-Well and Dot Infrared Photodetectors. *IEE Proc. Optoelectronic*, Vol. 153, No. 4, pp. 183-190

Nasr, A. (2007). Performance of Quantum Wire Infrared Photodetectors under Illumination Conditions. *Optics & Laser Technology, Elsevier*, Vol. 41, pp. 871–876

Nasr, A. (2008). Spectral Responsivity of the Quantum Wire Infrared Photodetectors. *Optics & Laser Technology, Elsevier,* Vol. 41, pp. 345– 350

Nasr, A. (2009). Theoretical Characteristics of Quantum Wire Infrared Photodetectors under Illumination Conditions. *J. of Optical Communications*, Vol. 30, Issue 3, pp. 126-132

Nasr, A. (2011). Detectivity Performance of Quantum Wire Infrared Photodetectors. *J. Opt. Commun.*, Vol. 32 No. 2 , pp. 101–106.

Nasr, A. (2011). Theoretical study of the photocurrent performance into quantum dot solar cells. *under publication*, 2011

National Nanotechnology Initiative Workshop. (2004). Nanotechnology in Space Exploration. *Report*

Norbert, K. & Jurgen, B. (2005). Guidelines for Future UV Observatories. *Institut für Astronomie und Astrophysik*, Abteilung Astronomie (IAAT), Universität Tübingen, pp. 1-6

Pauzauskie, P. J. & Yang, P. (2006). Nanowire Photonics: Review Feature. *Materials Today*, Vol. 9, No.10

Pernot et al. (1999). Improvement of Low-Intensity Ultraviolet Photodetectors. *Phys. Stat. Sol.(A)* Vol.176, pp. 147-151

Phillips et al. (1999). Self-Assembled InAs-GaAs Quantum-Dot Intersubband Detectors. *IEEE J. Quantum Electron*, Vol. 35, No. 6, pp. 936-943

Piotrowski, J. & Rogalski, A.(2007). High-Operating-Temperature Infrared Photodetectors. *SPIE Press Book*

Piotrowski, J. (2004). Uncooled Operation of IR photodetectors. *J. Optoelectronic Review*, Vol. 12, No. 1, pp. 111-122

Pucicki et al. (2007). Technology and Characterization of p-i-n Photodetectors with DQW (In,Ga)(As,N)/GaAs Active Region. *Optica Applicata*, Vol. XXXVII, No. 4

Ryzhii et al. (2000). Dark Current in Quantum Dot Infrared Photodetectors. *Jpn. J. Appl. Phys.* Vol. 39, pp. L 1283-L 1285

Ryzhii, V. & Khmyrova,I. (2001). On the Detectivty of Quantum-Dot Infrared Photodetectors. *Appl. Phys. Let.* Vol. 78, No. 22, pp. 3523-3525

Ryzhii, V. (2003). Intersubband Infrared Photodetectors. *World Scientific*, Vol. 27

Shen et al. (2000). Progress on Optimization of P-type GaAs/AlAs Quantum Well Infrared Photodetectors. *J. Vaac. Sci. Technol.* A, Vol. 18, pp. 601-604

Stillman, G. E. & Wolfe C. M.(1977). Avalanche Photodiodes. *Infrared Detectors II, R. K. Willardson and A. C. Beer, Eds.*, Vol. 12 of Semiconductors and Semimetals, pp. 291-393. Academic Press, New York

Sze, S. M. (1981). Physics of Semiconductor Devices. *Wiley*, New York

Tarif, L. E. (1991). Planar InP/InGaAs Avalanche Photodiodes with a Gain-Bandwidth Product Exceeding 100 GHz. *Proc. Optical Fiber Common. Conf.*, Optical Society of America, Washington, Dc, paper ThO3, Vol. 4

Tucker et al., (1986). Coaxially Mounted 67 GHz Bandwidth InGaAs PIN Photodiode. *Electron. Lett.*, Vol. 22, No. 17, pp. 917-918

URL http://www.its.caltech.edu/~feynman/

Waleker et al. (2007). The development of nitride base UV photodetectors. *J. Optoelectronic Review*, Vol. 8, No. 1, 2000.

www.nanoforum.org (2007). Nanotechnology and Civil Security Report

Zhao, X. (2006). Carrier Transport in High-Speed Photodetectors Based on Two-Dimensional-Gas. *Thesis*

6

Including Solar Radiation in Sensitivity and Uncertainty Analysis for Conjugate Heat Transfer Problems

Christian Rauch
Virtual Vehicle Research and Test Center (ViF)
Austria

1. Introduction

With the advancement of computer-aided engineering (CAE) and the development of coupling tools that enable the exchange of data from various codes specialized in their respective fields, the possibility of acquiring more accurate results rises, because some boundary conditions can be calculated. Rauch et al. (2008) stated various other reasons for making use of co-simulation. Hand in hand with performing multi-physics coupling the required knowledge of the underlying theories, model limitations, parameters, and numerical constraints increases. This can lead to involving specialists of various fields but it can also entice someone to make improper choices for parameters of models an engineer is not sufficiently acquainted with. Either way, the number of possible parameter settings usually rises with the number of models employed. To serve as an example, Rauch et al. (2008a) were involved with an underbody simulation of a car. The exhaust system transported the exhaust gas out of the engine and heat was released along the piping system. A commercial thermal radiation solver was used to calculate wall temperatures including conduction and convection. Convective heat transfer coefficients and fluid temperatures at the wall were imported from a commercial computational fluid dynamics (CFD) solver which, in turn, received the wall temperatures. Other parameters in the radiation solver to be set were conductivity, emissivity, and settings for the view factor calculation.

When confronted with deviations from measurements, the simulation engineer is left to choose those parameter changes that will improve the results based on a sound physical footing. This involves either expert knowledge, or a trial and error approach. Sensitivity analysis provides a systematic way for determining the most influential parameters. Global sensitivity analysis uses a statistical approach in understanding the total parameter space of the system. Saltelli et al. (2008) provide an overview on global methods which are suited for optimization. Local methods provide local information for the response of the system under investigation (Saltelli et al., 2008a).

Roy and Oberkampf (2011) give an overview of sources of uncertainties that influence the outcome of a simulation. They identified uncertainties that stem from model inputs, such as model parameters, geometry, boundary and initial conditions, and external influences such as forces. They do occur in the manufacturing process, in natural resources, and might change during the life-cycle. Another source of uncertainties is of a numerical nature. To this category belong discretizations in space and time, computational round-off errors, iterative convergence errors, or programming errors. Finally, the model itself might be based on uncertain footings such as assumptions or model simplifications.

In this work first-order local sensitivity analysis is being performed on a conjugate heat transfer case. This method derives the first-order derivatives of the model equations and calculates those derivatives with the results of a simulation. According to Saltelli and Annoni (2010), this method behaves similarly to one-at-a-time (OAT) screening in a high-dimensional parameter space. This method is more convenient to the engineer as it deals with results of the actual simulation without having to perform many simulations with parameters that are not of immediate interest. Starting from the simulation results small perturbations are excited in order to discover the surroundings of the solutions for all parameters of interest. In this way, the outcome of a perturbation can be ascribed to a particular parameter. Contrary to stochastic methods, non-influential parameters cannot falsely be identified as important. Furthermore, a model failure can be ascribed to one specific parameter. Finally, it helps the engineer to understand the behavior of the system in an easy to follow way. The draw-back of this method is its limited validity for extrapolation in non-linear systems.

To the author's knowledge, there are only a few papers devoted to sensitivity analysis for thermal cavity radiation including conduction and convection. Blackwell et al. (1998) investigated a control volume finite element system with conjugate heat transfer, but only simplified radiation to the environment without the use of view factors and reflections were considered. Taylor et al. (1993) gave a treatment on the impact of view factors on the outcome of the energy balances for cavity radiation. They were able to show by an academic example, that the accuracy of some form factors might influence the whole system, significantly. They also developed sensitivity balances for the emissivity and temperature for diffuse-gray radiation when a temperature is set as boundary condition. They further elaborated on this topic by expanding the sensitivities to changes in area and heat flux boundaries (Taylor et al., 1994, 1995). They found that enforcing reciprocity and closure brings great improvements to the energy balances. Taylor & Luck (1995) augmented those findings with the study of various methods in reciprocity and closure enforcement for view factors. They found that the weighted least squares correction for view factors gave the most promising result in producing meaningful configuration factors out of approximate view factors.

Korycki (2006) derived local sensitivity factors of the first order in the context of the finite element method (FEM). He derived expressions for conjugate heat transfer including the effect of changing view factors due to shape optimizations and heat transfer in participating media. He also offered an adjoint formulation for sensitivity analysis. Bhatia and Livne (2008) developed sensitivity equations for weak coupling. They employed two sets of meshes. On the coarser mesh, external boundaries were calculated using FEM whereas the finer mesh was used for radiation in a cavity. Both meshes were coupled by a connectivity matrix. Since thermal stresses and deformations were investigated, deformed view factors were approximated by a first order Taylor series expansion, which showed some descent results. In another paper (Bhatia & Livne, 2009) they also showed a way to put all temperature dependent parameters into the main diagonal of a matrix with consequent

Taylor series expansion to get the inverse of a matrix. Furthermore, transient behavior was introduced.

This article is a continuation of previously published papers (Rauch & Almbauer, 2010a; Rauch, 2011b) that dealt with steady-state uncertainty estimations of some parameters for conjugate heat transfer with cavity radiation. Later, the transient (Rauch, 2011a) case was investigated. The present author also treated view factors in conjugate heat transfer analysis for steady-state (Rauch & Almbauer, 2010) and transient (Rauch, 2011) cases.

In this treatment, this approach of steady considerations of sensitivities based on equations of an element-centered finite difference method (FDM) is extended for a solar heat source term. Therefore, the equation of first order local discrete sensitivities for wall temperature with respect to solar heat flux is being derived. The governing equation is first discretized followed by a first order derivation. Alternatively, the continuous approach would first do the derivation and than discretize (Lange & Anderson, 2010). Two formulations, namely with and without reciprocity relation for view factors, are being considered. The following section is devoted to introducing a simple case resembling a simplified flat plate solar collector. In section four a sensitivity and uncertainty analysis is performed. The numerical stability of the solution matrix for the sensitivity coefficients is investigated. The effect of partial sensitivity analysis on uncertainty coefficients is shown. The difference of incorporating reciprocity for uncertainty coefficients is demonstrated. A comparison to uncertainty factors of other parameters is made. This work aims to contribute to the enhancement of understanding the simulated model at hand by showing where input parameters have a large influence on the results.

2. Theory

In their work, Siegel & Howell (2002) show how to formulate the net-radiation method, which is based upon first principle thermodynamics in non-participating media.

2.1 Basic equations for conjugate heat transfer

To help understand the following treatment the geometry of two thermal nodes is shown in Fig. 1.

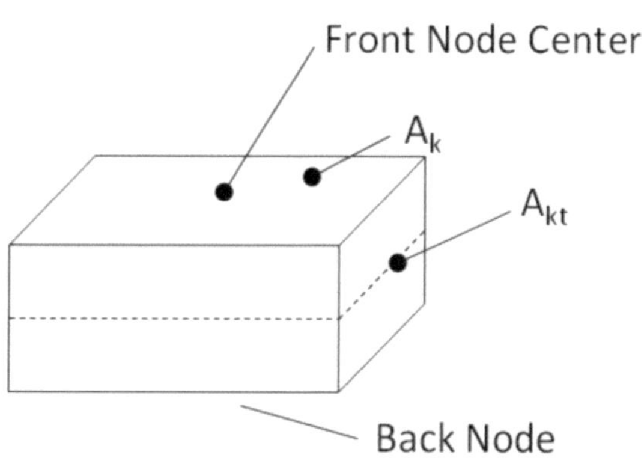

Fig. 1. Thermal node

A thermal node k consists of a center and a virtual extension. The node has an area A_k where it can exchange energy with the surroundings by convection and radiation. The dashed line in Fig. 1 marks the border to the thermal back node, if applicable. Through that border exchange from the surface node happens by conduction, only. The area A_{kt} marks the area to the neighboring node t, where heat is transferred from the surface node k to t by conduction.

First, the basic equations for conjugate heat transfer are introduced, starting with the relation for a given boundary heat flux b as a function of radiosity q_o, view factor F_{sk} between nodes s and k, and area A for an element k or s:

$$b_k = q_{o,k} - \frac{1}{A_k} \sum_{s=1}^{N} A_s q_{o,s} F_{sk} \tag{1}$$

The running index s spans all N thermal nodes of the geometry. This also includes the backside of a face, where applicable. The summation over all surface nodes sets thermal radiation apart from convective and conductive heat transfer. Where for the latter two heat transfer modes only neighboring elements for low order methods have an impact, in thermal cavity radiation all entities have to be included resulting in a dense solution matrix. The boundary heat flux in Eq. (1) consists of other heat transfer modes, source, and storage terms:

$$b_k = q_{solar,k} + h_k\left(T_{f,k} - T_k\right) + \frac{1}{A_k} \sum_{t=1}^{N_{Cond}} \left(-\frac{k_{kt} A_{kt}}{l_{kt}} (T_k - T_t) \right) \tag{2}$$

$T_{f,k}$ represents the fluid temperature for node k, T_k the wall temperature of node k, h_k the convective heat transfer coefficient, k_{kt} the thermal conductivity, which in this study is presumed to be independent of temperature, l_{kt} the distance between two thermal node centers, $q_{solar,k}$ the solar heat flux, N_{Cond} the number of thermally conducting neighbors, and T_t the wall temperature of the neighboring node t. The solar irradiance is attenuated by the atmosphere and strikes at the wall. There it can be reflected and in case of transparent material transmitted. The solar heat flux q_{solar}, with dots for heat fluxes being omitted in this treatment, is the actual heat absorbed by the wall. The radiation term of Eq. (1) is placed on the left-hand side (LHS) of Eq. (2) because the thermal radiation formulation does not follow thermodynamic sign convention.

With wall temperature and boundary heat fluxes as unknowns a second relationship is needed with emissivity ε_k:

$$b_k = q_{o,k} - (1 - \varepsilon_k) \frac{1}{A_k} \sum_{s=1}^{N} A_s q_{o,s} F_{sk} \tag{3}$$

Finally, the fourth relationship renders the set of equations solvable with the Stephan-Boltzmann constant σ:

$$b_k = \varepsilon_k \sigma T_k^4 \tag{4}$$

Local sensitivity coefficients of the first order are obtained by deriving the first order differentiation with respect to the solar heat flux. The order of the derivations follows the order of the four basic equations.

2.2 Sensitivity of solar heat flux

Starting with Eq. (1), total derivation yields:

$$\frac{db_k}{dq_{solar,i}} = \sum_{r=1}^{N} \frac{\partial b_k}{\partial q_{o,r}} \frac{\partial q_{o,r}}{\partial q_{solar,i}} \tag{5}$$

The underlying assumption is that a change in the solar heat flux is affecting the radiosity but not the geometry including view factors. Also, no temperature dependant properties are considered. The partial derivative of the boundary heat flux with respect to radiosity with δ as the Kronecker delta function yields:

$$\frac{\partial b_k}{\partial q_{o,r}} = \sum_{s=1}^{N} \left[\delta_{ks}(1 - F_{kk}) - (1 - \delta_{ks}) \frac{A_s}{A_k} F_{sk} \right] \tag{6}$$

At this point it is noteworthy no mention that in this treatment a derivation of a particular radiosity with respect to another radiosity is only considered when they are identical. In other words, a perturbation is kept local. The transportation of the perturbation information to the entire system is done by solving for the radiosity derivative with respect to the solar flux. Substitution of Eq. (6) into Eq. (5) results in:

$$\frac{db_k}{dq_{solar,i}} = \sum_{s=1}^{N} \left[\delta_{ks}(1 - F_{kk}) - (1 - \delta_{ks}) \frac{A_s}{A_k} F_{sk} \right] \frac{\partial q_{o,s}}{\partial q_{solar,i}} \tag{7}$$

The Kronecker delta function helps in the implementation and understanding of the equation as the radiosity $q_{o,k}$ in Eq. (1) had been drawn into the summation term. Next, the total derivative for Eq. (2) is formulated:

$$\frac{db_k}{dq_{solar,i}} = \frac{\partial b_k}{\partial T_k} \frac{\partial T_k}{\partial q_{solar,i}} + \sum_{r=1}^{N_{Cond}} \frac{\partial b_k}{\partial T_r} \frac{\partial T_r}{\partial q_{solar,i}} + \frac{\partial b_k}{\partial q_{solar,k}} \frac{\partial q_{solar,k}}{\partial q_{solar,i}} \tag{8}$$

The simplification of a local perturbation states:

$$\frac{\partial q_{solar,k}}{\partial q_{solar,i}} = 1 : i = k, \frac{\partial q_{solar,k}}{\partial q_{solar,i}} = 0 : i \neq k \tag{9}$$

The derivation of the boundary flow with respect to the solar heat flux is straightforward:

$$\frac{\partial b_k}{\partial q_{solar,k}} = 1 \tag{10}$$

Likewise, the derivation with respect to the wall temperature T_k gives:

$$\frac{\partial b_k}{\partial T_k} = -h_k - \frac{1}{A_k} \sum_{t=1}^{N_{Cond}} \frac{k_{kt}}{l_{kt}} A_{kt} \tag{11}$$

And for the neighboring wall temperature T_t:

$$\frac{\partial b_k}{\partial T_r} = \frac{1}{A_k} \sum_{t=1}^{N_{Cond}} \frac{k_{kt}}{l_{kt}} A_{kt} \tag{12}$$

Substituting Eq. (9) to (12) into Eq. (8) results in:

$$\frac{db_k}{dq_{solar,i}} = \left(-h_k - \frac{1}{A_k} \sum_{t=1}^{N_{Cond}} \frac{k_{kt}}{l_{kt}} A_{kt} \right) \frac{\partial T_k}{\partial q_{solar,i}} + \frac{1}{A_k} \sum_{t=1}^{N_{Cond}} \frac{k_{kt}}{l_{kt}} A_{kt} \frac{\partial T_t}{\partial q_{solar,i}} + \delta_{ki} \tag{13}$$

To bring the radiation term together with convection and conduction Eq. (13) is set equal to Eq. (7):

$$\begin{aligned} &\sum_{s=1}^{N} \left[\delta_{ks} (1 - F_{kk}) - (1 - \delta_{ks}) \frac{A_s}{A_k} F_{sk} \right] \frac{\partial q_{o,s}}{\partial q_{solar,i}} = \\ &= \left(-h_k - \frac{1}{A_k} \sum_{t=1}^{N_{Cond}} \frac{k_{kt}}{l_{kt}} A_{kt} \right) \frac{\partial T_k}{\partial q_{solar,i}} + \frac{1}{A_k} \sum_{t=1}^{N_{Cond}} \frac{k_{kt}}{l_{kt}} A_{kt} \frac{\partial T_t}{\partial q_{solar,i}} + \delta_{ki} \end{aligned} \tag{14}$$

This equation has two unknown gradients, namely radiosity and wall temperature with respect to solar heat flux.

Since the emissivity is not a function of temperature in this work, Eq. (3) is differentiated in the same way as in Eq. (5) which gives:

$$\frac{\partial b_k}{\partial q_{o,r}} = \sum_{s=1}^{N} \left[\delta_{sk} - \delta_{sk} (1 - \varepsilon_k) \frac{A_s}{A_k} F_{sk} - (1 - \delta_{sk})(1 - \varepsilon_k) \frac{A_s}{A_k} F_{sk} \right] \tag{15}$$

Substitution into Eq. (5) yields:

$$\frac{db_k}{dq_{solar,i}} = \sum_{s=1}^{N} \left[\delta_{ks} \left(1 - (1 - \varepsilon_k) F_{sk}\right) - (1 - \delta_{ks})(1 - \varepsilon_k) \frac{A_s}{A_k} F_{sk} \right] \frac{\partial q_{o,s}}{\partial q_{solar,i}} \tag{16}$$

Equation (4) is the last relation for total derivation:

$$\frac{db_k}{dq_{solar,i}} = \frac{\partial b_k}{\partial T_k} \frac{\partial T_k}{\partial q_{solar,i}} \tag{17}$$

$$\frac{\partial b_k}{\partial T_k} = 4 \varepsilon_k \sigma T_k^3 \tag{18}$$

$$\frac{db_k}{dq_{solar,i}} = 4 \varepsilon_k \sigma T_k^3 \frac{\partial T_k}{\partial q_{solar,i}} \tag{19}$$

Equation (19) is set equal with Eq. (16):

$$4 \varepsilon_k \sigma T_k^3 \frac{\partial T_k}{\partial q_{solar,i}} = \sum_{s=1}^{N} \left[\delta_{ks} \left(1 - (1 - \varepsilon_k) F_{sk}\right) - (1 - \delta_{ks})(1 - \varepsilon_k) \frac{A_s}{A_k} F_{sk} \right] \frac{\partial q_{o,s}}{\partial q_{solar,i}} \tag{20}$$

Equation (20) provides an expression for the temperature gradient when reformulated:

$$\frac{\partial T_k}{\partial q_{solar,i}} = \frac{1}{4\varepsilon_k \sigma T_k^3}\left(\sum_{s=1}^{N}\left[\delta_{ks}\left(1-(1-\varepsilon_k)F_{sk}\right)-(1-\delta_{ks})(1-\varepsilon_k)\frac{A_s}{A_k}F_{sk}\right]\frac{\partial q_{o,s}}{\partial q_{solar,i}}\right) \tag{21}$$

The same relationship will hold for a neighboring node. Therefore, index k in Eq. (21) is replaced with t:

$$\frac{\partial T_t}{\partial q_{solar,i}} = \frac{1}{4\varepsilon_t \sigma T_t^3}\left(\sum_{s=1}^{N}\left[\delta_{ts}\left(1-(1-\varepsilon_t)F_{st}\right)-(1-\delta_{ts})(1-\varepsilon_t)\frac{A_s}{A_t}F_{st}\right]\frac{\partial q_{o,s}}{\partial q_{solar,i}}\right) \tag{22}$$

These two equations can be substituted into Eq. (14):

$$\begin{aligned}
&\sum_{s=1}^{N}\left[\delta_{ks}(1-F_{sk})-(1-\delta_{ks})\frac{A_s}{A_k}F_{sk}\right]\frac{\partial q_{o,s}}{\partial q_{solar,i}} = \left(-h_k - \frac{1}{A_k}\sum_{t=1}^{N_{Cond}}\frac{k_{kt}}{l_{kt}}A_{kt}\right)\\
&*\frac{1}{4\varepsilon_k\sigma T_k^3}\left(\sum_{s=1}^{N}\left[\delta_{ks}\left(1-(1-\varepsilon_k)F_{sk}\right)-(1-\delta_{ks})(1-\varepsilon_k)\frac{A_s}{A_k}F_{sk}\right]\frac{\partial q_{o,s}}{\partial q_{solar,i}}\right)\\
&+\frac{1}{A_k}\sum_{t=1}^{N_{Cond}}\frac{k_{kt}}{l_{kt}}A_{kt}\frac{1}{4\varepsilon_t\sigma T_t^3}\\
&*\left(\sum_{s=1}^{N}\left[\delta_{ts}\left(1-(1-\varepsilon_t)F_{st}\right)-(1-\delta_{ts})(1-\varepsilon_t)\frac{A_s}{A_t}F_{st}\right]\frac{\partial q_{o,s}}{\partial q_{solar,i}}\right)+\delta_{ik}
\end{aligned} \tag{23}$$

When rearranging the terms with the unknown radiosity gradients to the left-hand side and the other terms to the right-hand side (RHS) the final relation is obtained:

$$\begin{aligned}
&\sum_{s=1}^{N}\left[\delta_{ks}(1-F_{sk})-(1-\delta_{ks})\frac{A_s}{A_k}F_{sk}\right]\frac{\partial q_{o,s}}{\partial q_{solar,i}} - \left(-h_k - \frac{1}{A_k}\sum_{t=1}^{N_{Cond}}\frac{k_{kt}}{l_{kt}}A_{kt}\right)\frac{1}{4\varepsilon_k\sigma T_k^3}\\
&*\sum_{s=1}^{N}\left[\delta_{ks}\left(1-(1-\varepsilon_k)F_{sk}\right)-(1-\delta_{ks})(1-\varepsilon_k)\frac{A_s}{A_k}F_{sk}\right]\frac{\partial q_{o,s}}{\partial q_{solar,i}}\\
&-\frac{1}{A_k}\sum_{t=1}^{N_{Cond}}\frac{k_{kt}}{l_{kt}}A_{kt}\frac{1}{4\varepsilon_t\sigma T_t^3}\\
&*\sum_{s=1}^{N}\left[\delta_{ts}\left(1-(1-\varepsilon_t)F_{st}\right)-(1-\delta_{ts})(1-\varepsilon_t)\frac{A_s}{A_t}F_{st}\right]\frac{\partial q_{o,s}}{\partial q_{solar,i}} = \delta_{ik}
\end{aligned} \tag{24}$$

2.3 Reciprocity

The equations of the previous chapter do not assume any interaction between view factors. But for physically meaningful view factors the reciprocity principle needs to be satisfied along with the closure relation. The investigation of closure is beyond the scope of this work. The interested reader may be referred to the work of Taylor & Luck (1995). The reciprocity relation is formulated as follows:

$$A_i F_{ij} = A_j F_{ji} \tag{25}$$

It states, that view factors can be transformed into each other by taking into account the size of the nodes. In this work this principle will be applied in reformulating the original equations. A difference of results for sensitivity formulations with and without reciprocity assumption will, apart from numerical issues, reflect the quality of the view factors.

The derivation for the solar flux with respect to radiosity with the reciprocity relation is identical with the one in the previous chapter. Consequently, it suffices to substitute Eq. (25) into Eq. (21) and Eq. (24).

$$\frac{\partial T_k}{\partial q_{solar,i}} = \frac{1}{4\varepsilon_k \sigma T_k^3}\left(\sum_{s=1}^{N}\left[\delta_{ks}\left(1-(1-\varepsilon_k)F_{ks}\right)-(1-\delta_{ks})(1-\varepsilon_k)F_{ks}\right]\frac{\partial q_{o,s}}{\partial q_{solar,i}}\right) \tag{26}$$

$$\begin{aligned}
&\sum_{s=1}^{N}\left[\delta_{ks}(1-F_{ks})-(1-\delta_{ks})F_{ks}\right]\frac{\partial q_{o,s}}{\partial q_{solar,i}}-\left(-h_k-\frac{1}{A_k}\sum_{t=1}^{N_{Cond}}\frac{k_{kt}}{l_{kt}}A_{kt}\right)\frac{1}{4\varepsilon_k\sigma T_k^3}\\
&{}^*\sum_{s=1}^{N}\left[\delta_{ks}\left(1-(1-\varepsilon_k)F_{ks}\right)-(1-\delta_{ks})(1-\varepsilon_k)F_{ks}\right]\frac{\partial q_{o,s}}{\partial q_{solar,i}}\\
&-\frac{1}{A_k}\sum_{t=1}^{N_{Cond}}\frac{k_{kt}}{l_{kt}}A_{kt}\frac{1}{4\varepsilon_t\sigma T_t^3}\\
&{}^*\sum_{s=1}^{N}\left[\delta_{ts}\left(1-(1-\varepsilon_t)F_{ts}\right)-(1-\delta_{ts})(1-\varepsilon_t)F_{ts}\right]\frac{\partial q_{o,s}}{\partial q_{solar,i}}=\delta_{ik}
\end{aligned} \tag{27}$$

2.4 Uncertainty analysis

The temperature or radiosity derivatives of the previous sections are called sensitivity coefficients. Once these are obtained they can be further processed to yield uncertainty coefficients by the following equation:

$$\frac{U_c}{T_k} = \sqrt{\sum_{s=1}^{N}\left(\frac{q_{solar,s}}{T_k}\frac{\partial T_k}{\partial q_{solar,s}}\right)^2}\frac{u_{q_{solar}}}{q_{solar,k}} + \sqrt{\sum_{s=1}^{N}\left(\frac{T_{f,s}}{T_k}\frac{\partial T_k}{\partial T_{f,s}}\right)^2}\frac{u_{T_f}}{T_{f,k}} + \sqrt{\sum_{s=1}^{N}\sum_{t=1}^{N_{Cond}}\left(\frac{C_{st}}{T_k}\frac{\partial T_k}{\partial C_{st}}\right)^2}\frac{u_C}{C_k} \tag{28}$$

U_c is the combined standard uncertainty and u is the standard uncertainty for the parameters q_{solar}, T_f, and conductance C which is defined as:

$$C_{kt} = \frac{k_{kt}}{l_{kt}}A_{kt} \tag{29}$$

The terms with the square root are the dimensionless uncertainty factors (UF) and reflect the changeability of the system to a parameter. The standard uncertainties can be supplied as difference for each parameter in form of a scalar or a probability density function. Provided the standard uncertainties for all parameters are the same, the uncertainty factors can be directly compared. Ideally, the standard uncertainty should be drawn into the square root but often a detailed knowledge of standard deviations, biases, or errors for each thermal node is not available. In this paper no distinction between errors, bias, or deviations is being made which would be beyond the scope of this work. A problem with Eq. (28) is when parameters become zero, because then the UF will lose some information. This can happen, when the effect of solar heat flux is investigated at nodes where no solar flux actually strikes. In order to circumvent this issue the following relationship is suggested:

$$U_c = \sqrt{\sum_{s=1}^{N}\left(\frac{\partial T_k}{\partial q_{solar,s}}\right)^2} u_{q_{solar}} + \sqrt{\sum_{s=1}^{N}\left(\frac{\partial T_k}{\partial T_{f,s}}\right)^2} u_{T_f} + \sqrt{\sum_{s=1}^{N}\sum_{t=1}^{N_{Cond}}\left(\frac{\partial T_k}{\partial C_{st}}\right)^2} u_C \tag{30}$$

A further advantage with this formulation is in its usability. The temperature information is already at hand and by multiplying with the standard uncertainties yields a range of temperatures or ranges of temperatures when using a probability density function.

2.5 Workflow

When performing an uncertainty analysis the following work-flow can be applied as described in Fig. 2.

The sensitivity and uncertainty analysis used here is performed a posteriori. First a conjugate heat transfer (CHT) simulation has to be run, be it a structural mechanics, a CFD, or a specialized heat transfer code. Then the radiosity gradients of Eq. (24) or Eq. (27) have to be calculated. This is followed by solving Eq. (21) or Eq. (26) for the temperature gradients. When uncertainty factors are desired, these temperature gradients need to be summed. This requires a lot of radiosity gradients. Wherefore, it is recommended to employ LU factorization for the LHS of the first set of equations, followed by matrix vector multiplication of the RHS of those relations. When using Eq. (28) or Eq. (30) a temperature range can be obtained.

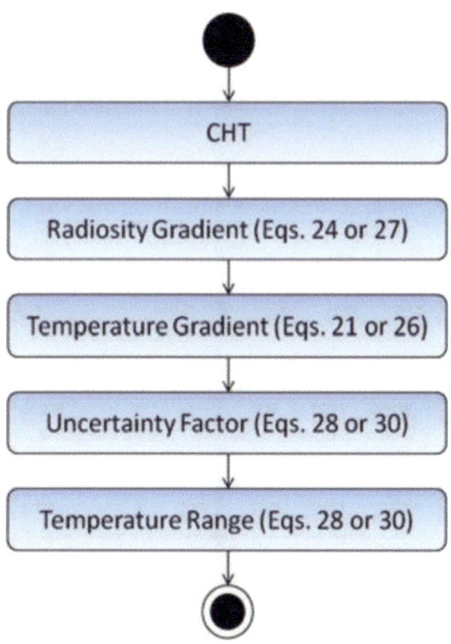

Fig. 2. Workflow

3. Example

For demonstration purposes a simplified test case has been set up as shown in Fig. 3. Some basic data has been obtained from Rodríguez-Hidalgo et. al (2011). The case represents a flat plate solar collector. It consists of a glass panel with a transmittance of 0.76 and a reflectance of 0.08. The absorber sheet is a crucial element. It consists of a thin copper sheet with a special surface coating with high solar absorbance but low emissivity. The heat absorbed is passed on the water pipe made of copper. The frame is made of aluminium. In Fig. 3 a small gap between the absorber sheet and the frame should signify, that those parts are actually insulated. The glass panel also is insulated from the front frame. Normally, between the absorber sheet, water pipe, and back frame there is a thick insulation layer. But because for ease of modelling the system in a finite difference solver and the steady-state calculation the insulation chamber is modelled as a vacuum. Basic parameters are to be found in tables 1a and 1b. As the connection between the absorber sheet and the water pipe would be a line the conduction is modelled by conduction bridges. The conductance as defined in Eq. (29) is estimated to be 77.1308 W/K. This value is based on the smaller nodal area between sheet and pipe element, the temperature-independent thermal conductivity of copper and the mean distance between the two nodes.

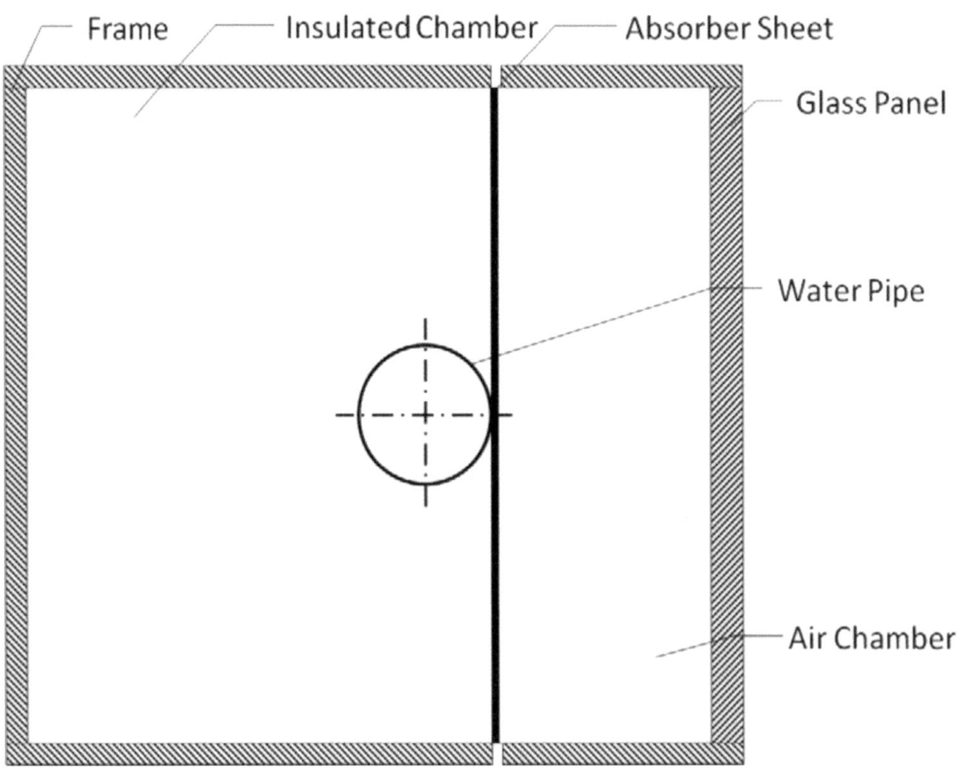

Fig. 3. Cross section through solar panel

Entity	*Unit*	*Glass Panel*		*Absorber Sheet*	
		Outside	*Inside*	*Outside*	*Inside*
Emissivity	[-]	0.1	0.1	0.1	0.1
Solar Absorbtivity	[-]	0.16	0.16	0.76	0.1
Thickness	[m]	0.002	0.002	0.000125	0.000125
Conductivity	[W/m/K]	1.1717		390	390
Wind Velocity	[m/s]	2.2	0		
Air Temperature	[K]	307.25	325.89	325.89	

Table 1a. Boundary conditions

Entity	*Unit*	*Water Pipe*		*Back Frame*		*Front Frame*	
		Outside	*Inside*	*Outside*	*Inside*	*Front*	*Back*
Emissivity	[-]	0.1	0.1	0.22	0.1	0.22	0.1
Solar Absorbtivity	[-]	0.1	0.2	0.49	0.1	0.49	0.1
Thickness	[m]	0.0005	0.0005	0.001	0.001	0.001	0.001
Conductivity	[W/m/K]	390	390	201.073	201.073	201.073	201.073
Wind Velocity	[m/s]			2.2		2.2	
Air Temperature	[K]			307.25		307.25	325.89

Table 1b. Boundary conditions continued

The environment is modelled as a desert in Arizona using environmental data from the commercial solver RadThermIR v.9.1. The solar irradiance is 950 W/m^2 and the zenith angle is 42°. The wind model of McAdam is used with the velocities in tables 1a and 1b which results in a heat transfer coefficient of 14.06 W/m^2/K for the exterior. The air chamber is modelled as a single air node and the resulting temperature is found in tables 1a and 1b.

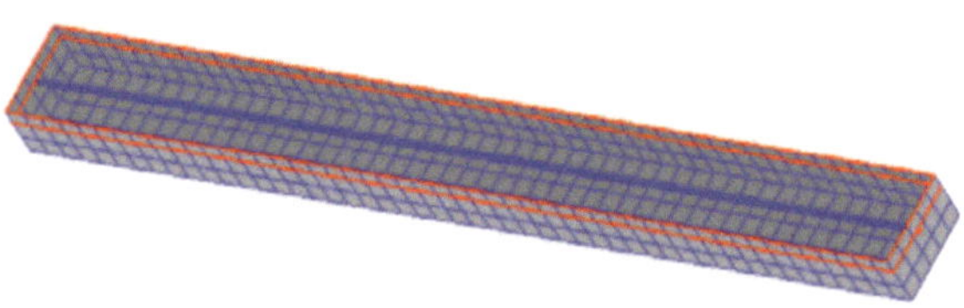

Fig. 4. Surface mesh of solar panel

Figure 4 shows the mesh of the panel with the glass panel and absorber sheet removed. The red lines indicate insulation to prevent conduction. Care is taken to use well shaped quadrilaterals. The outer dimensions are 2 x 0.19 x 0.082 m. The small width is caused because only one pass of the water serpentine is considered. The length of the quadrilaterals is about 0.04 m and the diameter of the pipe is 0.01 m. The water enters the pipe from left with an inlet temperature of 308 K and a flow rate of 0.3x10^{-4} m^{3}s^{-1}. The water is modelled with 50 fluid nodes along the pipe allowing for localized fluid temperatures.

Entity	*Unit*	*Back Frame*	*Absorber Sheet*	*Glass Panel*	*Water Pipe*
Node		*128*	*373*	*953*	*1324*
Conduction	[W/m²]	-13.74	-790.36	-0.003	6075.34
Convection	[W/m²]	-108.46	56.91	-217.82	-6085.27
Radiation	[W/m²]	28.58	9.76	-22.82	-0.51
Solar	[W/m²]	93.1	715.95	195.13	0
Temperature	[K]	314.96	311.54	322.74	311.41

Table 2. Results of radiation calculation

Table 2 shows the results of the thermal radiation solver with lowest view factor settings. The four front nodes are situated at the very centre of the respective part. The total thermal node count amounts to 3160.

4. Analysis

The results from the previous section form the basis for the following analysis.

4.1 Numerical stability

The infinity norm $\|M\|_\infty$ of the matrix *M* is one easily to be obtained indicator for stability. It is defined as follows with *max* as maximum operator and a_{ij} as a matrix entry of row *i* and column *j*:

$$\|\mathbf{M}\|_\infty = \max_{1<i<N} \sum_{j=1}^{N} |a_{ij}| \tag{31}$$

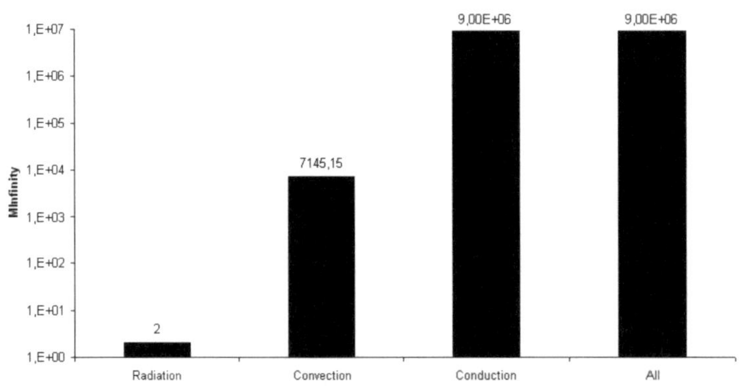

Fig. 5. Infinity norm of the solar panel

Figure 5 shows the infinity norm of the solution matrix for sensitivity coefficients for formulations with reciprocity of the LHS of Eq. (27). The set of bars denote whether only the radiation term (Radiation), the radiation and convection terms (Convection), radiation and conduction terms (Conduction), or all three terms (All) for the sensitivity coefficients had been used. It can be seen that for Radiation the infinity norm is quite low, because many nodes have view matrix entries as they cannot see neighboring nodes on a flat plane. The introduction of the conduction term lets the norm rise significantly, suggesting instability.

Another method for stability analysis is by using eigenvalues. The Gerschgorin circle theorem estimates eigenvalues λ and thus stability of a solver because it gives the maximum possible value of λ by the Gerschgorin radius r. The eigenvalue λ assures stability if it is less than or equal to 1. This is guaranteed when the radius r is less than or equal 0.5. In that case it can be expected, that the set of equations can be handled by an iterative solver without preconditioning.

$$r = \sum_{\substack{j=1 \\ j \neq i}}^{N} \left| a_{ij} \right| \tag{32}$$

Figure 6 shows the Gerschgorin radii r and the main diagonal entries for the nodes pertaining to the front side of the absorber sheet when only radiation and convection terms are considered. It reveals that although the radii are greater than 0.5 the diagonal entries are even greater. Thus, the absorber sheet behaves numerically stable because of its diagonal dominance.

Figure 7 shows the absorber sheet when all heat transfer modes are taken into account. The first thing to note, are the high values for both radii and diagonal entries as compared to Fig. 6. This supports the findings of the infinity norm in Fig. 5, that the introduction of conduction has a destabilizing effect.

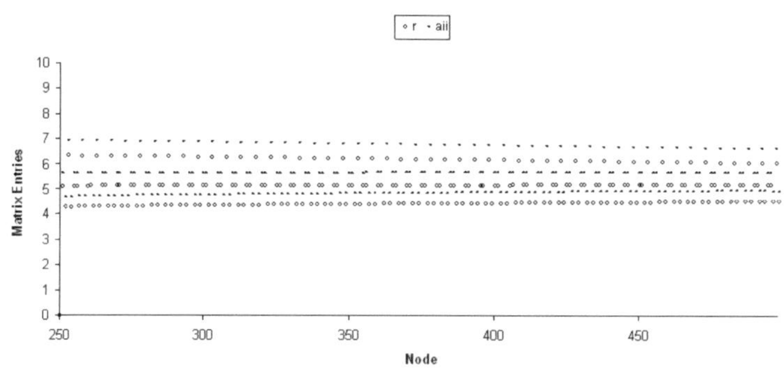

Fig. 6. Gerschgorin radii of the absorber sheet with convection and radiation term

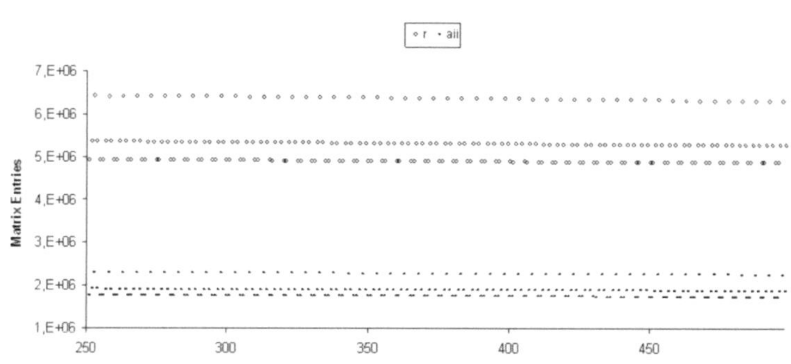

Fig. 7. Gerschgorin radii of the absorber sheet with all terms

The second finding is that now all radii are greater than their corresponding diagonal entry. This supports, along with the need of multiple solutions for uncertainty factors as mentioned in section 2.5, the use of a direct solver.

4.2 Partial uncertainty analysis

Seeing the stability issue when including all heat transfer terms there would be another incentive for omitting a heat transfer mode. For example, when considering conduction all neighboring nodes need to be processed. This cannot be felt while generating the solution matrix. But one needs to remember that the RHS needs to be recalculated for every sensitivity coefficient. In the present example this amounts to 3160 recalculations. This leads to the longer calculation time when estimating the uncertainty factor of, for example, conductance. In fact, the double summation of the last term in Eq. (30) represents together with a loop for the RHS calculations an N^3 time complexity.

Figure 8 shows uncertainty coefficients for the solar heat flux with the reciprocity relation, when incorporating various heat transfer mode. The back frame and the glass panel can handle the increase in solar flux quite well when radiation is the only heat transfer mode. But the absorber sheet and the water pipe show unphysical behavior. Introducing convection helps those nodes in passing the heat on to the water.

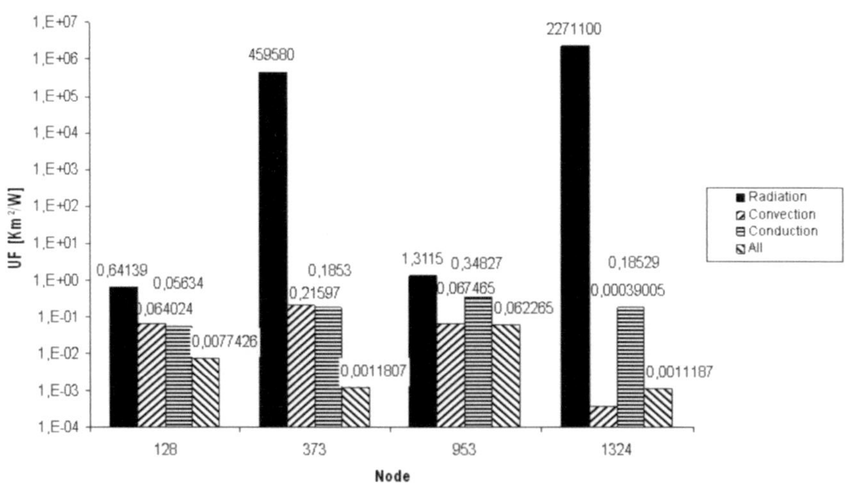

Fig. 8. Partial uncertainty factors

The changes in the order of magnitude emphasize the need to include all heat transfer modes into the calculations.

4.3 Reciprocity analysis

In section 3.2 the reciprocity relation was introduced together with the formulation of the sensitivity coefficient for solar flux.

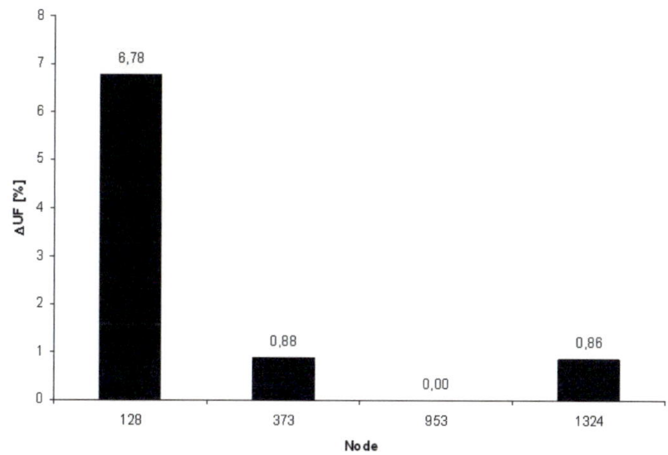

Fig. 9. Difference of standard uncertainty factors

The difference between the formulations with and without reciprocity in Fig. 9 show that the quality of the view factors behave quite satisfactorily with regards to reciprocity. The differences are not a matter of right or wrong with respect to uncertainties. Which formulation to use is a matter of whether the radiation solver employed uses the reciprocity relation. The solver used in this work does and for this reason all figures use the reciprocity formulation.

4.4 Results

The effect of a single perturbation at node 373 shows Fig. 10. At the node itself the impact is the largest and fades quickly with distance. But the information of the excitation is passed on to the whole absorber sheet.

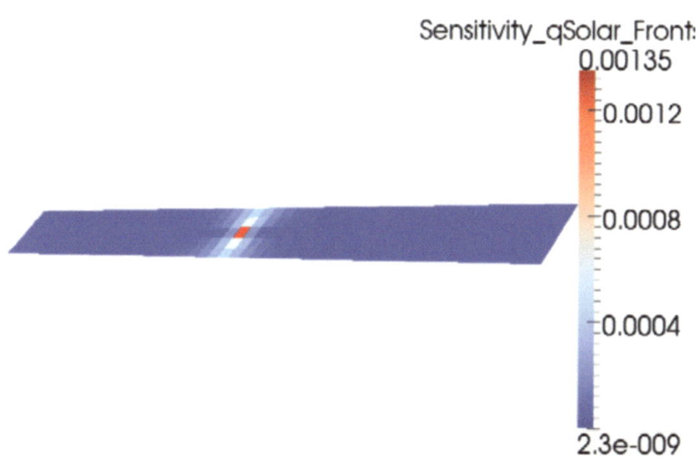

Fig. 10. Perturbation

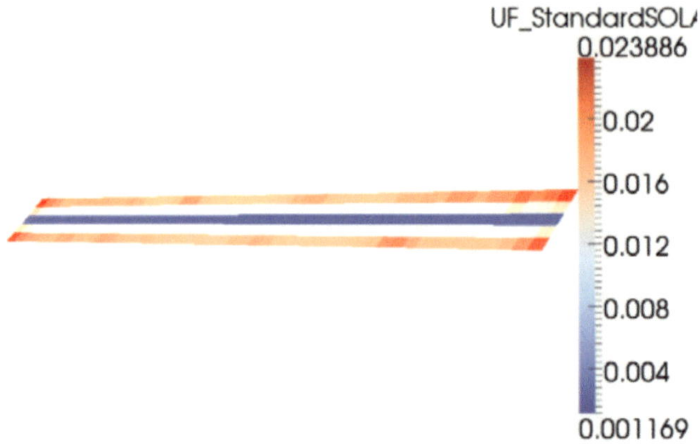

Fig. 11. UF q_{solar} at the absorber sheet

The dimensional UF for the solar heat flux at the absorber sheet in Fig. 11 reflects the inability of the absorber sheet at the border to pass on the heat to the center where it can be conducted to the water pipe due to the thinness of the sheet.

With the help of dimensionless UF one can see, whether uncertainties in the parameters are attenuated or amplified. In Fig. 12 dimensionless UF for q_{solar}, fluid temperature T_f, and conductance C are shown. The corresponding equations were published in Rauch & Almbauer (2010a) and Rauch (2011b). In none of the nodes investigated, exceeded the UF the value 1. Fluid temperature has the biggest value. In most cases, the solar flux has the smallest factor by one or two orders of magnitude.

In Fig. 13 the dimensional UF are shown with the dimensions Km²/W for the solar flux, K²/W for conductance and no units for the fluid temperature.

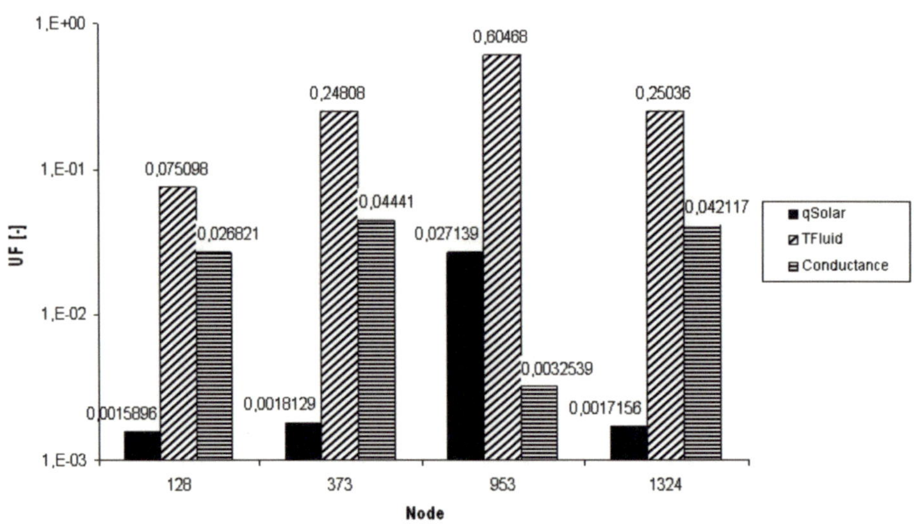

Fig. 12. Dimensionless UF for three parameters

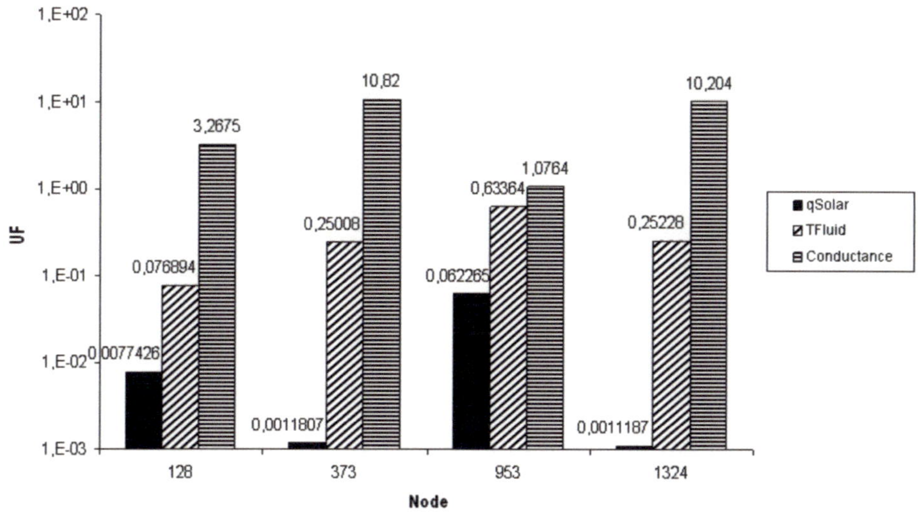

Fig. 13. Dimensional UF for three parameters

As the solar heat flux uncertainties have the smallest influence along the water pipe, the fluid temperature uncertainty has its biggest impact at the absorber sheet.

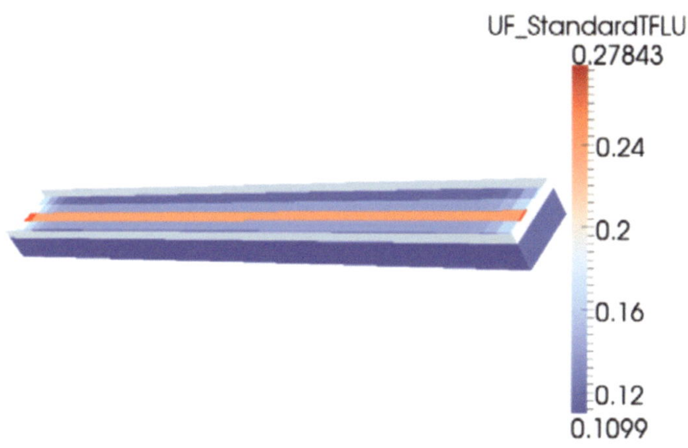

Fig. 14. Non-normalized UF for the fluid temperature

Finally, the obtained UFs from Fig. 13 shall be used to estimate the temperature range. Solar irradiation data that is available on a daily or monthly data for various locations can hold significant uncertainties as Gueymard & Thevenard (2009) has shown for the location Colorado, USA. To see whether the wall temperatures can be estimated when the solar fluxes change, a simulation is run with a solar irradiance of 1050 W/m². The obtained solar fluxes for the four nodes along with the UFs of Fig. 13 are put into Eq. (30). The positive combined uncertainties calculated for solar flux only is added to the temperatures in table 2 and expressed as percentage difference to the simulated temperatures with the case of solar irradiance of 1050 W/m².

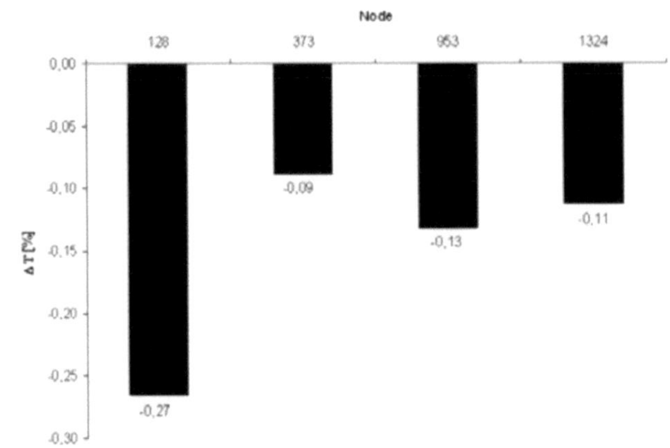

Fig. 15. Difference predicted to simulated temperatures

The predicted temperatures with UFs in Fig. 15 are in good agreement with the simulation.

5. Conclusion

Two formulations for solar heat flux uncertainties were derived. Partial uncertainties should not be used. Uncertainty factors can aid in understanding at what nodes which parameters need more attention. The combined standard uncertainty predicted a change in the solar heat flux of the test case very well. But this is just the beginning of the investigations. Questions of numerical stability, memory and time complexity need to be addressed. Also because of the non-linearity the range of predictability needs to be investigated. Very small values can cause high sensitivity coefficients. A draw-back of this formulation is that it can only be applied where radiosities have been calculated. This hampers the use of this method when systems are modeled without radiation in order to speed-up calculations or when contact between two surfaces is modeled where radiation cannot occur.

6. Acknowledgment

The author would like to acknowledge the financial support of the "COMET K2 - Competence Centres for Excellent Technologies Programme" of the Austrian Federal Ministry for Transport, Innovation and Technology (BMVIT), the Austrian Federal Ministry of Economy, Family and Youth (BMWFJ), the Austrian Research Promotion Agency (FFG), the Province of Styria and the Styrian Business Promotion Agency (SFG). The author would furthermore like to express his thanks to his supporting scientific partner Graz University of Technology.

7. References

Bhatia, M., & Livne, E. (2008). Design-Oriented Thermostructural Analysis with External and Internal Radiation, Part 1: Steady State. *AIAA Journal*, Vol. 46, No. 3, pp. 578-590, doi:10.2514/1.26236

Bhatia, M., & Livne, E. (2009). Design-Oriented Thermostructural Analysis with External and Internal Radiation, Part 2: Transient Response. *AIAA Journal*, Vol. 47, No. 5, pp. 1228-1240, doi:10.2514/1.40265

Blackwell, B., Dowding, K., Cochran, R., & Dobranich, D. (1998). Utilization of Sensitivity Coefficients to Guide the Design of a Thermal Battery, *ASME International Mechanical Engineering Congress and Exposition*, IMECE-981107, Anaheim, CA, USA

Gueymard, C., & Thevenard, D. (2009). Monthly average clear-sky broadband irradiance database for worldwide solar heat gain and building cooling load calculations. *Solar Energy*, Vol. 83, No. 11, pp. 1998-2018, doi.org/10.1016/j.solener.2009.07.011

Korycki, R. (2006). Sensitivity Analysis and Shape Optimization for Transient Heat Conduction with Radiation. *International Journal of Heat and Mass Transfer*, Vol. 49, No. 13, pp. 2033-2043, doi:10.1016/j.ijheatmasstransfer.2006.01.007

Lange, K., & Anderson, W., 2010, "Using sensitivity derivatives for design and parameter estimation in an atmospheric plasma discharge simulation," Journal of Computational Physics, Vol. 229, No. 17, pp. 6071-6083, doi.org/10.1016/j.jcp.2010.04.038

Rauch, C., Hoermann, T., Jagsch, S., & Almbauer, R. (2008). Advances in Automated Coupling of CFD and Radiation, SAE Paper 2008-01-0389

Rauch, C., Hoermann, T., Jagsch, S., & Almbauer, R. (2008a). An Efficient Software Architecture for Automated Coupling of Convection and Thermal Radiation Tools, *ASME Summer Heat Transfer Conference*, HT2008-56303, Jacksonville, FL, USA

Rauch, C., & Almbauer, R. (2010). Uncertainty and Local Sensitivity Analysis of View Factors for Steady-State Conjugate Heat Transfer Problems, *10th AIAA/ASME Joint Thermophysics Heat Transfer Conference*, AIAA-2010-4907, Chicago, IL, USA

Rauch, C., & Almbauer, R. (2010a). On Uncertainty and Local Sensitivity Analysis for Steady-State Conjugate Heat Transfer Problems, *14th International Heat Transfer Conference*, IHTC14-22859, Washington DC, USA

Rauch, C. (2011). Uncertainty and Local Sensitivity Analysis of View Factors for Transient Conjugate Heat Transfer Problems, *42nd AIAA Thermophysics Conference*, AIAA-2011-3942, Honolulu, HI, USA

Rauch, C. (2011a). On Uncertainty and Local Sensitivity Analysis for Transient Conjugate Heat Transfer Problems, *ASME-JSME-KSME Joint Fluids Engineering Conference*, AJK2011-23025, Hamamatsu, Shizuoka, Japan

Rauch, C. (2011b). On uncertainty and local sensitivity analysis for steady-state conjugate heat transfer problems part 1: Emissivity, fluid temperature, and conductance. *Frontiers in Heat and Mass Transfer*, Vol. 2, No. 3, p. 33006, doi.org/10.5098%2Fhmt.v2.3.3006

Rodríguez-Hidalgo, M.C., Rodríguez-Aumente, P.A., Lecuona, A., Gutiérrez-Urueta, G.L., & Ventas, R. (2011). Flat plate thermal solar collector efficiency: Transient behavior under working conditions. Part I: Model description and experimental validation. *Applied Thermal Engineering*, Vol. 31, No. 14, pp. 2394-2404, doi.org/10.1016/j.applthermaleng.2011.04.003

Roy, C., & Oberkampf, W. (2011). A comprehensive framework for verification, validation, and uncertainty quantification in scientific computing. *Computer Methods in Applied Mechanics and Engineering*, Vol. 200, No. 25, pp. 2131-2144, doi.org/10.1016%2Fj.cma.2011.03.016

Saltelli, A., Ratto, M., Andres, T., Campolongo, F., Cariboni, J., Gatelli, D., Saisana, M., & Tarantola, S. (2008). *Global Sensitivity Analysis The Primer*, Wiley, ISBN 978-0-470-05997-5, Chichester, UK

Saltelli, A., Chan, K., & Scott, E. (2008a). *Sensitivity Analysis*, Wiley, ISBN 978-0-470-74382-9, Chichester, UK

Saltelli, A., & Annoni, P. (2010). How to avoid a perfunctory sensitivity analysis. *Environmental Modelling & Software*, Vol. 25, No. 12, pp. 1508-1517, doi.org/10.1016%2Fj.envsoft.2010.04.012

Siegel, R., & Howell, J.R. (2002). *Thermal Radiation Heat Transfer* (4th ed.), Taylor & Francis, ISBN 1-56032-839-8, New York.

Taylor, R.P., Luck, R., Hodge, B.K., & Steele, W.G. (1993). Uncertainty Analysis of Diffuse-Gray Radiation Enclosure Problems - A Hypersensitive Case Study, *5th Annual Thermal and Fluids Analysis Workshop*, NASA, NASA CP 10122, Washington, DC, pp. 27-40

Taylor, R.P., Luck, R., Hodge, B.K., & Steele, W.G. (1994). Uncertainty Analysis of Diffuse-Gray Radiation Enclosure Problems, *32nd Aerospace Sciences Meeting and Exhibit*, AIAA 94-0132, Reno, NV, USA

Taylor, R.P., Luck, R., Hodge, B.K., & Steele, W.G. (1995). Uncertainty Analysis of Diffuse-Gray Radiation Enclosure Problems. *Journal of Thermophysics and Heat Transfer*, Vol. 9, No. 1, pp. 63-69, doi:10.2514/3.629

Taylor, R.P., & Luck, R. (1995). Comparison of Reciprocity and Closure Enforcement Methods for Radiation View Factors. *Journal of Thermophysics and Heat Transfer*, Vol. 9, No. 4, pp. 660-666, doi.org/10.2514/3.721

7

Infrared Radiation Detection Using Fiber Bragg Grating

Jean-Michel Renoirt, Christophe Caucheteur, Marjorie Olivier, Patrice Mégret and Marc Debliquy
University of Mons, Faculty of Engineering, Mons, Belgium

1. Introduction

Optical fiber sensors bring to measurement systems all the advantages offered by the optical fiber technology. In particular, their low weight and small dimensions yield non-intrusive capabilities while their immunity against electromagnetic interference leads to usefulness in harsh environments. Among the various optical fiber sensor configurations, fiber Bragg gratings (FBGs) are particularly interesting, especially thanks to their wavelength-encoded response and their inherent multiplexing capability. They can be used to measure static and dynamic perturbations such as temperature (Crunelle et al., 2009; Fernandez-Valdivielso et al., 2002), mechanical strain (Han, 2009; Ho, 2002) and pressure (Sheng et al., 2004). Their association with a sensitive layer converting the measurand into a local stress or a temperature elevation around the FBGs drastically increases the panel of applications.

Among others, gas sensing (Buric et al., 2007; Caucheteur et al., 2008), humidity sensing (Kronenberg et al., 2002; Yeo et al., 2005), salinity (Lu et al., 2008; Men et al., 2008), magnetic field (Davino et al., 2008; Yang et al., 2009) and pH monitoring (Corres et al., 2007) have been recently reported. Based on a similar principle, the association of an absorbing layer to a FBG can be realized for infrared (IR) radiation detection purposes as it has been demonstrated in (Caucheteur, 2010; Renoirt, 2010; Yüksel, 2011). The authors adapted the operating principle of a standard bolometer to the optical fiber in order to benefit from all the advantages of the latter.

Such a configuration is particularly interesting for early fire detection over long distances (Sivathanu, 1997). This kind of sensor presents significant advantages over smoke and heat detectors as it can monitor a wide area and can respond in less than a few seconds.

This chapter describes a fibered bolometer based on a pair of FBGs. One is covered with an IR-absorbing material while the other is uncoated and sheltered from the radiation in order to be used as a reference. Both FBGs act the same way as for the classical bolometer. The useful signal is the differential wavelength shift between the coated and the protected FBGs when exposed to IR radiation.

The remaining of the chapter is organized as follows. After a presentation of the usual existing technologies for infrared detection, the sensing principle is presented. Section 4 is dedicated to the sensor fabrication and the different designs, while section 5 presents the experimental results. In section 6, the modeling of the sensor is described for different configurations. The experimental results are compared to the modeling and discussed. Finally, section 7 draws some conclusions.

2. Existing systems

In this section, the main existing technologies for the infrared radiation detection are described (Gaussorgues et al., 1996). In Table 1, existing technologies are classified following their type. A short explanation of the working principle of each detector is given below.

Type	Detector	Spectral response
Thermal Thermal sensors are receivers in which the light flux is transformed into heat by an absorbing process and the useful signal is the temperature increase due to absorption. The temperature increase is transformed by different means into an electrical signal	Thermocouple – thermopile	Depends on window material
	Bolometer	
	Pneumatic cell	
	Pyroelectric detector	
	Pyromagnetic detector	
Quantum Detectors in which the signal corresponds to the measurement of the direct excitation of particles by the incident photons	Photoconductive type	PbS : 1 - 3.6 μm PbSe : 1.5 – 5.8 μm InSb : 2 – 6 μm HgCdTe : 2 – 16 μm
	Photovoltaic type	Ge : 0.8 – 1.8 μm InGaAs : 0.7 – 1.7 μm InSb : 1 – 5.5 μm
	Photoemissive type	AgOCs : 0.2 – 1 μm
	Quantum well type	GaAs/GaAlAs : 8 – 12 μm
Charge Coupled Devices Detectors based on coupled MOS capacitance structures that allow imaging	IRCCD	Si : 0.4 – 1μm PtSi : 3 – 5 μm $Hg_{(1-x)}Cd_xTe$: 8 – 12 μm

Fig. 1. Summary of the usual systems for infrared radiation detection

Working principles of the different systems

- ***Thermal***
  - *Thermocouple - thermopile*: A thermopile converts thermal energy into electrical energy. It is composed of several thermocouples usually connected in series or in parallel in contact with an absorbing material. Thermopiles do not generate a voltage proportional to absolute temperature, but to a local temperature difference or temperature gradient.
  - *Bolometer*: A bolometer measures the power of incident electromagnetic radiation via the heating of an absorbing material with a temperature-dependent electrical resistance.
  - *Pneumatic cell*: Pneumatic cells are based on the measurement of the pressure variation of a well-known mass of gas in a closed chamber. The radiation flux heats the sensitive element inside the chamber, increasing the gas pressure which induces a deformation of a membrane.

  - *Pyroelectric detector*: Pyroelectric detectors are made of pyroelectric films absorbing the radiation. For these materials, a temperature increase induces electrical charges at the surface. The surface charge density is proportional to the temperature increase. The charges are collected by electrodes transforming the temperature rise in an electric signal. These sensors can only measure variations and not steady radiations. For slowly changing radiation intensity, the pyroelectric film is periodically hidden by a mask inducing a rapidly changing exposure to the radiation.
  - *Pyromagnetic detector*: Pyromagnetic detectors use ferromagnetic materials with permanent magnetization. The magnetization decreases with temperature and becomes equal to zero at the Curie temperature. Below this temperature, a variation of temperature induces a reversible variation of magnetization. The magnetization is measured in a classical magnetic circuit.
- ***Quantum***
  - *Photoconductive type*: The observed signal is the conductivity increase of a crystal when exposed to radiation. The absorption of radiation, if energetic enough ($E > E_{gap}$ of the semiconductor), may induce the creation of electron-hole pairs. The free electrons and holes participate to the current flow in the material when biased. The variation of conductivity is measured using an external source of current.
  - *Photovoltaic type*: Incident photons of the impinging radiation with sufficient energy ($E > E_{gap}$ of the semiconductor) may be absorbed in the space charge zone of a p-n junction causing the formation of electron-hole pairs. The electrons and holes will migrate through the junction under the action of the junction field. The resulting current through the junction is proportional to the intensity of the radiation. Photovoltaic detectors behave as energy generators and can give a signal without polarization.
  - *Photoemissive type*: Incident photons of the impinging radiation with sufficient energy may extract electrons from a solid due to the joint effects of incident photons and a static polarization between the solid and a collecting electrode. The resulting current is proportional to the intensity of the radiation.
  - *Quantum well type*: This kind of detector consists in a stacking of materials between two highly doped materials for the electron collection. A bias voltage is applied to bend the energy bands. This is carried out at low temperature (typically 77K). Electrons are trapped in the fundamental state. An incident radiation can induce a transition and electrons go from emitter to collector, leading to a current that can be measured.
- ***Charge Coupled Devices***
  - *InfraRed Charge Coupled Devices (IRCCD)* : The basic principle of this kind of detector is the following: radiation induces electric charges in a semiconductor. The amount of charges is proportional to the intensity of the radiation. These charges are stored in an isolator element playing the role of a capacitor (MOS structure). These structures are placed in arrays. The generated charges are then transferred by an electric field applied successively to the next similar element. Collection of all the charges on the same array is made by a unique electrode and is treated as a video signal. A complete device is composed of a large number of such arrays defining a 2 dimensional structure consisting of pixels. This system is used for processing images.

All these systems give an electrical signal output and yield point detection while the presented system produces a purely optical signal and can provide quasi-distributed detection.

3. Sensing principle

3.1 Basic principle

The emission spectrum of a hot source, for instance a fire, can be approximated by Planck's law giving the spectral radiance of a black body as a function of the wavelength for a given temperature:

$$B_{\lambda,T} = \frac{2hc^2}{\lambda^5} \frac{1}{e^{\frac{hc}{\lambda kT}} - 1} \tag{1}$$

where $B_{\lambda,T}$ is the spectral radiance, h is Planck's constant (6.62×10^{-34} Js), c the speed of light in vacuum, k the Boltzmann constant (1.38×10^{-23} J/K) and T the absolute temperature of the black body. Fig. 2a depicts the radiance spectrum obtained from Equation 1 for different temperatures. To obtain a plot independent of the total radiance, each curve has been normalized (at each temperature, values are divided by the maximum radiance for this temperature). Therefore, it is easy to compare the effect of a higher temperature on the spectral distribution.

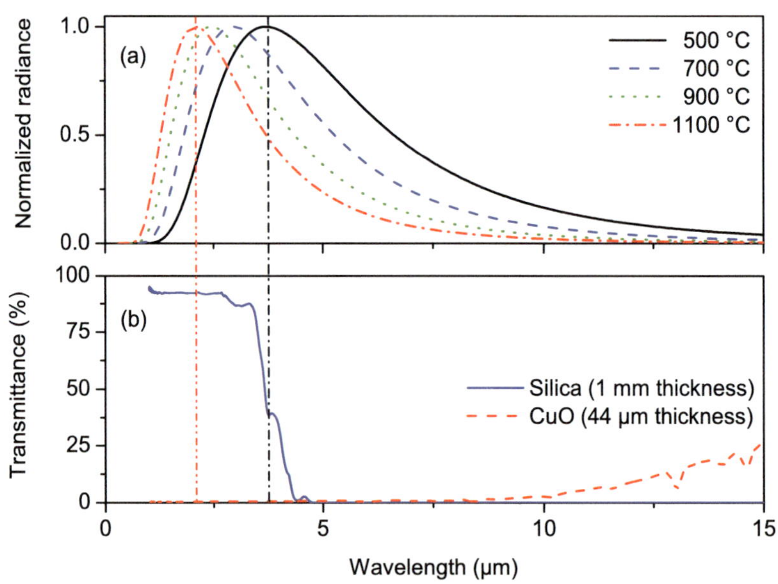

Fig. 2. (a) Spectral radiance at different temperatures, (b) silica – CuO transmittance spectra.

On this graph, a maximum is observed at around 5 µm for 500 °C while the maximum is observed at 2 µm for 1100 °C. A higher temperature has a narrower distribution centered on a lower wavelength. The major part of the spectrum emitted by hot spots and fires lies in the range from 0.8 µm to 15 µm.

Since silica absorbs part of the IR radiation, an uncoated FBG could be used as an IR sensor. However, it will have a weak response because silica absorbs only for wavelengths longer than 4 µm. The transmittance spectrum of silica (thickness: 1 mm) is shown on Fig. 2b. Silica can be seen as a low-pass filter in wavelength. Below 4 µm, silica is almost transparent (transmittance is not equal to 100 % because part of the incident beam is reflected by the material) and very absorbing above 4 μm. It means that for high temperature sources the signal will be underestimated. Therefore, in order to cover the whole wavelength range of the radiation and to yield a sensitivity independent of the source temperature, an additional coating is needed to absorb radiation at shorter wavelengths. The chosen coating is a layer of copper(II) oxide. Copper(II) oxide is a black compound and is well-known for its good absorbing properties in the near IR-domain (this material is used in solar panels). The transmittance spectrum for a layer of 44 μm CuO encapsulated in a polymer is displayed in Fig. 2b.

3.2 Operating principle of FBGs sensors

Fiber Bragg gratings sensors are often used for temperature and strain measurements (Othonos, 1999). A FBG is a periodic and permanent modulation of the core refractive index along the fiber axis (Fig. 3). In its simplest form, it acts as a mirror, selective in wavelength around a resonance wavelength given by the following relationship:

$$\lambda_B = 2\Lambda n_{\text{eff}} \tag{2}$$

where n_{eff} is the effective refractive index of the fiber core mode at the Bragg wavelength and Λ the grating period. To obtain Bragg wavelengths centered in the telecommunication window around 1.55 μm, the order of magnitude of Λ is 500 nm.

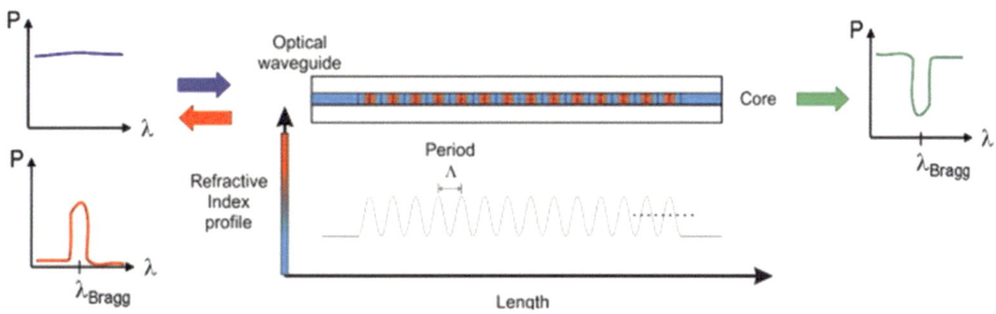

Fig. 3. Principle of a fiber Bragg grating.

Temperature and strain modify both n_{eff} and Λ and as a result, change the Bragg wavelength. A specific packaging can be used to isolate the strain effect in a way that the sensor is only sensitive to temperature. The average Bragg wavelength-shift at 1550 nm due to a temperature change is about 10 pm/°C. In the presented sensor configuration, a pair of gratings is used, one coated and the other one protected from the outside radiation. The temperature change is measured by the shift in the Bragg wavelengths. The wavelength of the protected grating depends on the room temperature change while the wavelength of the coated grating depends on the infrared radiation and the room temperature fluctuations. The measurement of the differential shift between the two grating wavelengths provides the signal corresponding to the temperature rise due to the infrared radiation. Typical spectra for both cases (with and without IR radiation) are sketched in Fig. 4.

The useful signal is the differential Bragg wavelength shift between the coated and the protected gratings:

$$\Delta\lambda_B = \lambda_{B_2'} - \lambda_{B_1'} - \Delta \tag{3}$$

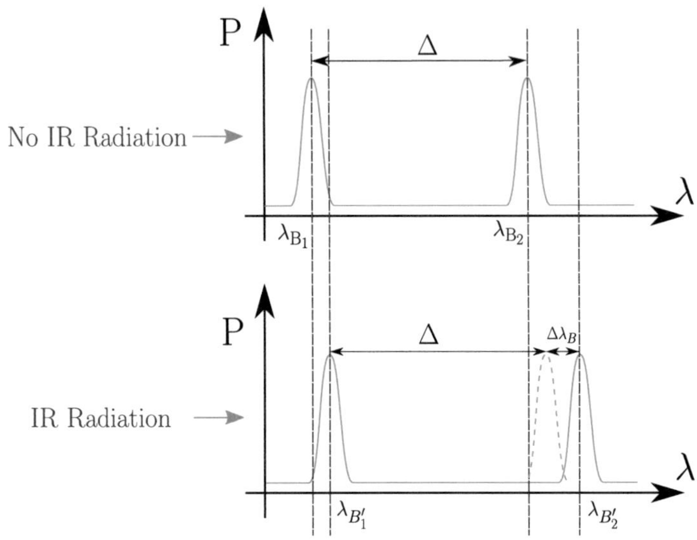

Fig. 4. Optical reflected spectra of a pair of gratings.

where Δ is the difference in Bragg wavelengths between the two gratings when there is no radiation ($\Delta = \lambda_{B_2} - \lambda_{B_1}$).

3.3 Sensitivity

The sensitivity Σ_T of the sensor is defined in terms of temperature rise for a given radiation flux f:

$$\Sigma_T = \frac{\Delta T}{f} \tag{4}$$

where ΔT is the difference between fiber and room temperature (°C) and f the total infrared flux on the sensor (W/m^2). The sensitivity Σ_λ can also be defined in terms of wavelength shift for a given heat flux and is then given by:

$$\Sigma_\lambda = \frac{\Delta \lambda_B}{f} \tag{5}$$

3.4 Sensitivity enhancement

The sensitivity for this sensor can be easily enhanced using a glass capillary around the coated grating. It creates a greenhouse effect around the fiber and therefore, for a given radiation power, the temperature rise is higher, resulting in an increased sensitivity. Indeed, infrared radiation below 4 μm passes through the glass tube and are absorbed by the CuO-layer. The CuO-coated fiber will reemit at lower wavelengths but this radiation is confined in the capillary because glass absorbs radiation above 4 μm. As a result, the glass tube absorbs radiation with wavelengths above 4 μm not only coming from the source but also from the fiber. The glass tube exchanges heat with the coated grating by a radiation process and increases the fiber temperature for a given radiating flux. It has a strong effect on the sensitivity of the sensor.

4. Sensor fabrication

4.1 Bragg grating and coating

The gratings were inscribed on a standard germanium-doped silica fiber according to classical procedures. The two gratings were 10 cm separated, the reference grating was centered at 1528 nm and the coated grating at 1533 nm. The fiber was coated with a CuO loaded polymer by dip coating followed by drying at 110 °C during 5 min and UV-curing for 30 s. This method gives a good absorbing layer (thickness: 44 μm) (Fig. 2b) with good adhesion on the fiber. The polymer matrix leads to a thicker layer with a very short deposition time (< 5 min) compared to other deposition processes (sputtering ...). Moreover the coating has a higher roughness, decreasing reflection and as a consequence, increasing absorption.

4.2 Glass capillary and packaging

To improve the sensor response, a thin glass tube (outer diameter $\approx$ 1 mm) was put around the CuO-coated FBG. To ensure good measurements and good performances, care was taken to place the fiber perfectly in the center of the capillary. If the sensor would be in contact with the capillary, a thermal bridge would be established between the capillary and the fiber, considerably modifying the response of the sensor. A packaging is needed to protect the uncoated grating from the outside radiation. It is fundamental to have a stable reference in order to eliminate room temperature fluctuations. This packaging can be advantageously tailored around the double FBG to improve the sensitivity (Fig. 5). The packaging is metallic (aluminum) and acts as a mirror. Furthermore, by concentrating the radiation on the fiber, it allows to reduce the minimal incident intensity which can be detected. Although the parabolic shape is the best one to concentrate the radiation on the sensor, experiments have been done with a circular aluminum tube. Aluminum is a good reflector for infrared radiation (for wavelengths > 2 μm the reflectivity is greater than 97 %); it isolates the second grating from the outside infrared radiation and concentrates radiation on the copper oxide coated grating. Figure 5 shows an example of such a packaging (aluminum tube of 20 mm outer diameter and 17 mm inner diameter). A rectangular aperture is made to form a cylindrical mirror (Fig. 5). As in the case of the glass tube, the aluminum tube diameter can be modified. A larger tube can harvest and concentrate more radiation but the advantages (low weight, small dimensions) of optical fiber are lost.

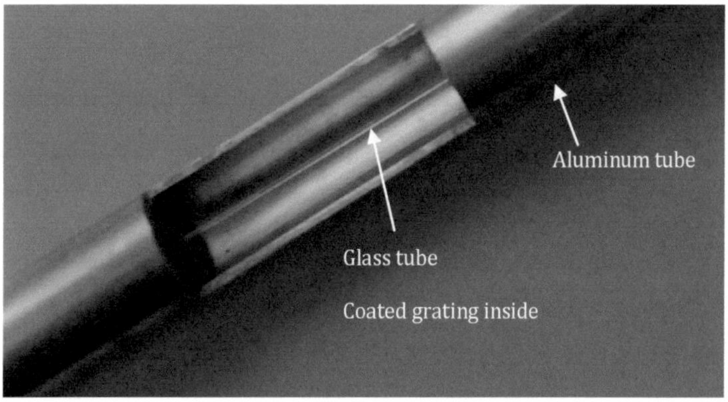

Fig. 5. Packaging made of an aluminum tube and the coated FBG inside the capillary tubing.

5. Sensitivity measurements

5.1 Experimental set-up

In this section, the experimental set-up and the measurement processes are explained. Fig. 6 presents a sketch of the experimental set-up. An ASE (Amplified Spontaneous Emission) source covering the C+L bands (1525 nm - 1610 nm) and an optical spectrum analyzer (OSA) ANDO AQ6317C are used to collect the reflected spectrum of the double grating configuration. The heat source is provided by a furnace (Edouard Defrance, inner volume 3.6 l (T_{room}-1100 °C)). In order to measure the incident flux, a radiant flux sensor (Captec Enterprise, 10x10 mm^2 , 0.438 $\mu V/(W/m^2)$) was put close to the fiber. A simple way to modify the incident power is to put the sensor at different distances from the heat source. The infrared radiation flux decreases with the square of the distance to the source. The temperature has also an important effect on the flux value. When increasing the furnace temperature, the spectrum of the incident radiation is changed. From the reflected spectrum measurement, the two Bragg wavelengths can be obtained.

The difference between both wavelengths is nearly proportional to the IR radiation absorbed by the sensor. All the gratings were coated with copper oxide encapsulated into polymer deposited on a length of 5 cm. Other parameters can significantly influence the sensitivity: diameter and thickness of the capillary tubing and thickness of the copper oxide coating. Modeling of the sensor surrounded by the glass capillary was carried out in order to understand the complete mechanism of the sensor response and optimize it by changing relevant parameters (see section 6). Another key point, for all experiments, is to ensure that the FBG sensor is not strained. The strain can indeed vary with time, leading to a wrong measurement of the resonance shift. FBGs sensors are very sensitive to strain. Therefore, the sensor and capillary must be carefully designed to allow natural expansion of the fiber without supplementary strain. For each measurement, the temperature of the furnace was stabilized at the target value. The furnace door was opened and the measurement time was between 15 and 45 s.

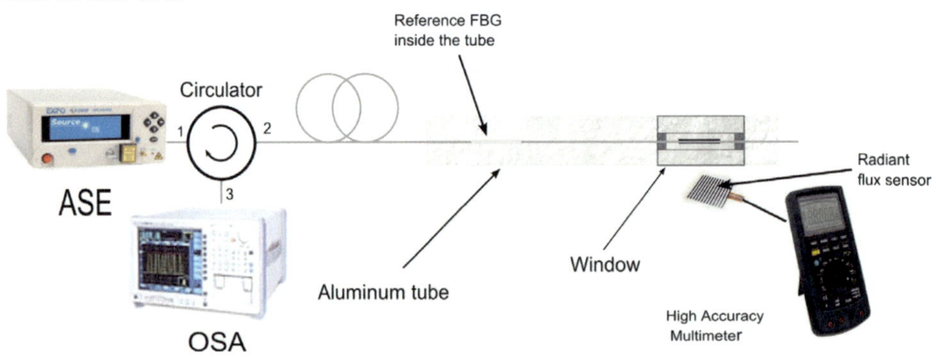

Fig. 6. Sketch of the experimental set-up.

5.2 Results

The following paragraph presents the experimental results conducted with three sensor configurations: a CuO coated fiber, a fiber inserted into a glass capillary and finally, fiber and capillary put in a reflecting packaging. In order to follow the influence of the source temperature, the sensitivities were measured at 4 temperatures: 500, 700, 900 and 1100 °C.

In the case of the first configuration (fiber without capillary), Fig. 7 shows the response (Bragg wavelength shift) as a function of the incident flux when the furnace temperature is kept at 700 and 900 °C. The response is linear with the incident flux. The sensitivity at 700 °C is about 8×10^{-3} pm/(W/m^2) which corresponds to 8×10^{-4} °C/(W/m^2). The resolution of the OSA used for the experiments is 10 pm.

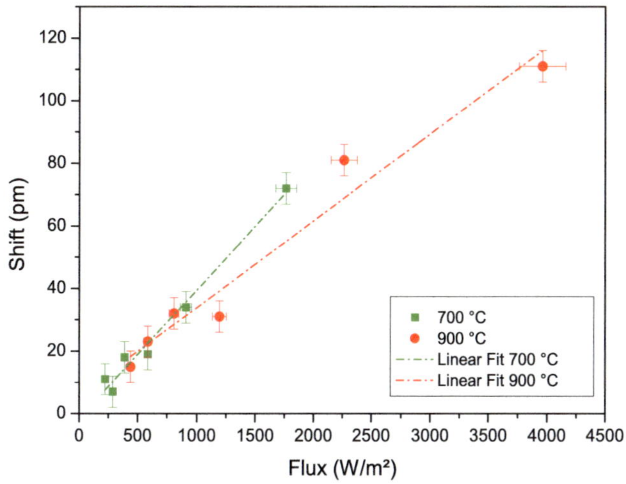

Fig. 7. Shift of the Bragg wavelength for a furnace temperature of 700 and 900 °C, without capillary tubing on the grating.

Results at 500, 900 and 1100 °C are similar, with respectively a slope of 12×10^{-3}, 6×10^{-3} and 4×10^{-3} pm/(W/m^2).

For the second configuration, with gratings surrounded by a glass capillary tube, the sensitivity is, as expected, higher but the influence of the furnace temperature is still substantial. Results for the grating covered with capillary tubing are shown in Fig. 8.

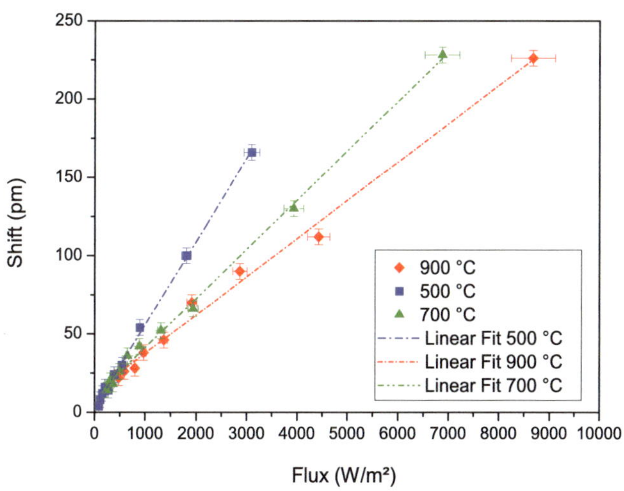

Fig. 8. Shift of the Bragg wavelength for a fiber surrounded by a glass capillary for different temperatures.

Finally, in the case of the whole packaging (Fig. 9), the sensitivity Σ_λ decreases again when the source temperature rises. At 500 °C, Σ_λ is around 0.11 pm/(W/m^2) and is about 0.05 pm/(W/m^2) at 1100 °C. The influence of the source temperature on the sensitivity can be reduced by adding a CuO layer on the capillary tube itself.

For the sake of comparison, table 1 presents an overview of the results for the different tested temperatures and sensor configurations.

10^{-3} pm/(W/^{2}m)	500 °C	700 °C	900 °C	1100 °C
Without packaging	12	8	6	4
With capillary	50	31	25	20
Whole packaging	110	100	75	50

Table 1. Sensitivity for different furnace temperatures and sensors configurations

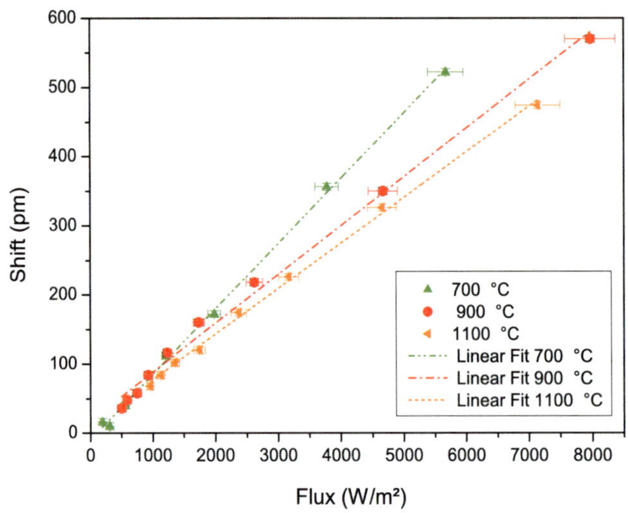

Fig. 9. Shift of the Bragg wavelength for different temperatures, with glass and aluminum tubes.

The experimental results show that it is possible to achieve an IR detection system with a good sensitivity with fiber Bragg gratings. For a given source temperature, the sensor shows a linear behavior according to the flux intensity. However, the sensitivity of the sensor is still influenced by the source temperature. The sensitivity of the sensor is higher at lower temperature. The most important property in the perspective of early fire detection is a low threshold for alarm. The threshold is in the range of the kW/m^2.

The first two configurations described above are theoretically modeled in the next section.

6. Theoretical response of the sensor

Considering the fact that a grating has a temperature sensitivity of the order of 10 pm/°C, it is possible to compute and model the behavior of the sensor under a given infrared radiation flux. Indeed, the temperature of the fiber under the infrared radiation and so the Bragg wavelength resonance shift is the result of the balance between the thermal losses and incident energy absorbed energy.

Firstly, the simple case of a fiber covered with CuO without protection is discussed.

6.1 Case 1: fiber with absorbing coating alone

In Fig. 10, the different losses and energy fluxes are displayed in the case of a fiber without protection. Let us consider L the length of the exposed zone of the fiber (L is typically a few cm). The total energy absorbed by the section of length L (Φ_{abs}) is equal to the losses of the same section due to convection (Φ_{conv}), radiation with the ambient (Φ_{rad}) and conductivity along the fiber (Φ_{cond}). The temperature of the exposed section is supposed to be uniform. At thermal equilibrium, the fiber will stabilize at a temperature higher than the room temperature according to the following thermal balance (Equation 6):

$$\Phi_{abs} = \Phi_{conv} + \Phi_{rad} + \Phi_{cond} \tag{6}$$

The positive contribution to the balance (Φ_{abs}) comes from the incident infrared flux of the hot spot. The absorption properties of the materials constituting the sensor (which depend on the wavelengths of the incident radiation) must be taken into account. The losses can be divided into 3 parts:

- Φ_{conv} results from the exchanges with ambient air due to convection.
- Φ_{rad} comes from heat exchanges due to radiation between the fiber and the ambient.
- Φ_{cond} is due to thermal conduction along the axis of the fiber. This term is small compared to the first two because thermal conductivity of silica is small.

All these terms depend on the temperature of the fiber.

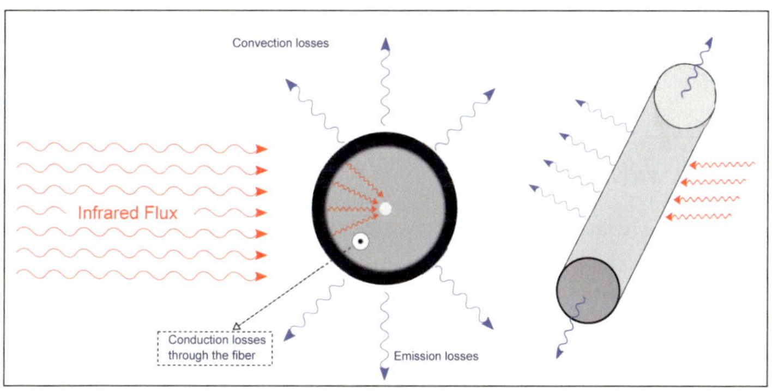

Fig. 10. Thermal exchanges existing on the optical fiber.

As the diameter of the fiber is very small, it can be shown that there is no radial significant gradient inside the fiber and the temperature is considered as uniform in the exposed zone. Indeed, Biot's number is defined by:

$$Bi = \frac{K \cdot D}{\kappa_{silica}} \tag{7}$$

where K is the exchange coefficient resulting from convection and radiation (typically $50\ W/(m^2K)$), D is the diameter of the fiber including the coating (200 μm) and κ_{silica} the thermal conductivity of silica ($\approx 1\ W/(mK)$). In our case, Biot's number is approximately equal to 10^{-2}. For a solid in contact with a fluid and exchanging heat with the fluid, Biot's number represents the ratio of the heat transfer resistance through the solid and the heat transfer resistance through the surrounding fluid (Perry, 1984). When Biot's number is small (<0.1), it means that the main resistance is due to the exchange with the fluid and so the radial temperature gradient in the solid can be neglected.

The different terms in Equation 6 are detailed in the following paragraphs.

6.1.1 Absorption of the fiber

The fraction of energy absorbed by the fiber and its coating is given by :

$$\Phi_{\text{abs}} = K_{\text{abs}} R_s L f \tag{8}$$

K_{abs} represents the global absorption coefficient. In practice, $0 \leq K_{\text{abs}} \leq 2$. The case $K_{\text{abs}} = 0$ corresponds to a totally transparent material, while the case $K_{\text{abs}} = 2$ describes a completely absorbed radiation, because the latter comes from one direction.

R_s is the radius of the fiber including the coating (in our case, $R_s = R_f + e_{\text{CuO}}$) where e_{CuO} is the thickness of the coating.

K_{abs} can be computed by a simple model using the Beer-Lambert law and assuming the spectral intensity to follow Planck's law. Transmission of the material is given by the Beer-Lambert law:

$$\mathrm{T}_\lambda = \frac{\mathrm{I}_\lambda}{\mathrm{I}_{0\lambda}} = \mathrm{e}^{-\alpha_\lambda d} \tag{9}$$

where $\mathrm{I}_{0\lambda}$ is the intensity of the radiation at wavelength λ, I_λ is the intensity transmitted through the absorbing material, d is the distance travelled in the material and α_λ is the absorption coefficient that depends on the wavelength for a given material.

The fraction of absorbed energy after traveling a length d is given for each wavelength by :

$$\mathrm{A}_\lambda = (1 - \mathrm{R}_\lambda)(1 - \mathrm{e}^{-\alpha_\lambda d}) \tag{10}$$

with R_λ being the reflection coefficient of the material. The total absorbed energy is obtained by integration over all wavelengths.

The optical properties of the absorbing material (CuO dispersed in a polymer matrix) were experimentally determined by spectroscopy as it presents a complex behavior. For fused silica, data from the literature (Desvignes (1991)) were used. From these data, the absorption coefficients as a function of the wavelength were extracted (cf. Fig. 11 and 12).

To determine K_{abs}, the fiber is mathematically cut in slices (cf. Fig. 13) and Equation 10 is used on each slice. Absorption of the coating and absorption of the fiber are then added.

Using Equation 10 for coating and fiber, the absorbed radiation is given by Equation 11 for the portion of the coating/fiber between θ and $\theta + d\theta$. By integrating from $-\pi/2$ to $+\pi/2$, we obtain the whole absorption of the fiber (Fig. 11).

Remark: As the absorption coefficient of the material is quite large, the secondary reflections at the back are neglected in this model.

$$\Phi_{\text{abs}} = R_f L f \int_{\lambda=0}^{\lambda=\infty} \int_{\theta=-\frac{\pi}{2}}^{\theta=\frac{\pi}{2}} \cos(\theta) \frac{f_\lambda}{f} \Big(\mathrm{e}^{-\alpha_\lambda^c e(\theta)} \Big[1 - \mathrm{e}^{-2\alpha_\lambda^v R_f \cos(\theta)} + (1 - \mathrm{e}^{-\alpha_\lambda^c e(\theta)}) \mathrm{e}^{-2\alpha_\lambda^v R_f \cos(\theta)}\Big] + 1 - \mathrm{e}^{-\alpha_\lambda^c e(\theta)} \Big) \, \mathrm{d}\lambda \mathrm{d}\theta \tag{11}$$

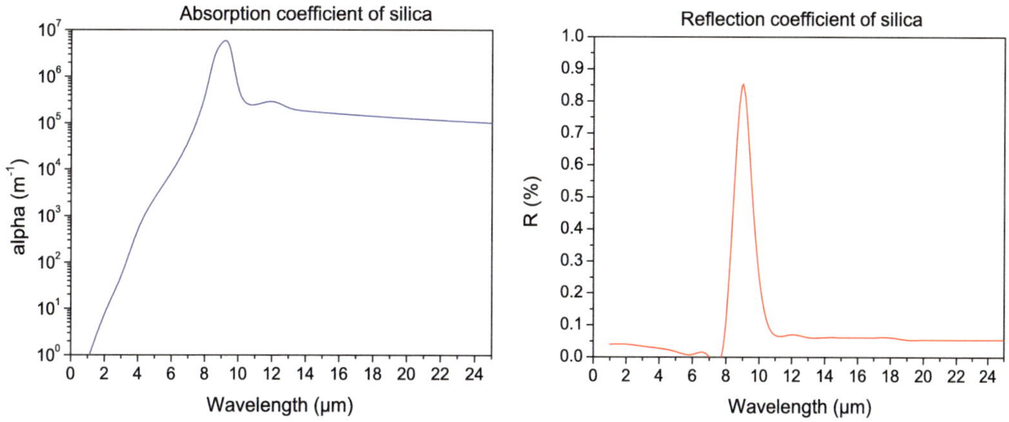

(a) Absorption (b) Reflection

Fig. 11. Optical coefficients of silica

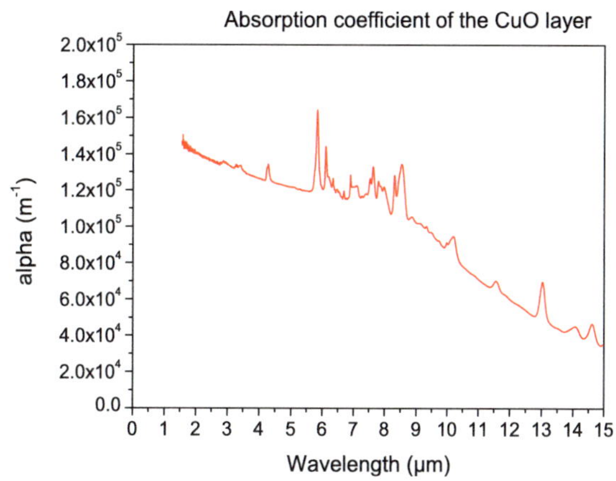

Fig. 12. Absorption coefficient of CuO in its polymer matrix.

where

- α_λ^c is the absorption coefficient of CuO
- α_λ^v is the absorption coefficient of fused silica
- f_λ is the incident spectral density and is proportional to $B_{\lambda,T}$ given by Planck's law (cf. Equation 1)
- f is the total incident spectral density $f = \int_{\lambda=0}^{\lambda=\infty} f_\lambda \mathrm{d}\lambda$
- $e(\theta)$ is the effective thickness defined by:

$$e(\theta) = \sqrt{(R_f + e_{\text{CuO}})^2 - R_f^2 \sin^2\theta} - R_f^2 \cos\theta \tag{12}$$

This equation represents the projection of the thickness of the coating for an angle θ. The part H corresponding to the heads must be added.

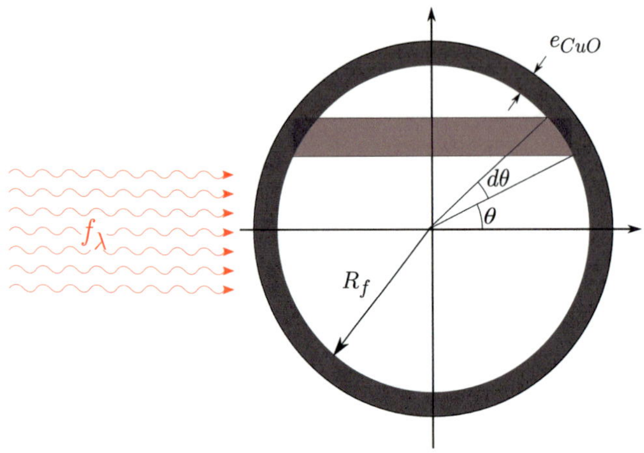

Fig. 13. Scheme for the computation of the fiber absorption.

If $R_s = R_f + e_{\text{CuO}}$ and $\theta^* = \arcsin \frac{R_f}{R_s}$

$$H = 2R_sL \int_{\lambda=0}^{\lambda=\infty} \int_{\theta=\theta^*}^{\theta=\frac{\pi}{2}} \left(1 - \mathrm{e}^{-2\alpha_\lambda^c R_s \cos\theta}\right) \frac{f_\lambda}{f} \cos\theta \mathrm{d}\theta \mathrm{d}\lambda \tag{13}$$

The total absorption coefficient of the fiber becomes:

$$K_{\text{abs}} = \frac{R_f}{R_s} \int_{\lambda=0}^{\lambda=\infty} \int_{\theta=-\frac{\pi}{2}}^{\theta=\frac{\pi}{2}} \cos(\theta) \frac{f_\lambda}{f}$$
$$\left(\mathrm{e}^{-\alpha_\lambda^c . e(\theta)} \left[1 - \mathrm{e}^{-2\alpha_\lambda^v . R_f \cos(\theta)} + (1 - \mathrm{e}^{-\alpha_\lambda^c e(\theta)}) \mathrm{e}^{-2\alpha_\lambda^v . R_f \cos(\theta)}\right] + 1 - \mathrm{e}^{-\alpha_\lambda^c e(\theta)}\right) \mathrm{d}\lambda \mathrm{d}\theta$$
$$+ 2 \int_{\lambda=0}^{\lambda=\infty} \int_{\theta=\theta^*}^{\theta=\frac{\pi}{2}} \left(1 - \mathrm{e}^{-2\alpha_\lambda^c R_s \cos\theta}\right) \frac{f_\lambda}{f} \cos\theta \mathrm{d}\theta \mathrm{d}\lambda \tag{14}$$

6.1.2 Convection losses

The fiber exchanges heat with the ambient air by a natural convection mechanism. The convection losses are given by the classical Equation 15

$$\Phi_{\text{conv}} = K_c 2\pi R_s L(T - T_a) \tag{15}$$

where

- K_c is the exchange coefficient with the fluid around the fiber (here = air)
- R_s is the radius of the fiber with its coating
- L is the exposed length
- T is the temperature of the fiber (what is calculated)
- T_a is the ambient temperature

Due to the cylindrical geometry of the problem, K_c is calculated using the adimensional Nüsselt number defined by :

$$\text{Nü} = \frac{K_c D}{\kappa} \tag{16}$$

where

- D is the diameter of the fiber with its coating ($D = 2R_s$)
- κ is the thermal conductivity of the fluid (air here)

The Nüsselt number depends on the fluid properties (viscosity, thermal conductivity, density), on the geometry and on the type of convection mechanism (natural or forced). In this case, natural convection in air is considered and the fiber is a horizontal tube of diameter D at uniform temperature in the considered zone. Under these conditions, the Nüsselt number is given by (Churchill, 1983; Padet, 2005) :

$$\text{Nü} = A\text{Ra}^n \tag{17}$$

where A and n are two empirical parameters and Ra is the Rayleigh number. It is the product of Grashof and Prandl numbers given by Equation 18 and Equation 19 respectively.

$$\text{Gr} = \frac{g\beta D^3 \Delta T}{\nu^2} \tag{18}$$

$$\text{Pr} = \frac{\nu}{a} \tag{19}$$

where

- g is the acceleration of gravity in m/s^2
- β is the buoyancy coefficient (for perfect gases $\beta = \frac{1}{T}$) in K^{-1}
- ΔT is the temperature difference between the surface and the fluid in K
- a is the effusivity of air ($= \frac{\kappa}{\rho C}$) with κ the thermal conductivity in $\text{Wm}^{-1}\text{K}^{-1}$, ρ the density in kg/m^3 and C the specific heat in $\text{Jkg}^{-1}\text{K}^{-1}$
- ν is the cinematic viscosity in m^2/s ($\nu_{air} = 1.2 \times 10^{-5}\ \text{m}^2/\text{s}$)

For gases, Pr is 0.7. Due to the fact that n is small, the Nüsselt number will not vary a lot in the considered temperature range. In the range of fiber diameter used in this study, Nü ≈ 0.6. The order of magnitude of K_c is 50 $\text{W}/(\text{m}^2\text{K})$. Equation 17 is valid for $10^{-4} < \text{Ra} < 10^{-2}$.

$$\text{Ra} = \frac{g\beta D^3 \Delta T}{a\nu} \tag{20}$$

Replacing in Equation 17 the values of the physical constants for air taken at $T_a = 20$ °C, this equation turns into Equation 21 in the range of diameters used in this study with A =1.02 and n =0.148. As n is small, the temperature dependence on *Nü* is small and this coefficient is almost constant and so is coefficient K_c.

$$\text{Nü} = A(10^8 D^3 \Delta T)^n \tag{21}$$

6.1.3 Radiation losses

The radiation losses with the ambient are given by Equation 22

$$\Phi_{\text{rad}} = \epsilon 2\pi R_s L\sigma(T^4 - T_a^4) \tag{22}$$

where

- R_s is the radius of the fiber with its coating
- ϵ is the emissivity of the surface of the material (here CuO or silica)
- σ is Stefan's constant (according to Stefan's law of emission of hot bodies, $\sigma = 5.67 \times 10^{-8}$ W/(m^{2}K^4))

As the difference $T - T_a$ is generally small, Equation 22 can be linearized

$$\Phi_{\text{rad}} = \epsilon 2\pi R_s L\sigma(T^4 - T_a^4) = K_r 2\pi R_s L(T - T_a) \tag{23}$$

with

$$K_r = \epsilon\sigma\frac{(T^4 - T_a^4)}{(T - T_a)} \approx 4\epsilon\sigma T_a^3 \tag{24}$$

The order of magnitude for K_r is 8 W/(m^{2}K).

6.1.4 Conduction losses

The losses due to the thermal conduction along the fiber are given by the following relationship:

$$\Phi_{\text{cond}} = K_l 2\pi R_s^2(T - T_a) \tag{25}$$

K_l is a coefficient relative to the conduction through the fiber along its axis. The conduction through the thin layer of absorbing material is neglected. K_l can be calculated by assuming that the rest of the fiber behaves as 2 cylindrical winglets. The conductivity of silica is small and this term has little influence on the computed result ($K_l \approx 65$ W/(m^{2}K)).

Now that the different terms are detailed, the temperature of the fiber at equilibrium can be computed and by using equations 12 , 22 and 25, Equation 6 becomes:

$$K_{abs} R_s L f = K_r 2\pi R_s L(T - T_a) + K_c 2\pi R_s L(T - T_a) + K_l 2\pi R_s^2(T - T_a) \tag{26}$$

Or dividing by $R_s L$

$$K_{abs} f = 2\pi(K_r + K_c + K_l')\,(T - T_a) \tag{27}$$

with $K_l' = K_l \frac{R_s}{L}$ (≈ 0.5 W/(m^{2}K))

The sensitivity Σ_T is then given by Equation 28

$$\Sigma_T = \frac{(T - T_a)}{f} = \frac{K_{\text{abs}}}{2\pi(K_r + K_c + K_l')} \tag{28}$$

K_{abs} may depend on the source temperature T if the absorption depends on the wavelength. Notice that for large temperature variations (high radiation fluxes), K_c and K_r depend on temperature (increase with temperature) and the sensitivity is no longer constant. As in the ideal absorbing case $K_{\text{abs}} = 2$, this equation shows that the sensitivity depends essentially on the losses through the coefficients K_c, K_l' and K_r. As the length of the exposed zone plays only a role in the coefficient K_l' which is small compared to the other coefficients, it may be concluded that the length has very little influence on the sensitivity.

The model was compared to the experimental results. Figure 14 shows the calculated results compared to the experimental values of the fiber coated with CuO for different source temperatures. The model curve was obtained directly from the physical and spectral properties of the materials. The parameters used in the model are listed in table 2.

Absorption coefficient of CuO coating	$\alpha_{\text{CuO}}(\lambda) = \frac{1.5\times10^8 e^{-0.038\lambda}}{1000+(e^{\lambda})^{0.5}}$
Absorption coefficient of fused silica	$\alpha_v(\lambda) = \frac{10^5}{1+e^{-1.8(\lambda-7)}}$
Reflection coefficient of fused silica	$R_v(\lambda) = 0.06 + 0.8\, e^{-\frac{(7-\lambda)^2}{0.8}}$
Thermal conductivity of air	$\kappa_{air} = 0.024 + 7\times10^{-5}(T_a - 273.15)$
Emissivity of CuO coating	$\epsilon_{\text{CuO}} = 0.78$
Emissivity of glass	$\epsilon_{\text{glass}} = 0.94$
Radius of the fiber	$R_f = 62.5$ µm
Thickness of CuO coating	$e_{\text{CuO}} = 40$ µm
Room temperature	$T_a = 293$ K
Nüsselt number	$\text{Nü} = 1.02\ (8\times10^8 R_s^3 \Delta T)^{0.148}$
ΔT used for the Rayleigh number	$\Delta T = 1$ K
Conductivity along the fiber	$K_l = 65$ W/(m^{2}K)
Exposed length	$L = 1$ cm

Table 2. Table of parameters injected into the model

The obtained results confirm that the model fits reasonably well with the experimental data. Small discrepancies are attributed to the fact that, during the measurements, the source temperature varies with ± 5 °C and the sensitivity of the sensor is measured with ± 5 % relative error.

6.2 Case 2: fiber with absorbing coating and capillary

As explained in section 4, the sensitivity of the sensor can be improved by protecting the fiber with a glass capillary. This case can be modeled and the model is explained in the next paragraph.

The structure is schematically represented in Figure 15.

The fiber and the capillary may both be covered with an absorbing coating or not. The system is more complicated and different energy exchanges take place through 3 surfaces:

- surface 1 (S_1) of the fiber covered with the absorbing coating at temperature T_1

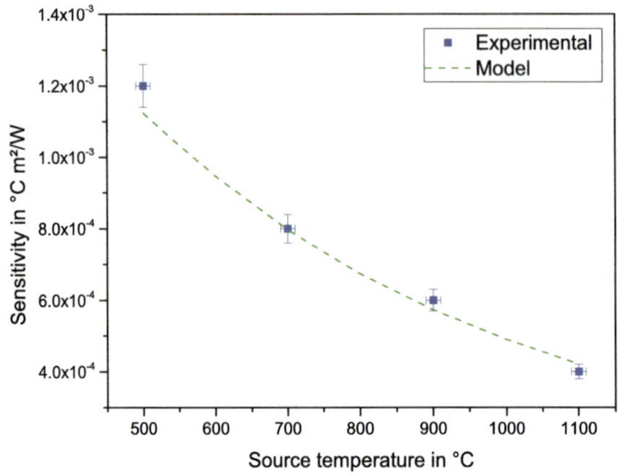

Fig. 14. Sensitivity Σ_T of the bolometer for different source temperatures (solid line = model, dots = experimental results).

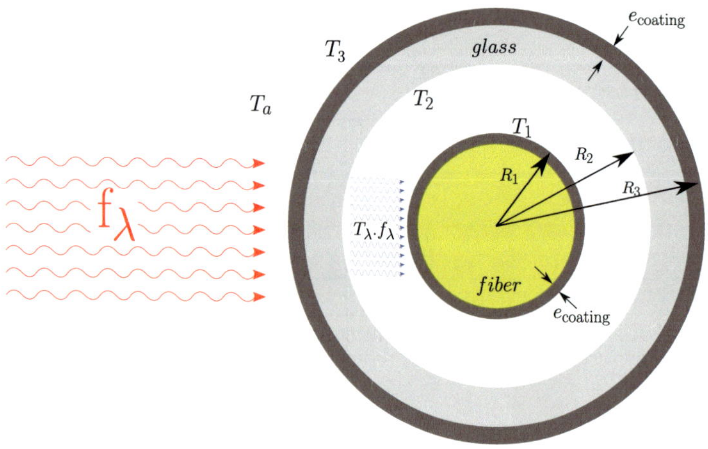

Fig. 15. Scheme of the fiber with a capillary.

- surface 2 (S_2) at temperature T_2 inside surface of the capillary
- surface 3 (S_3) at temperature T_3 in contact with ambient air

Surface 1 will be the reference surface for the calculations ($R_1 = R_f + e_{\text{coating}}$).

The thickness of the capillary is small and the temperature gradient between the walls is weak, thus it is assumed that $T_2 = T_3$.

Again, as developed in the previous paragraph, considering L the length of the exposed zone of the fiber, the energy balance of the various elements of the sensor can be computed.

For S_1, the balance between the different energy fluxes can be expressed (Equation 29):

$$\Phi^f_{\text{abs}} = \Phi_{\text{r12}} + \Phi_{\text{c12}} + \Phi_{\text{long}} \tag{29}$$

- Φ^f_{abs} is the absorbed energy after partial absorption by the fiber of the incident radiation through the capillary

- Φ_{r12} is the exchange (loss) by radiation with surface S_2
- Φ_{c12} is the heat transfer by conduction through the filling gas with S_2
- Φ_{long} is the heat transfer by conduction along the fiber

For S_2, heat is exchanged by radiation with S_1 and by conduction through the filling gas. The heat is transferred by thermal conduction through the capillary to S_3.

$$\Phi_{c23} = \Phi_{r12} + \Phi_{c12} \tag{30}$$

- Φ_{c23} is the heat transfer by conduction through the capillary with S_2

For S_3, the outer surface of the capillary, the energy balance may be expressed by Equation 31:

$$\Phi_{loss} = \Phi_{c23} + \Phi^{c}_{abs} \tag{31}$$

- Φ^{c}_{abs} is the partial absorption of the incident flux by the capillary
- Φ_{loss} represents the losses with the ambient (conduction along the capillary is neglected) $\Phi_{loss} = \Phi_{conv_{ext}} + \Phi_{rad_{ext}}$

Then

$$\Phi_{loss} = \Phi_{r12} + \Phi^{c}_{abs} + \Phi_{c12} \tag{32}$$

6.2.1 Absorption

The absorbed energy by the fiber of the transmitted radiation through the capillary is given by Eq. 33.

$$\Phi^{f}_{abs} = T_c K^{f}_{abs} R_1 f L \tag{33}$$

where T_c is the mean transmission coefficient of the capillary K^{f}_{abs} is computed in the same way as explained in the first case but absorption of the capillary is taken into account now

$$T_c K^{f}_{abs} = \frac{1}{f}\int_{\lambda=0}^{\lambda=\infty}\int_{\theta=-\frac{\pi}{2}}^{\theta=\frac{\pi}{2}} (1-R^{v}_{\lambda}) f'_{\lambda}(1-\mathrm{e}^{-2\alpha^{v}_{\lambda}e(\theta)})\sqrt{1-\left(\frac{R_f}{R_3}\right)^2 \sin^2\theta}\, d\theta d\lambda$$
$$+2\int_{\lambda=0}^{\lambda=\infty}\int_{\theta=\theta^*}^{\theta=\frac{\pi}{2}} \left(1-\mathrm{e}^{(-2\alpha_{\lambda}R_3\cos\theta)}\right) f'_{\lambda}\cos\theta\, d\theta d\lambda \tag{34}$$

where $f'_{\lambda} = T_{\lambda}.f_{\lambda}$ with

$$T_{\lambda} = \frac{(1-R_{\lambda})\mathrm{e}^{-(\alpha^{c}_{\lambda}e_{\mathrm{CuO}}+\alpha^{v}_{\lambda}e_{capillary})}}{1-R^2_{\lambda}\mathrm{e}^{-(\alpha^{c}_{\lambda}e_{\mathrm{CuO}}+\alpha^{v}_{\lambda}e_{capillary})}} \tag{35}$$

with $e_{\mathrm{capillary}} = \frac{\text{outer diameter}-\text{inner diameter}}{2}$ and $\theta^* = \arcsin\frac{R_2}{R_3}$.

The curvature of the capillary is neglected because the fiber diameter is small in comparison with the capillary dimension.

In equation 35, $R_{\lambda} \approx 0$ because the roughness of the CuO coating used is high. K^{c}_{abs} is obtained using Equation 36

$$K^{c}_{abs} = \frac{R_3}{R_1}\frac{1}{f}\int_{\lambda=0}^{\lambda=\infty}\int_{\theta=-\frac{\pi}{2}}^{\theta=\frac{\pi}{2}} (1-R^{v}_{\lambda}) f_{\lambda}(1-\mathrm{e}^{-2\alpha^{v}_{\lambda}e(\theta)})\sqrt{1-\frac{R_2}{R_3}^2 \sin^2\theta}\, d\theta d\lambda$$
$$+2\int_{\lambda=0}^{\lambda=\infty}\int_{\theta=\theta^*}^{\theta=\frac{\pi}{2}} \left(1-\mathrm{e}^{(-2\alpha^{c}_{\lambda}R_3\cos\theta)}\right) f_{\lambda}\cos\theta\, d\theta d\lambda \tag{36}$$

6.2.2 Convection losses with the ambient

The capillary exchanges heat with the ambient air by a convection mechanism like already presented before. The losses are expressed by Equation 37.

$$\Phi_{\text{conv}_{\text{ext}}} = K_{c_{\text{ext}}} 2\pi R_3 L(T_3 - T_a) \tag{37}$$

This term has the same meaning as in the first case and $K_{c_{\text{ext}}}$ is the exchange coefficient to be calculated with Nü using radius equal to R_3.

6.2.3 Radiation losses with the ambient

The radiation losses with the ambient can be calculated as in the first case following relation 38

$$\Phi_{\text{rad}_{\text{ext}}} = K_{r_{\text{ext}}} 2\pi R_3 L(T_3 - T_a) \tag{38}$$

with

$$K_{r_{\text{ext}}} = \epsilon\sigma \frac{(T^4 - T_a^4)}{(T - T_a)} \approx 4\epsilon.\sigma T_a^3 \tag{39}$$

This term has the same meaning as before and $K_{r_{ext}}$ is the exchange coefficient to be calculated with radius $= R_3$ and $\epsilon = 0.78$ for CuO or 0.94 for glass.

6.2.4 Conduction losses along the fiber

This term is the same as in the first case without the glass capillary (Eq. 25).

6.2.5 Exchange by radiation between 1 and 2

Surface 1 exchanges heat by radiation with surface 2. The heat exchange between surface 1 and surface 2 is given by

$$\Phi_{\text{r}_{12}} = K_{\text{r}_{12}} 2\pi R_1 L(T_1 - T_2) \tag{40}$$

$K_{\text{r}_{12}}$ is given by the classical formula for radiation heat exchange between two coaxial cylinders

$$\Phi_{\text{r}_{12}} = \frac{2\pi R_1 L 4\sigma(T_1^4 - T_2^4)}{\frac{1}{\epsilon_1} + \frac{\rho_2}{\epsilon_2}\left(\frac{R_1}{R_2}\right)^2} \tag{41}$$

or

$$\Phi_{\text{r}_{12}} \cong \frac{2\pi R_1 L 4\sigma T_1^3(T_1 - T_2)}{\frac{1}{\epsilon_1} + \frac{\rho_2}{\epsilon_2}\left(\frac{R_1}{R_2}\right)^2} \tag{42}$$

where $\epsilon_1 = \epsilon_{\text{CuO}}$ and $\epsilon_2 = \epsilon_{\text{glass}}$ and

$$K_{\text{r}_{12}} = \frac{4\sigma T_1^3}{\frac{1}{\epsilon_1} + \frac{\rho_2}{\epsilon_2}\left(\frac{R_1}{R_2}\right)^2} \tag{43}$$

6.2.6 Exchange by conduction between 1 and 2

Surfaces 1 and 2 exchange heat through the filling gas. As the gap between the fiber and the capillary is small, no natural convection takes place and the heat transfer is a pure conductivity phenomenon. The conduction through the air layer is given by the classical relationship corresponding to the cylindrical geometry of the transfer:

$$\Phi_{c12} = \frac{2\pi\kappa L}{\ln\left(\frac{R_2}{R_1}\right)}(T_1 - T_2) \tag{44}$$

where κ is the thermal conductivity of air.

Or

$$\Phi_{c12} = K_{c_{12}} 2\pi R_1 L(T_1 - T_2) \tag{45}$$

with $K_{c_{12}} = \frac{\kappa}{R_1 \ln \frac{R_2}{R_1}}$

Using Equations 29 and 31 and assuming $T_2 = T_3$, the following system of equations can be obtained:

$$\begin{aligned} K^f_{\text{abs}} R_1 L T_c f &= K_{r_{12}} 2\pi R_1 L(T_1 - T_2) + K_{c_{12}} 2\pi R_1 L(T_1 - T_2) + K_l 2\pi R_f^2 (T_1 - T_a) \\ K^c_{\text{abs}} R_3 L f + K_{r_{12}} 2\pi R_1 L(T_1 - T_2) + K_{c_{12}} 2\pi R_1 L(T_1 - T_2) &= 2\pi R_3 L(K_{c_{\text{ext}}} + K_{r_{\text{ext}}})(T_3 - T_a) \end{aligned} \tag{46}$$

The solution of this system of equations gives the temperature of the fiber T_1 and from this, the sensitivity Σ_T can be calculated:

$$\begin{aligned} \Sigma_T &= \frac{(T_1 - T_a)}{f} \\ &= \frac{\left[(K_{r_{12}} + K_{c_{12}})\frac{R_1}{R_3} + (K_{c_{\text{ext}}} + K_{r_{\text{ext}}})\right] T_c K^f_{\text{abs}} + (K_{r_{12}} + K_{c_{12}}) K^c_{\text{abs}}}{2\pi\left[(K_{r_{12}} + K_{c_{12}})\frac{R_1}{R_3} + (K_{c_{\text{ext}}} + K_{r_{\text{ext}}})\right](K_{r_{12}} + K_{c_{12}} + K'_l) - 2\pi\frac{R_1}{R_3}(K_{r_{12}} + K_{c_{12}})^2} \end{aligned} \tag{47}$$

As K'_l can be neglected, equation 47 becomes :

$$\Sigma_T = \frac{\left[\frac{R_1}{R_3} + \frac{(K_{c_{\text{ext}}} + K_{r_{\text{ext}}})}{(K_{r_{12}} + K_{c_{12}})}\right] T K^f_{\text{abs}} + K^c_{\text{abs}}}{2\pi(K_{c_{\text{ext}}} + K_{r_{\text{ext}}})} \tag{48}$$

Again, it is clearly put in evidence that the sensitivity depends essentially on the losses of the capillary with the ambient, the diameter of the fiber is of little importance. The most important parameter is the diameter of the capillary.

This model can now be confronted with experimental results. Fig. 16 and 17 show the comparison of the model and experimental data for different source temperatures and different capillary diameters. The values of the different parameters are given in table 2.

Fiber protected by a capillary

The fiber is coated with CuO (44 μm) and surrounded by a glass capillary. The influence of the capillary diameter was experimentally tested and modeled as shown in Fig.16. The dots are the experimental points.

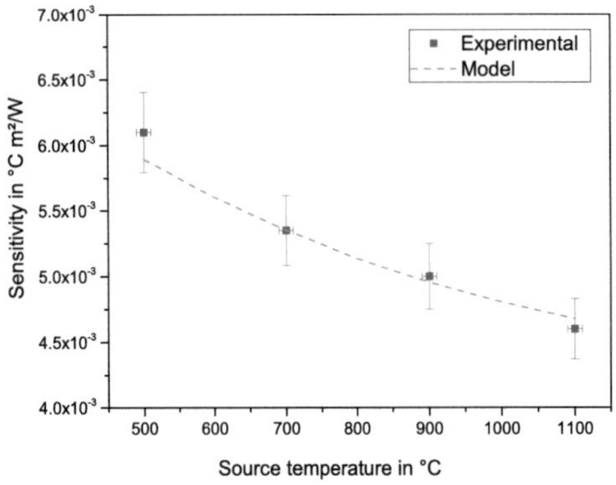

(a) **Capillary 1** outer: 635 - inner: 560 μm

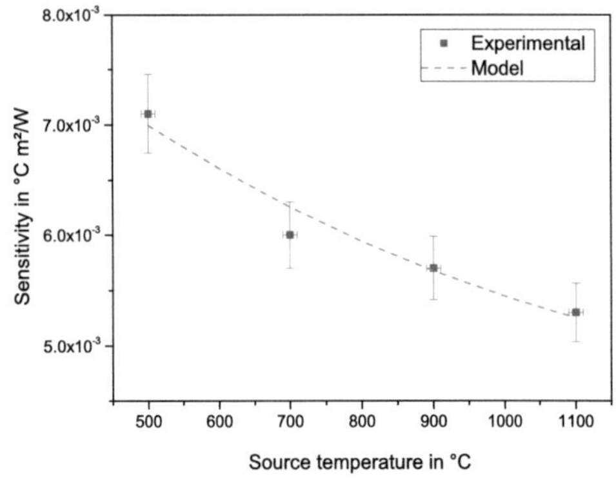

(b) **Capillary 2** outer: 950 - inner: 750 μm

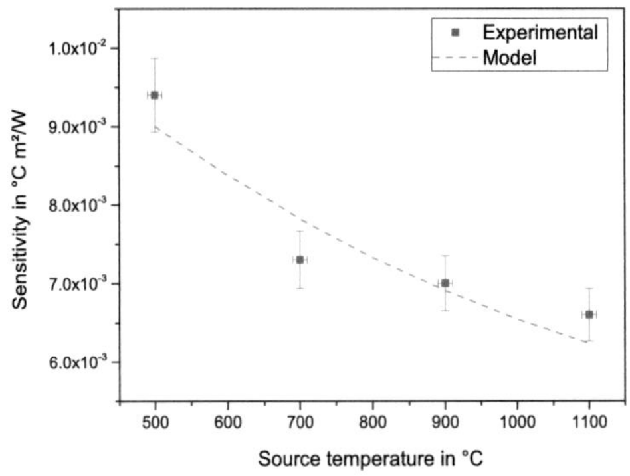

(c) **Capillary 3** outer: 1900 - inner: 1660 μm

Fig. 16. Influence of the source temperature on the sensitivity of the sensor for a different capillary diameters.

Despite of the simplifications, the model predicts well the behavior of the sensor. These results show that the diameter of the capillary has a strong influence on the sensitivity of the system. In the following paragraph, the theoretical evolution of the sensor sensitivity as a function of the capillary diameter will be presented.

Fiber protected by a capillary covered with CuO.

Fig. 17 shows the result of the modelling for the case when the capillary is covered with CuO. The dots are the experimental points.

In fact, as the dimensions remain small, the exchanges between the capillary and the sensing fiber are very good and, as a consequence, the temperature at the surface of the fiber is close to the one at the surface of the capillary. The system behaves as if the diameter of the fiber was increased. The consequence is then a decrease in the convection losses (K_c decreases with increasing diameter) and thus an increase in sensitivity.

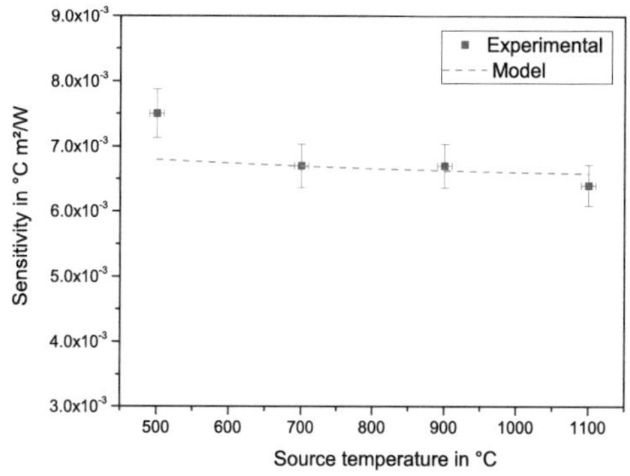

Fig. 17. Influence of the source temperature on the sensitivity of the sensor for a fiber protected by a capillary coated with CuO (diameter outer: 635 - inner: 560 μm).

6.3 Limit cases

The model can be predictively used to study the influence of some important parameters and limit cases. If the coating is perfectly absorbing, the sensitivity does not depend on the source temperature as it can be seen in Figure 18. In this example, the outer diameter of the capillary is 1000 μm and inner diameter 800 μm. Thickness of the CuO layer is equal to 80 μm.

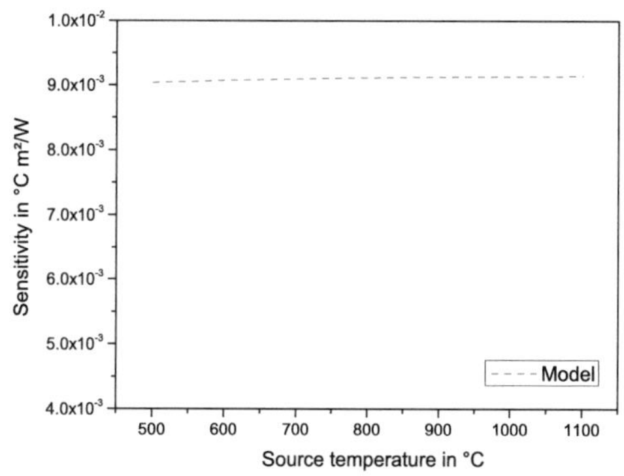

Fig. 18. Simulation of the source temperature dependence when the coating is perfectly absorbing.

The sensitivity is of the order of 9×10^{-3} °C/(W/m^2). As explained before, the outer diameter of the capillary is an important parameter. Fig. 19 shows the influence of the capillary diameter when the absorption is complete. The thickness of the capillary is kept at 200 μm. The point for diameter = 0 means the fiber without capillary.

As expected, the increase of the diameter increases the sensitivity until a saturation is reached. It can be seen that using a capillary thicker than 5 mm is not interesting because the sensitivity is not substantially increased and the sensor becomes cumbersome which precisely reduces the advantage of using a fiber. The maximum sensitivity Σ_T is about 2.5×10^{-2} °C/(W/m^2) or expressed in Bragg wavelength shift $\Sigma_\lambda = 0.25$ pm/(W/m^2). This sensitivity is the maximum that could be reasonably expected for a sensor without concentrating packaging.

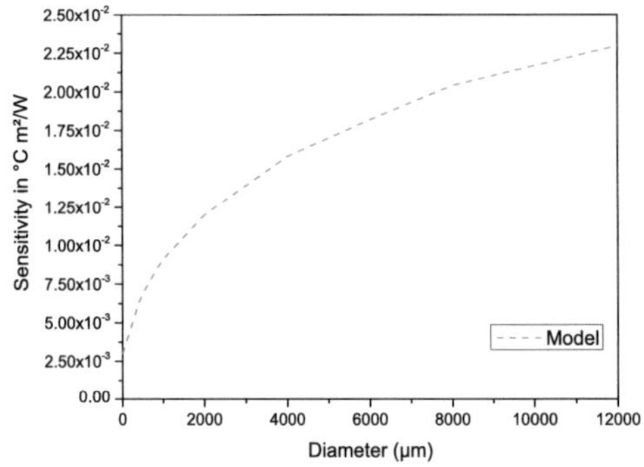

Fig. 19. Simulation of the capillary diameter dependence when the coating is perfectly absorbing.

7. Conclusions

In this chapter, a simple IR detection system with a pair of fiber Bragg gratings based on the working principle of the bolometer has been described. The sensor shows a linear behavior according to the flux intensity, a good sensitivity and a short response time (estimated to be less than 1 s). The behavior of the sensor and its different configurations can be modeled and the predictions fit well the experimental results.

This kind of sensor possesses all the advantages of the optical fiber sensors : low weight, small size, immunity to electromagnetic interferences and remote operation. Indeed, quasi-distributed measurements based on this system can be made without difficulty using classical techniques (cascaded gratings, reflectometric techniques,...). The deposition method used here to coat the grating (encapsulation in a polymer matrix) requires only a few minutes, is cheap, and easily scaled up for mass production, which is important for commercial use of this device.

One of the most important applications would be in the perspective of early fire detection. The simple design presented above is sufficient for the detection of car fires in tunnels because the combustion of hydrocarbons and polymers releases hundreds of kW/m^2 of heat and this sensor can easily detect a few kW/m^2. With dedicated packaging and a good equipment for the measurement of the Bragg wavelength, a lower threshold can be achieved. The temperature dependence can be explained by a non-uniform absorption on the whole infrared wavelength range. Considering emission spectra, it can be deduced that silica contributes a lot to the absorption at low temperature while copper oxide is not as efficient as expected for the wavelengths below 4 μm and so the absorption is not complete. The model shows that a better sensitivity and a lower temperature dependence can be achieved using a better absorbing layer (increasing the thickness for example). A second observation is that the shielding with a capillary tube is very efficient (sensor sensitivity enhanced by 5 to 8 times). A glass capillary also reduces the sensitivity to air streams. Finally, the packaging increases the shift of the Bragg wavelength by a factor of 2 to 3. This simple packaging is cylindric (simple commercial tube cover with aluminium) and is not optimal. Using a parabolic shape should result in a better focalization and, as a consequence, a still higher sensitivity.

8. Acknowledgments

This study was performed in the framework of the Opti^{2}Mat project financially supported by the Wallonia in Belgium. C. Caucheteur is supported by the Fonds National de la Recherche Scientifique (F.R.S.-FNRS). The authors take part to the Attraction Pole Program IAP 6/10 (photonics@be) of the Belgian Science Policy.

9. References

Buric, M., Chen, K. P., Bhattarai, M., Swinehart, P. R. & Maklad, M. (2007). Active Fiber Bragg Grating Hydrogen Sensors for All-Temperature Operation, *IEEE Photonics Technology Letters* 19(5): 255–257.

Caucheteur, C., Debliquy, M., Lahem, D. & Megret, P. (2008). Hybrid fiber gratings coated with a catalytic sensitive layer for hydrogen sensing in air, *Optics Express* 16(21): 16854.

Caucheteur, C., R. J. D. M. M. P. (2010). Infrared radiation detection with matched fiber bragg gratings, *IEEE Photonics Technology Letters* 22(23): 1732–1734.

Churchill, S. (1983). *Heat Exchanger Design Handbook*, Hemisphere.

Corres, J. M., del Villar, I., Matias, I. R. & Arregui, F. J. (2007). Fiber-optic pH-sensors in long-period fiber gratings using electrostatic self-assembly, *Optics Letters* 32(1): 29.

Crunelle, C., Wuilpart, M., Caucheteur, C. & Mégret, P. (2009). Original interrogation system for quasi-distributed FBG-based temperature sensor with fast demodulation technique, *Sensors and Actuators A: Physical* 150(2): 192–198.

Davino, D., Visone, C., Ambrosino, C., Campopiano, S., Cusano, a. & Cutolo, a. (2008). Compensation of hysteresis in magnetic field sensors employing Fiber Bragg Grating and magneto-elastic materials, *Sensors and Actuators A: Physical* 147(1): 127–136.

Desvignes, F. (1991). *Rayonnements optiques : Radiometrie - Photometrie*, Editions Masson.

Fernandez-Valdivielso, C., Matias, I. R. & Arregui, F. J. (2002). Simultaneous measurement of strain and temperature using a fiber Bragg grating and a thermochromic material, *2002 15th Optical Fiber Sensors Conference Technical Digest. OFS 2002(Cat. No.02EX533)* 101: 203–206.

Gaussorgues, G., Micheron, F. & Pocholle, J. (1996). Détecteurs infrarouges, *Techniques de l'Ingenieur;* pp. 1–42.

Han, Y.-G. (2009). Directional bending sensor with temperature insensitivity using a sampled chirped fiber Bragg grating, *Journal of Applied Physics* 105(6): 063103.

Ho, H. (2002). A fiber Bragg grating sensor for static and dynamic measurands, *Sensors and Actuators A: Physical* 96(1): 21–24.

Kronenberg, P., Rastogi, P. K., Giaccari, P. & Limberger, H. G. (2002). Relative humidity sensor with optical fiber Bragg gratings, *Optics Letters* 27(16): 1385.

Lu, P., Men, L. & Chen, Q. (2008). Tuning the sensing responses of polymer-coated fiber Bragg gratings, *Journal of Applied Physics* 104(11): 116110.

Men, L., Lu, P. & Chen, Q. (2008). A multiplexed fiber Bragg grating sensor for simultaneous salinity and temperature measurement, *Journal of Applied Physics* 103(5): 053107.

Othonos, A., K. K. (1999). *Fiber Bragg gratings: fundamentals and applications in telecommunications and sensing*, Artech House.

Padet, J. (2005). *Techniques de l'Ingenieur* BE8206.

Perry, R., G. D. (1984). *Perry's Chemical Engineers' Handbook*, 6th edition edn, Mc Graw Hill.

Renoirt, J.-M., C. C. M. P. D. M. (2010). Infrared radiation detector using a pair of fiber bragg gratings, *Proceedings of SPIE - The International Society for Optical Engineering* 7726.

Sheng, H., Fu, M., Chen, T., Liu, W. & Bor, S. (2004). A lateral pressure sensor using a fiber Bragg grating, *IEEE PHOTONICS TECHNOLOGY LETTERS* 16(4): 1146–1148.

Sivathanu, Y. (1997). Fire detection using time series analysis of source temperatures, *Fire Safety Journal* 29(4): 301–315.

Yang, M., Dai, J., Zhou, C. & Jiang, D. (2009). Optical fiber magnetic field sensors with TbDyFe magnetostrictive thin films as sensing materials, *OPTICS EXPRESS* 17(23): 20777–20782.

Yeo, T., Sun, T., Grattan, K., Parry, D., Lade, R. & Powell, B. (2005). Characterisation of a polymer-coated fibre Bragg grating sensor for relative humidity sensing, *Sensors and Actuators B: Chemical* 110(1): 148–156.

Yüksel, K, C. C. R. J. M. P. D. M. W. M. (2011). Infrared radiation detector interrogated by optical frequency domain reflectometer (ofdr), *Optical Sensors, OSA Technical Digest (CD) (Optical Society of America, 2011)* (SMB6).

8

Wave Optics of Synchrotron Radiation

Nikolay Smolyakov
National Research Center "Kurchatov Institute", Moscow, Russia

1. Introduction

Synchrotron radiation is a unique source of infrared radiation being highly polarized, pulsed, with the broad emission band and about thousand times brighter than standard thermal source. All just mentioned synchrotron radiation adventures apply to a large choice of experimental techniques and investigations. Among them are high-pressure studies, earth science and biology, microspectroscopy, reflectance and absorption spectroscopy for surface study, time-resolved spectroscopy and ellipsometry. Interest in infrared synchrotron radiation goes back to the 1980s (Duncan & Williams, 1983). At present numerous infrared beamlines have been developed at synchrotron radiation facilities throughout the world, see, e. g., (Bocci et al., 2008; Carr & Dumas, 1999; Guidi et al., 2005; H. Kimura et al., 2001; S. Kimura et al., 2001; S. Kimura et al., 2006; Roy et al., 2006; Williams & Dumas, 1997). Efforts to improve the radiation beam characteristics lead to elaboration more and more sophisticated beamline optics. To achieve this goal, we need to know all characteristics of emitted radiation, its intensity distribution, polarization and phase distribution. In particular, the synchrotron radiation wave properties play a major role in conventional diagnostics of electron beams in storage rings (Andersson et al., 2006; Elleaume et al., 1995; Fang et al., 1996; Flanagan et al., 1999; Hs & Huang, 1993; Weitkamp et al., 2001). In this case, the image of the electron beam is formed by an optical lens. The synchrotron radiation diffraction on the lens iris aperture restricts the resolution of the beam profile measurements. Infrared synchrotron radiation was used for longitudinal beam diagnostics at FLASH free electron laser (Grimm et al., 2008; Paech et al., 2006, 2007). Recent trends show an increased usage of synchrotron radiation interferometers for high precision measurements of the electron beam sizes (Artemiev et al., 1996; Chubar, 1995; Hiramatsu et al., 1999; Katoh & Mitsuhashi, 1999; Naito & Mitsuhashi, 2010). For proper interpretation of observed data, we also need to know synchrotron radiation phase distributions.

As far as we know, the paper (Sabersky, 1973) was one of the first papers to pioneer the investigation of geometrical-optical properties of synchrotron radiation coming from the curving relativistic electron beam. The phase-space techniques, commonly applied to charged beam optics, was used for analysis. The focusing of synchrotron radiation by a convex lens within the framework of geometrical optics was considered in (Green, 1976; Ogata, 1987). The qualitative agreement with the experimentally observed data was found, although the quantitative discussion needs taking into account of the diffraction effects. The synchrotron radiation treatment as a laser-like Gaussian beam with a small opening angle was performed in (Coisson & Marchesini, 1997; Kim, 1986; Ogata, 1991; Takayama et al.,

1998, 1999). The important benefit to the use of this approximation is that the Gaussian beams have been much studied. It has been shown however that the Gaussian approximation has some limitations when calculating the synchrotron radiation coherence. At the same time the Gaussian beam approximation can reasonably be applied to the horizontally polarized component of synchrotron radiation only. It has been also found that the Gaussian beam approximation is poor for undulator radiation (Kim, 1986). In addition, as it was noted (Miyahara & Kagoshima, 1995), the brightness function defined by Kim can be negative (that has no physical meaning) and should therefore be modified.

The most-used scheme of diffraction phenomena implementation into synchrotron radiation theory was suggested by A. Hofmann and F. Meot (Hofmann & Meot, 1982). This technique involves the following steps. First, the synchrotron radiation from a single electron has supposedly the phase distribution of a point source that is spherically symmetrical one, while its amplitude distributions (for both horizontal and vertical polarizations) should be identical to that of an electron in a homogeneous magnetic field. Under these assumptions and supposing small observation angles, for such source we can calculate the Fraunhofer diffraction pattern. Then, assuming that we have a transverse Gaussian distribution of such point-like sources, and that they are not coherent (it corresponds to incoherent radiation of different electrons in the beam), we can calculate the image distribution by a convolution of the single-electron radiation diffraction pattern with the point sources distribution. Assume further that, instead of moving relativistic beam, we have a uniform longitudinal distribution of such immobile point sources along the circular trajectory of the beam, which are again incoherent. Then we can calculate the resultant image distribution from the electron beam by straightforward integration along longitudinal coordinate. This scheme is relatively simple, easy-to-interpret physically and allows easily estimating the optical resolution for the beam cross-section measurements. It was employed by a number of groups for optimization of the electron beam profile monitor systems and experimental date interpretations, e.g. (Andersson & Tagger, 1995; Arp, 2001; Clarke, 1994; Hs & Huang, 1993). The main weakness of this method is that some steps in this scheme are not deduced from the fundamental principles of radiation theory that is to say from the Maxwell equations. Close inspection of this model shows that some assumptions seem not sufficiently self-consistent. Thus, considering the synchrotron radiation as the radiation from the immobile point sources distributed uniformly along electron's trajectory, we in fact presume that the radiation, generated by the electron at neighboring sections of its path, are incoherent, but this is not the case. The simplest counter example is undulator radiation, where the radiation, emitted by the electron at different periods (5-10 and more cm in length), is coherent. Accordingly, there is no reason to expect the opposite for synchrotron radiation. Rather, we should expect that the radiation from different parts of the electron trajectory is coherent in the case of synchrotron radiation as well.

In terms of wave optics a consistent definition of the problem is as follows. Let a physical device consists of multiple optical parts. Each part has characteristics known beforehand. It is sufficient to know the radiation wave amplitude and phase at every point of the device entrance window in order to take advantage of the Helmholtz - Kirchhoff integral theorem (Born & Wolf, 1986). If these values are known, one can calculate the distribution of the radiation intensity on the device output screen.

The analysis of synchrotron radiation phase distribution based on the fundamental radiation principles was presented at the Eleventh Russian Conference on the usage of Synchrotron Radiation (Novosibirsk, 1996) (Smolyakov, 1998a, 1998b). Using exact solutions of the Maxwell equations, the expansion of the synchrotron radiation phase in powers of small observation angles was obtained. A leading quadratic term of this expansion shows that synchrotron radiation by its phase distribution nature is much closer to the radiation emitted by an immobile point source rather than to the longitudinally extended source according to the geometrical-optical approach mentioned above. At the same time the correction cubic term produces a self-aberration effect of synchrotron radiation. The synchrotron radiation exhibits a searchlight effect which manifests itself in the amplitude distribution non-homogeneity in the vertical direction. Using the known amplitude and phase distributions, the intensity distribution of the focused synchrotron radiation from one electron (single electron image) was computed (Smolyakov, 1998c). The effect of the lens aperture on the optical resolution of the electron beam profile measurement system by means of synchrotron radiation was also analyzed. Later the similar results were obtained in the paper (Bosch, 1999), in the part which deals with the synchrotron radiation focusing by a reflecting sphere with an aperture.

An accurate simulation of synchrotron radiation propagation through the beamline optical system inevitably needs the application of specialized computer codes. Nowadays two codes are mostly in use: the code SRW (Chubar & Elleaume, 1998) and the code PHASE (Bahrdt, 1997, 2007). Comparison studies of these codes may be found in (Bowler et al., 2008). Some distributions of focused synchrotron radiation with and without phase corrections were simulated with the help of SRW (Chubar et al., 1999, 2001).

The codes, which are based on the wave propagation, are precise but time consuming. It should be mentioned that simpler and hence faster approaches are also used for analysis of beamline optical system properties (Ferrero et al., 2008).

In this chapter we will consider solely the case of standard synchrotron radiation, namely, the radiation generated by highly relativistic electrons while they pass through the uniform magnetic field of bending magnets in storage rings. Intensity distributions of synchrotron radiation, as well as its polarization properties, have been studied quite intensively and are widely covered in the literature. The amplitudes of horizontally (σ) and vertically (π) polarized wave components of synchrotron radiation are expressed in terms of modified Bessel functions. Although the first formulas for the wave amplitudes were derived more than 60 years ago, regular study of the synchrotron radiation phase distributions started relatively recently, about dozen years ago and not yet analyzed in full measure. Here an exact expression for the phase distribution in synchrotron radiation wave will be derived. Surprisingly, the resultant exact formula is written in terms of elementary functions only, though this formula is rather cumbersome. The lens aperture effect on the focused radiation intensity distribution is analyzed.

We do not consider edge radiation in this chapter. A comprehensive list of papers on edge radiation can be found in (Geloni et al. 2009a, 2009b; Korchuganov & Smolyakov, 2009; Smolyakov & Hiraya, 2005).

2. Qualitative analysis of synchrotron radiation wave properties

2.1 Wave optics of convex lens

To gain greater insight into physics of synchrotron radiation wave optics, let us consider a standard case of the point source radiation focusing by a refractive lens, see Fig. 1. Within the framework of geometrical optics, the set of rays (R_1 and R_2) from the point source P, located at a distance $D>0$ ahead of the lens, is focused to an image point S aft of the lens at the distance L so that: $1/D+1/L=1/f$, where f is the focal length of the lens.

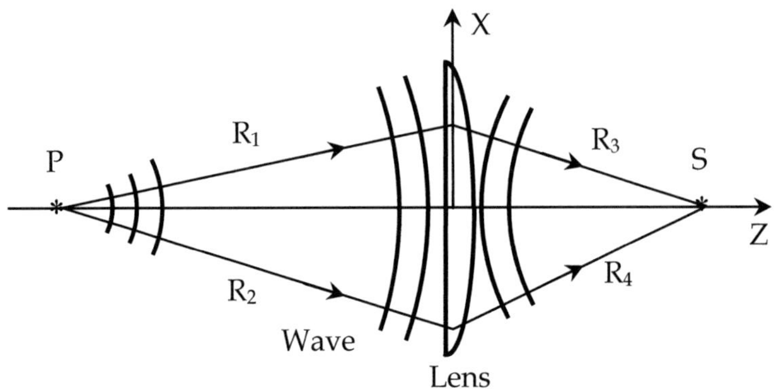

Fig. 1. Point source radiation focusing by a lens: P - point source, S - its image, R_1 - R_4 - radiation rays.

In terms of wave optics, a point source emits a divergent spherical wave $\vec{E}(x,y,z)=\frac{\vec{E}_0}{r}e^{ikr}$, with the spherically symmetric amplitude and phase distributions. Here $k=2\pi/\lambda$, $r=\sqrt{x^2+y^2+(z+D)^2}$. In its passing through the ideal lens, the wave does not change its amplitude distribution (no absorption) while the phase distribution is transformed from the divergent-type one into the convergent-type wave as it is shown in Fig. 1. This is due to the fact that the lens, with its spherical shape and correspondently with smoothly varied thickness, imparts the proper different phase delays to different parts of the incident wave. Supposing that the lens sizes are far less than the distance to the source D, we have for the incident wave at $z=0$:

$$\vec{E}_{in}(x,y,z=0)=\frac{\vec{E}_0}{D}\exp\left(ikD+ik(x^2+y^2)/(2D)\right). \qquad (1)$$

After passing the ideal infinitely thin lens with the focal length f, the correspondent distributions of the outgoing wave $\vec{E}_{out}(x,y,z=0)$ at $z=0$ will be equal to:

$$\vec{E}_{out}(x,y,z=0)=\vec{E}_{in}(x,y,z=0)\cdot\exp\left(-ik(x^2+y^2)/(2f)\right). \qquad (2)$$

Here the phase factor $\exp\left(-ik(x^2+y^2)/(2f)\right)$ describes the action of the lens on the radiation passing through it. With a knowledge of the radiation distributions $\vec{E}_{out}(x,y,z=0)$ aft of the lens and applying the Kirchhoff integral theorem (Born & Wolf, 1986), we can calculate the radiation intensity at any point of the observation screen $\{x_s,y_s,z_s\}$ at $z_s>0$.

$$\vec{E}(x_s, y_s, z_s) = \frac{-i}{\lambda} \int_A \vec{E}_{out}(x, y, z=0) \frac{\exp(ikr_s)}{r_s} dxdy, \tag{3}$$

where $r_s = \sqrt{(x-x_s)^2 + (y-y_s)^2 + z_s^2} \cong z_s + \frac{(x-x_s)^2 + (y-y_s)^2}{2z_s}$ is the distance of the lens element $dxdy$ from the point $\{x_s, y_s, z_s\}$ at the observation screen and the integral is taken over the lens aperture A. The standard simplification of the Kirchhoff integral was made by neglecting the small terms of the order of λ/D, λ/L and λ/f. The expression (3) displays the well-known Huygens-Fresnel Principle: every point of a wave front $\vec{E}_{out}(x, y, z=0)$ may be considered as a center of a secondary disturbance which gives rise to spherical waves $\frac{\exp(ikr_s)}{r_s}$. The total disturbance at the observation point $\{x_s, y_s, z_s\}$ is the result of all these secondary waves interference.

Substituting Eq. (1) and Eq. (2) into Eq. (3), we get:

$$\vec{E}(x_s, y_s, z_s) = \frac{-i\vec{E}_0}{\lambda D z_s} \int_A \exp\left((ik/2)\left(\frac{1}{D} + \frac{1}{z_s} - \frac{1}{f} \right)(x^2 + y^2) \right) \cdot \exp((-ik/z_s)(xx_s + yy_s)) dxdy \tag{4}$$

Assume for simplicity that the lens is rectangular in shape: $|x| \le l_x$, $|y| \le l_y$, where $2l_x$ and $2l_y$ are the lens horizontal and vertical sizes respectively. Considering the integral (4) in the image plane $z_s = L$, where $1/D + 1/L - 1/f = 0$, we get:

$$\vec{E}(x_s, y_s, z_s = L) = \frac{-4i\vec{E}_0}{\lambda DL} l_x l_y \frac{\sin(\xi_s)}{\xi_s} \cdot \frac{\sin(\eta_s)}{\eta_s}, \tag{5}$$

where $\xi_s = \frac{2\pi l_x}{\lambda L} x_s$, $\eta_s = \frac{2\pi l_y}{\lambda L} y_s$. The radiation intensity is proportional to $\left|\vec{E}(x_s, y_s, z_s)\right|^2$. Notice that the focused spot has horizontal and vertical sizes $\lambda L / l_{x,y}$ respectively, which are in inverse proportion to the lens aperture sizes $2l_{x,y}$. Second, the radiation intensity at the spot center $x_s = 0$, $y_s = 0$, is proportional to $l_x^2 l_y^2$. It is physically clear that with the lens horizontal aperture $2l_x$ increasing, the photon flux through the lens varies proportionally and the horizontal size of the spot is correspondently diminished. So, the density is quadratic in $2l_x$. Similar to the vertical aperture $2l_y$.

2.2 Time-domain analysis of synchrotron radiation

Here, we will consider synchrotron radiation in the time domain and describe it in terms of electric field of the emitted wave and its arrival time to an observer. A relativistic electron generates synchrotron radiation in the uniform magnetic field of a storage ring bending magnet. Let us consider a physical experiment with a geometry shown in Fig. 2. The electron rotates anticlockwise.

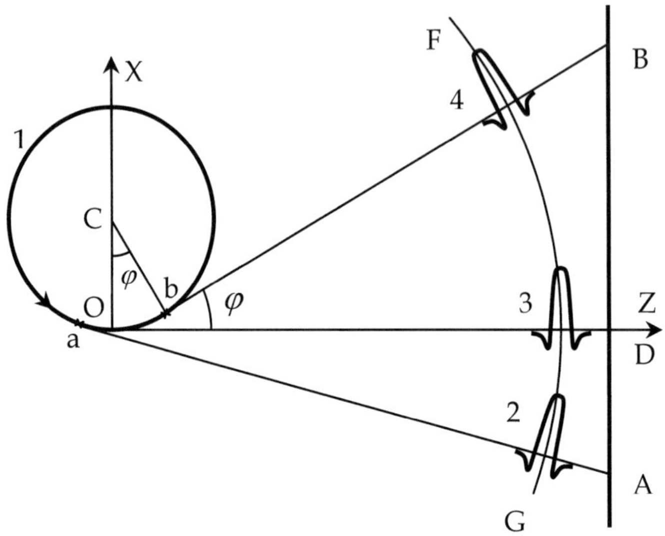

Fig. 2. Typical layout of synchrotron radiation observation: 1 – electron trajectory; 2, 3, 4 – synchrotron radiation field pulses; A-D-B – device entrance window.

For simplicity, we will restrict here our qualitative analysis to the case of radiation in the storage ring median plane XOZ . Briefly, the idea of analysis is as following.

In uniform magnetic field, an electron generates synchrotron radiation homogeneously along its orbit. In the median plane on which the electron moves, the wave shape of the emitted radiation (the electric field temporal structure, curves 2, 3 and 4 in Fig.2) is the same at any observer positions in this plane (e.g., points A , D and B) provided the distances from the emitting points a , O and b respectively to the observers are the same. In the strict sense the distances aA , OD and bB are slightly different from each other. Here, however, this difference can be neglected being negligibly small in comparison with the distance itself. The difference in the field between observers A , D and B is in the relative arrival time τ_A , τ_D and τ_B of the radiation field. The distinctions $\Delta\tau$ between arrival time τ_A , τ_D and τ_B cannot be neglected since the quantity $c \cdot \Delta\tau$ is not negligibly small as compared with the radiation wavelength, where c is the speed of light. This arrival time difference gives rise to the phase difference in the frequency domain.

Let a relativistic electron moves along its circular orbit in a horizontal plane XOZ (the magnetic field is aligned with the vertical Y-axis). It is convenient to choose a frame of reference in such a way that at the initial moment $t = 0$ the electron was at the origin of the coordinates with its velocity vector directed along the Z- axis. Then the equations of the electron motion are:

$$
\begin{aligned}
r_x(t) &= R - R\cos(\Omega t), \\
r_y(t) &= 0, \\
r_z(t) &= R\sin(\Omega t).
\end{aligned}
\tag{6}
$$

$$\begin{aligned}\beta_x(t) &= \beta \sin(\Omega t),\\ \beta_y(t) &= 0,\\ \beta_z(t) &= \beta \cos(\Omega t).\end{aligned} \tag{7}$$

Here, R is the electron orbit radius, Ω is the angular velocity of the electron $\Omega = c\beta/R$, β is the electron reduced velocity $\vec{\beta}(t) = \frac{1}{c}\cdot\frac{d\vec{r}(t)}{dt}$, $\beta = \left|\vec{\beta}(t)\right| = const$, c is the speed of light.

Let us consider a points at the observation screen (plane A-D-B, device entrance window), which is transversal to the Z-axis. The observer, who is at the point $\{0,0,D\}$, will detect the peak in the electric field distribution of the radiation wave (synchrotron radiation field pulse) at the moment $\tau = D/c$. This is obvious because, at time $t = 0$ the electron was at the origin of coordinates, and the vector of its velocity pointed to the observer with coordinates $\{0,0,D\}$. If there are other observers located at some different points A or B of the observation screen, they will detect a similar synchrotron radiation pulse. Nevertheless, these observers will detect the signal peak at different instants. Obviously, the time detected by each observer consists of two components. The first part $t_e(x_s)$ is the time, when the electron was at such point of its orbit $\vec{r}(t_e)$ from where the velocity vector pointed to the observer with coordinates $\vec{X}_s = \{x_s, 0, D\}$ (points *a* or *b* in Fig.2). The other part is the time of radiation propagation from position of the electron $\vec{r}(t_e(x_s))$ to the point at the observation screen $\vec{X}_s = \{x_s, 0, D\}$, that is the quantity $\frac{1}{c}\left|\vec{X}_s - \vec{r}(t_e(x_s))\right|$.

At some moment t the electron's position was $\vec{r}(t)$ and its velocity was pointed to the observer with the coordinates $\{x_s(t), 0, D\}$. It can easily be shown from Fig. 2 that:

$$x_s(t) = r_x(t) + (D - r_z(t))\cdot tg\varphi\,, \tag{8}$$

where $\varphi = \Omega t$.

It follows from Eq. (6) and Eq. (8) that:

$$x_s(t) = R + D\cdot tg(\Omega t) - \frac{R}{\cos(\Omega t)}\,, \tag{9}$$

The case $\beta_z(t) > 0$ (i.e. $\cos(\Omega t) > 0$) is of interest for our analysis. By using the relation $1 + tg^2(\varphi) = 1/\cos^2(\varphi)$, we can get from Eq. (9):

$$\left(D^2 - R^2\right)tg^2(\Omega t) + 2D\left(R - x_s(t)\right)tg(\Omega t) + x_s^2(t) - 2Rx_s(t) = 0\,. \tag{10}$$

This equation is quadratic in terms of $tg(\Omega t)$ with the solution:

$$tg(\Omega t_e(x_s)) = \frac{R\sqrt{(R - x_s)^2 + D^2 - R^2} - D(R - x_s)}{D^2 - R^2}\,. \tag{11}$$

This is precisely the solution of the two solutions of the quadratic equation (10) which gives $t_e(x_s=0)=0$. Here we consider the quantity x_s as an independent variable. Thus, $t_e(x_s)$ is the moment when the electron, having the position $\vec{r}(t_e(x_s))$, points to the observer with the coordinates $\{x_s,0,D\}$. As a result, we have the following expression for the arriving time $\tau(x_s)$ of synchrotron radiation pulse:

$$\tau(x_s)=t_e(x_s)+\frac{1}{c}\left|\vec{X}_s-\vec{r}(t_e(x_s))\right|. \tag{12}$$

It is convenient to define the function $\varphi(x_s)$:

$$\varphi(x_s)=\arctan\left(\frac{R\sqrt{(R-x_s)^2+D^2-R^2}-D(R-x_s)}{D^2-R^2}\right), \tag{13}$$

with $\varphi(x_s=0)=0$ and $t_e(x_s)=\frac{1}{\Omega}\varphi(x_s)=\frac{R}{c\beta}\varphi(x_s)$. In essence the function $\varphi(x_s)$ is the angle $\varphi=\Omega t$, at which the electron velocity $\vec{\beta}(t)$ points to the observer $\vec{X}_s=\{x_s,0,D\}$, see Fig. 2.

As a result, we will get the following expression for the arriving time $\tau(x_s)$:

$$\tau(x_s)=\frac{R}{c\beta}\varphi(x_s)+\frac{1}{c}\sqrt{(x_s-R+R\cos(\varphi(x_s)))^2+(D-R\sin(\varphi(x_s)))^2}. \tag{14}$$

Looking ahead, we note that the value of $2\pi c\tau(x_s)/\lambda$ is a phase of the radiation with the wavelength λ.

The x_s-dependence of the function $c\Delta\tau(x_s)=c(\tau(x_s)-\tau(0))$ is shown in Fig. 3a. The parameters of Siberia-2 storage ring were used for this simulation: electron beam energy 2.5 GeV (electron reduced energy is equal to γ =4892), electron orbit radius R =4905.4 mm, synchrotron radiation critical energy 7.2 keV (Korchuganov et al., 2005; Korchuganov & Smolyakov, 2009). The distance of the observation plane from the emission point was taken to be D = 20000 mm.

The quadratic behaviour of retardation $c\Delta\tau(x_s)$ suggests that the synchrotron radiation phase distribution is very close to that of an immobile point source. Closer examination of the derived above expressions shows that the correspondent equivalent point source, which produces the similar phase distribution, should be placed at the point with coordinates $\left\{\frac{-R}{2\gamma^2},0,0\right\}$. Practically the value $\frac{-R}{2\gamma^2}$ is very small; in our case it is equal to 10^{-4} mm. The front position for this point source is described by the following expression:

$$\Delta d(x_s)=\sqrt{\left(x_s+R/(2\gamma^2)\right)^2+D^2}-\sqrt{\left(R/(2\gamma^2)\right)^2+D^2}. \tag{15}$$

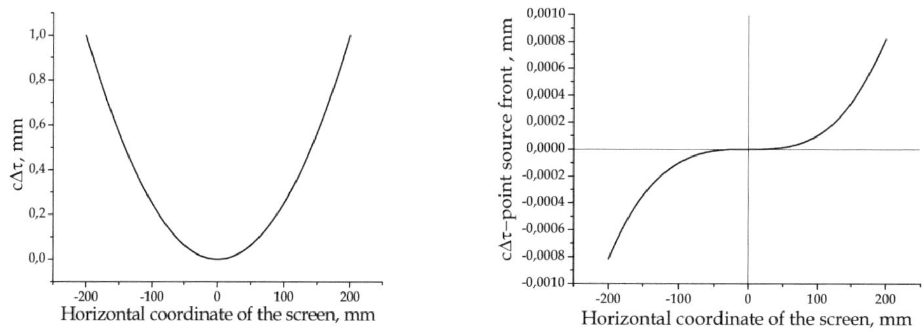

a b

Fig. 3. (a) Synchrotron radiation retardation $c\Delta\tau(x_s)$ versus observer horizontal position x_s. (b) Difference in positions between synchrotron radiation front and point source front.

The difference between the synchrotron radiation front position $c\Delta\tau(x_s)$ and the point source front position $\Delta d(x_s)$ is shown in Fig. 3b. It should be pointed out that this difference is small, less to 10^{-3} mm at the horizontal aperture boundaries ± 200 mm in our example. It means that the synchrotron radiation phase can be considered at zero approximation as the spherically symmetric phase of the correspondent immobile point source. Expanding Eqs. (13) - (15) in powers of x_s we get the following approximation for the difference in front positions of synchrotron radiation and point source front (Fig. 3b):

$$c\Delta\tau(x_s)-\Delta d(x_s)\cong \left(Rx_s^3\right)/\left(6D^3\right). \tag{16}$$

It describes the curve in Fig. 3b with a very good accuracy. Though this term is cubic in x_s and is very small at first glance, it plays an important part in synchrotron radiation imaging, as we will see below.

Finally we will give another well-known example, which on closer examination also shows the phase distribution of synchrotron radiation. Let an electron moves anticlockwise along the circle trajectory $\vec{r}(t)$, see Fig. 4. The radiation pulse (the maximum of the generated electric field) moves along the velocity vector $\vec{\beta}(t)$. At time τ it will reach the point $\vec{X}(\tau)=\vec{r}(t)+\frac{\vec{\beta}(t)}{\beta}c(\tau-t)$. Let us substitute Eqs. (6) and (7) into this relation, fix an observation point in time τ and consider the set of points in time t when the radiation was emitted: $t<\tau$. We will get the simultaneous distribution of the radiation pulses in the space, the well-known spiral radiation pattern, see Fig. 4. Some comments to this pattern can be found in (Jackson, 1999). The numbers of computed diagrams of electric field lines of radiating electron are published by (Tsien, 1972), see also (Shintake, 2003).

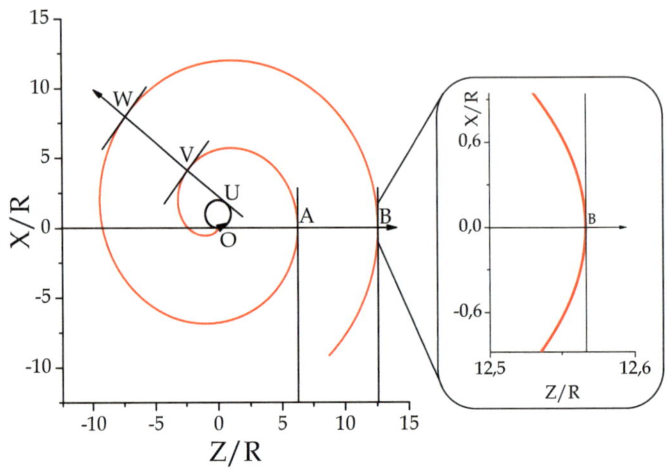

Fig. 4. The left side: space distribution of synchrotron radiation in the median plane. Small black circle is the electron trajectory which rotates anticlockwise. The right side: a magnified element of the spiral pattern.

Let us draw a line along the electron velocity $\vec{\beta}(t)$ tangentially to the electron trajectory (the direction of the radiation propagation) and consider the observation screen transversal to this line as it is shown on Fig.4, left side. The radiation at the point A was emitted at the point O one electron turn ago and the radiation at the point B was emitted two electron turns ago. The radiation at the point A moves along the line $O-A-B$ towards the point B. Similarly, we can say about the radiation which is generated at the point U and propagates along the line $U-V-W$. What is important for us in this picture is the fact that the spiral pattern is tangent to the screen and locally has the behaviour of quadratic function relative to horizontal (X) coordinates as it is shown at the right side of Fig.4. So, this spiral pattern also displays the phase distribution similar to the point source phase behaviour.

3. Synchrotron radiation phase distribution

In this section we will derive the exact expressions for synchrotron radiation phase. Let us consider an electron moving along trajectory $\vec{r}(t)$ with reduced energy $\gamma >> 1$. Electric field in the emitted wave which is observed at the point $\vec{X} = \{x, y, D\}$ and time τ is equal to:

$$\vec{E}(\tau,\vec{X}) = \frac{e}{c\left|\vec{X}-\vec{r}(t)\right|}\frac{[\vec{n}\times[(\vec{n}-\vec{\beta}(t))\times\dot{\vec{\beta}}(t)]]}{(1-\vec{n}\cdot\vec{\beta}(t))^3} + \frac{e}{\left|\vec{X}-\vec{r}(t)\right|^2}\frac{(\vec{n}-\vec{\beta}(t))(1-\beta^2(t))}{(1-\vec{n}\cdot\vec{\beta}(t))^3}, \tag{17}$$

where c is the speed of light, e is the electron charge, $\vec{\beta}$ is its reduced velocity, $\gamma = \frac{1}{\sqrt{1-\beta^2}}$, $\vec{n} = \frac{\vec{X}-\vec{r}(t)}{\left|\vec{X}-\vec{r}(t)\right|}$, and t is emission time:

$$\tau = t + \frac{1}{c}\left|\vec{X} - \vec{r}(t)\right| . \tag{18}$$

The radiation spectrum with wavelength λ is defined by the Fourier transform:

$$\tilde{\vec{E}}(\lambda, \vec{X}) = \int_{-\infty}^{\infty} \exp(i\frac{2\pi c}{\lambda}\tau)\vec{E}(\tau, \vec{X})d\tau , \tag{19}$$

The unit vectors of polarization $\vec{q}_\sigma$ and $\vec{q}_\pi$ are directed transversally to the radiation propagation and describe horizontally and vertically polarized radiation correspondently. The number of photons per unit time per unit area per unit relative spectral interval emitted by an electron beam with current I is equal to:

$$\frac{dN_{\sigma,\pi}}{d\tau\, ds(d\lambda/\lambda)} = \frac{c}{4\pi^2\hbar}\frac{I}{e}\left|\left(\vec{q}_{\sigma,\pi} \cdot \tilde{\vec{E}}(\lambda, \vec{X})\right)\right|^2 \tag{20}$$

It can easily be shown by the direct differentiation of Eq. (18) that:

$$\frac{\partial\tau(t)}{\partial t} = 1 - (\vec{n} \cdot \vec{\beta}(t)) . \tag{21}$$

Using Eq. (21) we can write: $\tau(t) = \tau(T) + \int_T^t (1 - \vec{n}\cdot\vec{\beta}(t'))dt'$, or:

$$\tau(t) = T + \frac{1}{c}\left|\vec{X} - \vec{r}(T)\right| + f(T,t) , \tag{22}$$

where

$$f(T,t) = \int_T^t (1 - \vec{n}\cdot\vec{\beta}(t'))dt' . \tag{23}$$

Here T is an arbitrary quantity to be defined below. It is important to keep in mind that T does not depend on time t .

Substituting Eqs. (17) and (18) into Eq. (19) and changing the integration from the independent variable τ to the variable t with the help of Eq. (21), we get:

$$\tilde{\vec{E}}(\lambda, \vec{X}) = \exp\left(i\Phi(T, \vec{X})\right)\tilde{\vec{E}}_0(\lambda, \vec{X}) , \tag{24}$$

where

$$\Phi(T, \vec{X}) = \frac{2\pi}{\lambda}\left\{cT + \left|\vec{X} - \vec{r}(T)\right|\right\} \tag{25}$$

$$\tilde{\vec{E}}_0(\lambda, \vec{X}) = \frac{e}{c}\int_{-\infty}^{\infty} \exp\left(i\frac{2\pi c}{\lambda}f(T,t)\right)\left\{\frac{[\vec{n}\times[(\vec{n}-\vec{\beta})\times\dot{\vec{\beta}}]]}{\left|\vec{X}-\vec{r}(t)\right|(1-\vec{n}\cdot\vec{\beta})^2} + \frac{c}{\left|\vec{X}-\vec{r}(t)\right|^2}\frac{(\vec{n}-\vec{\beta})(1-\beta^2)}{(1-\vec{n}\cdot\vec{\beta})^2}\right\}dt \tag{26}$$

Once the radiation intensity is calculated, the Eq. (24) should be substituted into Eq. (20) and the phase $\Phi(T,\vec{X})$ is of no concern. But the applications of optical elements generate a need for the radiation phase $\Phi(T,\vec{X})$ accurate calculations.

It is a matter of direct verification to prove the following exact relation:

$$\frac{[\vec{n}\times[(\vec{n}-\vec{\beta})\times\dot{\vec{\beta}}]]}{\left|\vec{X}-\vec{r}(t)\right|(1-\vec{n}\cdot\vec{\beta})^2}+\frac{c}{\left|\vec{X}-\vec{r}(t)\right|^2}\frac{(\vec{n}-\vec{\beta})(1-\beta^2)}{(1-\vec{n}\cdot\vec{\beta})^2}=\frac{d}{dt}\left\{\frac{\vec{n}-\vec{\beta}}{\left|\vec{X}-\vec{r}(t)\right|(1-\vec{n}\cdot\vec{\beta})}\right\}+\frac{c\vec{n}}{\left|\vec{X}-\vec{r}(t)\right|^2} \quad (27)$$

Substituting Eq. (27) into Eq. (26) and integrating by parts with using Eq. (23), we get:

$$\tilde{\vec{E}}_0(\lambda,\vec{X})=-i\frac{2\pi e}{\lambda}\int_{-\infty}^{\infty}\exp\left(i\frac{2\pi c}{\lambda}f(T,t)\right)\left\{\vec{n}(t)\left(1+\frac{i\lambda}{2\pi\left|\vec{X}-\vec{r}(t)\right|}\right)-\vec{\beta}(t)\right\}\frac{dt}{\left|\vec{X}-\vec{r}(t)\right|} \quad (28)$$

It should be noted that so far the exact formulas (17) - (28) were used and no approximations were employed. Particular attention should be paid to the phase as its scale of variation is very small, the phase changes by 2π along the radiation wavelength. Furthermore, we will derive the exact expression for synchrotron radiation phase distribution which is given by Eq. (25). At the same time we will apply the standard far-field approximation in the radiation amplitudes calculations, see Eq. (28).

We detect the radiation at the point $\vec{X}=\{x,y,D\}$, where D is considered to be much greater than the size of the emission region. Then we can write $\left|\vec{X}-\vec{r}(t)\right|=D$, $n_x=x/D$, $n_y=y/D$ and certainly we can neglect by the small term $i\lambda/\left(2\pi\left|\vec{X}-\vec{r}(t)\right|\right)$ in Eq. (28). The polarization vectors are

$$\begin{aligned}\vec{q}_\sigma&=[\vec{j}\times\vec{n}]/\left|[\vec{j}\times\vec{n}]\right|,\\ \vec{q}_\pi&=[\vec{n}\times\vec{q}_\sigma],\end{aligned} \quad (29)$$

where $\vec{j}$ is the unit vector along vertical Y-axis. Let us consider the radiation at small observation angles: $\left|n_{x,y}\right|<<1$ so that $n_z=\sqrt{1-n_x^2-n_y^2}$ is approximately equal to:

$$n_z=1-0.5\cdot(n_x^2+n_y^2). \quad (30)$$

Similarly, the expansion for $\beta_z(t)=\sqrt{\beta^2-\beta_x^2(t)}$ is equal to:

$$\beta_z(t)=1-0.5\cdot(\gamma^{-2}+\beta_x^2(t)). \quad (31)$$

It can be easily derived from Eq. (30) and Eq. (31) that:

$$1-(\vec{n}\cdot\vec{\beta})=0.5\cdot\left(\gamma^{-2}+(n_x-\beta_x)^2+n_y^2\right). \tag{32}$$

Then we obtain from Eq. (28):

$$\widetilde{E}_{0\sigma,\pi}(\lambda,\vec{X})=-i\frac{2\pi e}{\lambda D}\int_{-\infty}^{\infty}\exp\left(i\frac{2\pi c}{\lambda}f(T,t))\right)\left(\vec{n}-\vec{\beta}(t)\right)_{x,y}dt \tag{33}$$

and

$$f(T,t)=\frac{1}{2\gamma^2}\int_T^t\left(1+\gamma^2(n_x-\beta_x(t'))^2+\gamma^2n_y^2\right)dt'. \tag{34}$$

Usually, calculating synchrotron radiation intensity, the observation point is chosen in such a way that $n_x=0$ thus substantially simplifying the resultant calculations. It is possible for the amplitudes calculations because they are axially uniform and depends on vertical angle only. In this case we obtain $\left(\vec{q}_\sigma\cdot\tilde{\vec{E}}(\lambda,\vec{X})\right)$ as a purely real function and $\left(\vec{q}_\pi\cdot\tilde{\vec{E}}(\lambda,\vec{X})\right)$ as a purely imaginary function. But this simplification is not applicable when calculating the phase distribution of synchrotron radiation and we need to apply more accurate calculating procedure.

Now we determine the parameter T by making it equal to $t_e(x)$, where x is the horizontal coordinate of the observation point $\vec{X}=\{x,y,D\}$ and the function $t_e(x)$ is given by Eq. (11). By using the relation $tg\varphi=\dfrac{\beta_x(t)}{\beta_z(t)}$, see Eqs. (7), we get from Eq. (8):

$$\frac{x-r_x(t_e(x))}{D-r_z(t_e(x))}=\frac{n_x}{n_z}=\frac{\beta_x(t_e(x))}{\beta_z(t_e(x))}. \tag{35}$$

Saving the linear terms of Eq. (35), which is to say that $n_z\cong 1$ and $\beta_z\cong 1$, we get: $\beta_x(t_e(x))=n_x=\dfrac{x}{D}$ and hence in the frame of linear approximation:

$\beta_x(t)=\beta\sin(\Omega t_e(x))\cos(\Omega\Delta t)+\beta\cos(\Omega t_e(x))\sin(\Omega\Delta t)\cong n_x+\Omega\Delta t$, where $t=t_e(x)+\Delta t$.

Substituting this relation into Eq. (34) and denoting $\chi=\gamma\Omega\Delta t$, where $\Omega=\beta c/R$, we obtain:

$$f(T=t_e(x),t)=\frac{R}{2\gamma^3c}\left\{\left(1+\gamma^2n_y^2\right)\chi+\frac{1}{3}\chi^3\right\}. \tag{36}$$

Using Eq. (36), we can easily derive from Eq. (33) the well-known expressions for synchrotron radiation fields:

$$\widetilde{E}_{0\sigma}(\lambda,\vec{X})=\frac{4\pi e\gamma}{Dc}\sqrt{\Lambda}Ai'\left(\Lambda\left(1+\gamma^2n_y^2\right)\right), \tag{37}$$

$$\widetilde{E}_{0\pi}(\lambda,\vec{X}) = -i\frac{4\pi e\gamma^2}{Dc}\Lambda n_y Ai\left(\Lambda\left(1+\gamma^2 n_y^2\right)\right) \tag{38}$$

where $\Lambda = \left(\frac{\pi R}{\lambda\gamma^3}\right)^{2/3} = \left(\frac{3\lambda_c}{4\lambda}\right)^{2/3}$, $\lambda_c = \frac{4\pi R}{3\gamma^3}$ is the synchrotron radiation critical wavelength, $Ai(x)$ and $Ai'(x)$ are the Airy function and its derivative.

We have obtained standard formulas, describing the synchrotron radiation amplitudes $\widetilde{E}_{0\sigma}$ and $\widetilde{E}_{0\pi}$ in the far-field approximation. The phases of the amplitudes $\widetilde{E}_{0\sigma}$ and $\widetilde{E}_{0\pi}$ are constant (independent of the observation point position), so the phase dependence of the synchrotron radiation is described by the function $\Phi(T = t_e(x), \vec{X})$, see Eq. (25):

$$\Phi(\vec{X}) \equiv \Phi(T = t_e(x), \vec{X}) = \frac{2\pi c}{\lambda}\left\{t_e(x) + \frac{1}{c}\left|\vec{X} - \vec{r}(t_e(x))\right|\right\} \tag{39}$$

Notice that $t_e(x)$ is the moment when the electron, having the position $\vec{r}(t_e(x))$, points to the observer with the coordinates $\{x,0,D\}$. The second term $\frac{1}{c}\left|\vec{X} - \vec{r}(t_e(x))\right|$ is the time interval required for the light passing from the position of the electron $\vec{r}(t_e(x))$ to the observer position $\vec{X} = \{x,y,D\}$. Thus, we have obtained a physically reasonable result that the synchrotron radiation phase is proportional to the arrival time of the maximum of function $E_\sigma(\tau,\vec{X})$: the arrival time is equal to $\lambda\Phi(\vec{X})/(2\pi c)$.

Let us determine the following variables: $\rho = \frac{R}{D}$, $\theta_x = \frac{x}{D}$, $\theta_y = \frac{y}{D}$ and a function $\delta(\theta_x)$:

$$\delta(\theta_x) = \arctan\frac{\rho\sqrt{(\rho-\theta_x)^2 + 1 - \rho^2} - (\rho - \theta_x)}{1-\rho^2}. \tag{40}$$

Notice that $\delta(\theta_x = 0) = 0$. Then the phase distribution of synchrotron radiation is described by the following exact expression:

$$\Phi(\vec{X}) = \frac{2\pi D}{\beta\lambda}\left\{\rho\delta + \beta\sqrt{(1-\rho\sin\delta)^2 + (\theta_x + \rho\cos\delta - \rho)^2 + \theta_y^2}\right\}. \tag{41}$$

It is worth noting that we have obtained, as a result, the exact analytic expression involving only elementary functions. It is important to outline also that the phase difference, rather than the phase value $\Phi(\vec{X})$, has a physical sense. In other words, the phase is defined up to a constant value identical for all the points of the optical device entrance window.

4. Synchrotron radiation regarded as a point-source radiation

Consider now the phase $\Phi(\vec{X})$ at the observation point $\vec{X} = \{x,y,D\}$, where $|x| \ll D$, $|y| \ll D$. Analysis shows that the function $\Phi(\vec{X})$ has a minimum in the vicinity of the point

$\{0,0,D\}$. Let us expand this phase in a power series with respect to the small transverse angles $\theta_x = \frac{x}{D}$ and $\theta_y = \frac{y}{D}$ up to the third power inclusive. This expansion has the form:

$$\Phi(\vec{X}) = \Phi_0 + \Phi_2(\vec{X}) + \Phi_3(\vec{X}), \tag{42}$$

$$\Phi_0 = \frac{2\pi D}{\lambda}\left\{1 - R^2/(8D^2\gamma^4)\right\}, \tag{43}$$

$$\Phi_2(\vec{X}) = \frac{\pi D}{\lambda}\left\{\left(\theta_x + \frac{\rho}{2\gamma^2}\right)^2 + \theta_y^2\right\}, \tag{44}$$

$$\Phi_3(\vec{X}) = \frac{\pi R}{\lambda}\left\{\frac{1}{3}\theta_x^3 + \theta_x\theta_y^2\right\}. \tag{45}$$

If both angles θ_x and θ_y are sufficiently small, one can neglect the term $\Phi_3(\vec{X})$, which is of the third order of smallness. The size of the corresponding region can be evaluated from condition $\Phi_3(\vec{X}) < 0.5\pi$. As a result, one finds that the term $\Phi_3(\vec{X})$ can be neglected if:

$$\left|\theta_{x,y}\right| < \frac{1}{\gamma}\sqrt[3]{\frac{\lambda}{\lambda_c}}. \tag{46}$$

In this case $\Phi(\vec{X}) = \Phi_2(\vec{X})$ and the distribution of the synchrotron radiation phases coincides with the phase distribution of an immobile point source located at the point $\left\{-\frac{R}{2\gamma^2},0,0\right\}$. The quantity $\frac{R}{2\gamma^2}$ usually is very small and can be neglected. Then it is seen from Eq. (44) that the equivalent immobile point source is located at the origin of the coordinates. In this case the synchrotron radiation properties are rather well understood. The spatial distributions of the synchrotron radiation amplitudes are described by Eqs. (37) and (38). The synchrotron radiation phase distribution inside the angle limited by Eqs. (46) coincides with that of an equivalent spherical wave propagating from the origin of the coordinates and is described by Eq. (44). For the Siberia-2 storage ring (γ =4892, λ_c =0.175 nm, λ =6000 nm) the region given by Eq. (46) is rather large: $\left|\theta_{x,y}\right| < 6.6$ mrad. Any decrease in the radiation wavelength reduces the size of this region.

All the above stated concerned a single relativistic electron moving along the circular trajectory with zero initial conditions see Eqs. (6) and (7) and Fig. 5. This electron passed through the origin of the coordinates, where its velocity was directed along the Z-axis. The immobile point source, which is equivalent in phase distribution to the synchrotron radiation generated by this electron, is at the origin of the coordinates. However, in practice the electron beam has always some spread in both positions and angles. It is clear that for electrons with different orbits the locations of the equivalent immobile point sources would also be different. The explicit calculation of the position of the equivalent point source for an arbitrary electron involves rather cumbersome formulas and is beyond the scope of this chapter. However, for an electron with a trajectory lying in the storage ring median plane, the solution is reasonably simple.

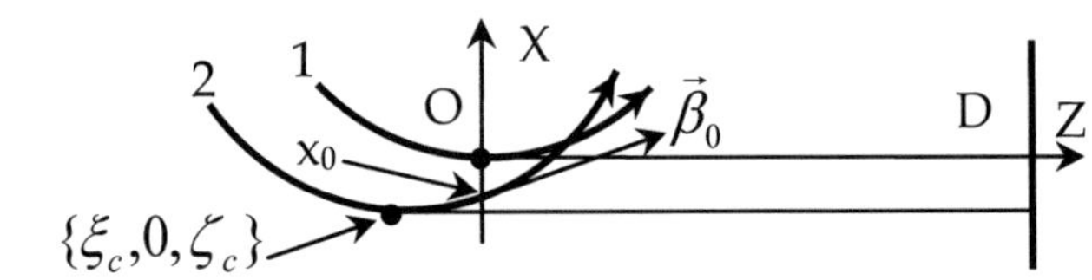

Fig. 5. Equivalent immobile point sources positions sketch: the point source $\left\{-R/(2\gamma^2), 0, 0\right\}$ for the electron 1 with zero initial conditions and the point source $\{\xi_c, 0, \zeta_c\}$ for the electron 2 with nonzero initial conditions.

Let us consider now an electron moving along a circular trajectory which lies in the XOZ-plane and intersects the X-axis at some point x_0 so that the electron initial position is $\{x_0, 0, 0\}$. Let the vector $\{\beta_{x0}, 0, \beta_{z0}\}$ be the reduced velocity of the electron at this initial point and β_{x0} be small. According to the analysis, the phase distribution of the synchrotron radiation in this case coincides with the phase distribution of an equivalent spherical wave outgoing from a point source with the coordinates $\{\xi_c, 0, \zeta_c\}$, where:

$$\xi_c = x_0 - \frac{R}{2\gamma^2} - 0.5R\beta_{x0}^2, \zeta_c = -R\beta_{x0}. \tag{47}$$

Up to $R/(2\gamma^2)$, this point source lies on the electron orbit. Namely, it is the point of tangency of a normal to the screen with the electron circular trajectory, see Fig.5. Notice that the transverse (X) coordinate ξ_c depends mainly on the electron initial position x_0, while its longitudinal (Z) coordinate ζ_c is determined by the electron initial angle β_{x0}.

Let us consider a relativistic beam, in which all the electron trajectories lie in the XOZ-plane (zero vertical emittance condition). Nevertheless, these electrons are spread over horizontal positions and angles. In the vicinity of the optical device window (as defined by Eq. (46)) the synchrotron radiation of such a beam can be regarded as a radiation from a set of immobile point sources, whose coordinates are defined by Eqs. (47). The phases of the synchrotron radiation emitted by each of the electrons coincide with the phases of the radiation emitted by the corresponding immobile point sources. However, according to Eqs. (37) and (38), the radiation amplitudes of the immobile point sources are modulated in the vertical direction. This approach permits us to substitute of a determination of a pattern produced by the distributed immobile point sources for the calculation of the image created by the synchrotron radiation. The solution of this problem has been comprehensively examined by the wave optics (Born & Wolf, 1986). It is significant that the length along Z-axis of this immobile point sources distribution is determined by the electron beam angular spread rather than the orbit curvature and a focusing lens aperture as implied in the geometrical optics approach (Hofmann & Meot, 1982).

It should be said in closing that the term $\Phi_3(\vec{X})$, which was outside our analysis here, plays an important role in synchrotron radiation optics, determining its self-aberration effect.

5. Focusing of synchrotron radiation

Let us consider the standard experimental layout for synchrotron radiation imaging shown in Fig. 6. For simplicity sake the geometry with an ideal refractive lens is considered. The distance between the origin of coordinates and lens is equal to D, the distance from the lens to the observation screen is equal to L.

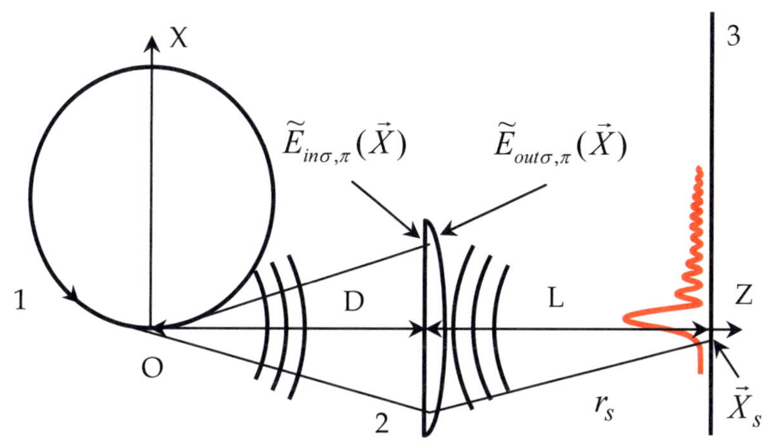

Fig. 6. Synchrotron radiation imaging experiment geometry: 1 - electron trajectory; 2 - lens, 3 - observation screen.

The ideal lens with focal length f does not change the synchrotron radiation amplitudes but adds an extra shift in the radiation phase distribution:

$$\vec{E}_{out}(x,y,z=D)=\vec{E}_{in}(x,y,z=D)\cdot\exp\left(-i\pi(x^2+y^2)/(\lambda f)\right). \tag{48}$$

Here $\vec{E}_{in}(x,y,z=D)=\exp\left(i\Phi(\vec{X})\right)\widetilde{\vec{E}}_0(\lambda,\vec{X})$ is the synchrotron radiation field. Its amplitudes $\widetilde{\vec{E}}_0$ are given by Eqs. (37) and (38) and the phase $\Phi(\vec{X})=\Phi_2(\vec{X})+\Phi_3(\vec{X})$ is given by Eqs. (42) - (45). Notice that the cubic term $\Phi_3(\vec{X})$ will also be taken here into account.

The radiation field at the point $\vec{X}_s=\{x_s,y_s,z_s=D+L\}$ of the observation screen can be found by applying the Kirchhoff integral theorem (Born & Wolf, 1986):

$$\vec{E}(x_s,y_s,z_s)=\frac{-i}{\lambda}\int_A \vec{E}_{out}(x,y,z=D)\frac{\exp(ikr_s)}{r_s}dxdy\ , \tag{49}$$

where $r_s = \sqrt{(x-x_s)^2+(y-y_s)^2+L^2} \cong L+\frac{(x-x_s)^2+(y-y_s)^2}{2L}$ is the distance of the lens element $dxdy$ from the observer $\vec{X}_s = \{x_s, y_s, z_s = D+L\}$ and A is the lens aperture, which is assumed rectangular in shape here. Calculating numerically the Kirchhoff integral (49), we can find with Eq. (20) the intensity distribution of the focused synchrotron radiation.

The computed images of horizontally and vertically polarized synchrotron radiation are shown in Fig. 7. The simulations were performed at the following conditions: orbit bending radius 4905.4 mm; electron beam energy of 2.5 GeV; beam current 0.1 A; beam emittance is zero (one-electron approximation); lens focal length 5 m; distance from tangential source point to the lens 10 m; distance from the lens to the observation screen 10 m; horizontal and vertical sizes of the lens 600 mm (angular aperture 60 mrad); radiation wavelength 6000 nm. The lens focuses the synchrotron radiation emitted into all vertical angles, see Fig. 8.

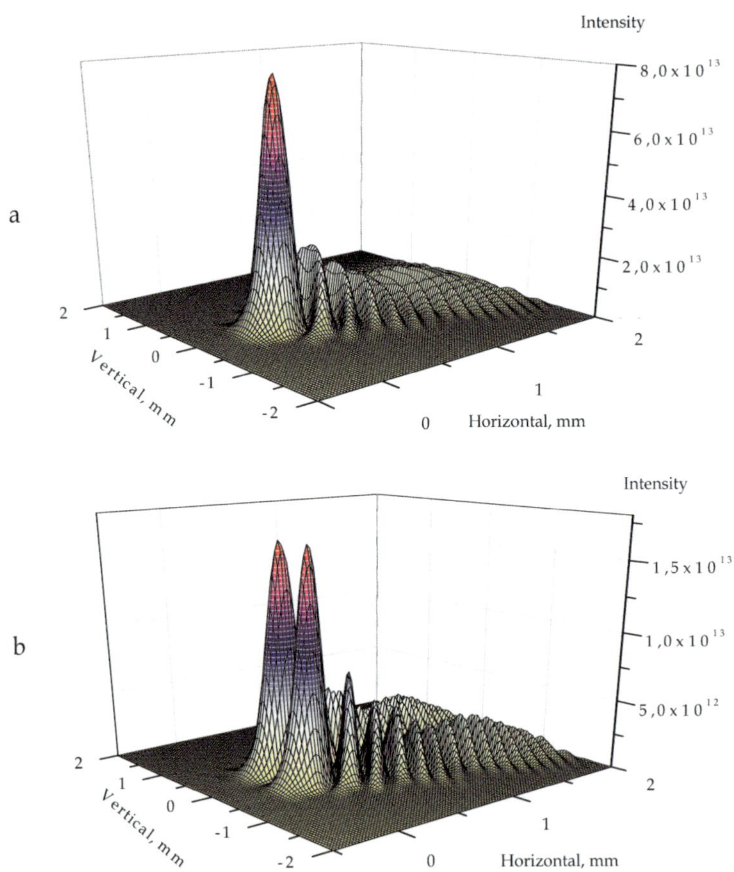

Fig. 7. Intensity distributions of the focused synchrotron radiation: a – horizontally (σ) polarized component of radiation; b – vertically (π) polarized component of radiation. Intensity distributions are given in photons/s/mm²/(0.1%bandwidth).

It is significant that the intensity of focused vertically polarized synchrotron radiation is equal to zero in the electron orbit plane ($y = 0$), see Fig. 7b. This is a result of the anti-symmetry of the vertically polarized electric field with respect to the electron orbit plane, see Fig. 8 and Eq. (38): $\widetilde{E}_{0\pi}(\lambda, \vec{X})$ is proportional to $n_y = y/D$. Thus two rays with the opposite signs of n_y and respectively with the opposite electric field directions, when focused to the point $\{x_s, 0, z_s = D + L\}$, will cancel each other.

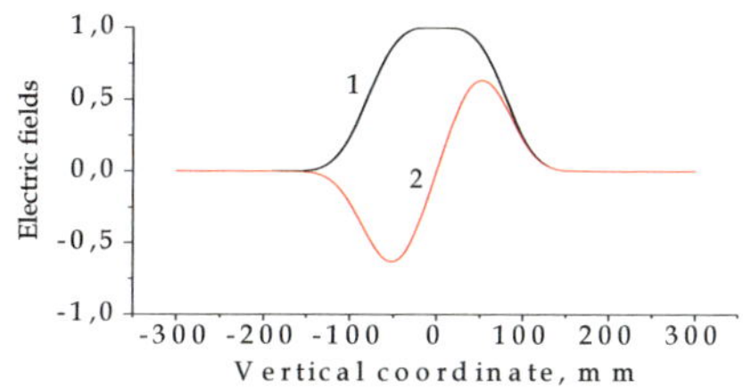

Fig. 8. Vertical distributions of synchrotron radiation electric fields $\widetilde{E}_{0\sigma,\pi}(\lambda, \vec{X})$ on the lens surface: 1 – horizontally polarized radiation, 2 – vertically polarized radiation.

At large horizontal aperture of the lens (600 mm) the cubic term $\Phi_3(\vec{X})$ in the radiation phase manifests itself as additional asymmetric fringes in the radiation image (self-aberration effect). Real electron beam emittance will smooth out these fringes. Nevertheless, such asymmetry was observed in 100 - 20000 cm^{-1} spectral range at BL43IR infrared beamline of Spring-8 (Ikemoto et al., 2003), see Fig. 9.

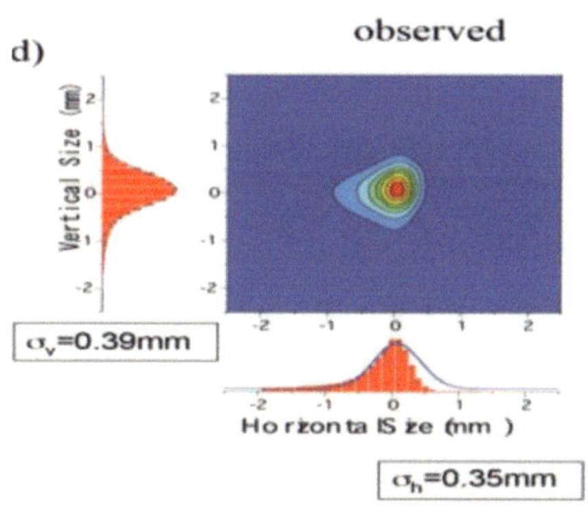

Fig. 9. Experimentally observed at Spring-8 image of the electron beam.
(Published with the kind permission of Prof. T. Nanba)

Let us consider now the influence of the lens horizontal aperture on the synchrotron radiation imaging. It follows from Eq. (46) that the natural scale for the angular aperture is

$$\Delta\theta_{SR} = \frac{2}{\gamma} \cdot \sqrt[3]{\frac{\lambda}{\lambda_c}} \cong 1.24 \cdot \sqrt[3]{\frac{\lambda}{R}} . \tag{50}$$

For the Siberia-2 storage ring with the orbit bending radius $R = 4905.4$ mm and for $\lambda = 6000$ nm this value is equal to $\Delta\theta_{SR} = 13.3$ mrad. Fig. 10 shows the computed radiation intensity distributions in the electron orbit plane $y = 0$. The simulations were performed under the same conditions as above except for the lens horizontal size. The lens horizontal aperture $2l_x$ was taken equal to 66 mm, 132 mm, 264 mm and 528 mm, that is the lens horizontal angular aperture $\theta_{lens} = 2l_x/D$ is equal to $0.5\Delta\theta_{SR}$, $\Delta\theta_{SR}$, $2\Delta\theta_{SR}$ and $4\Delta\theta_{SR}$ respectively.

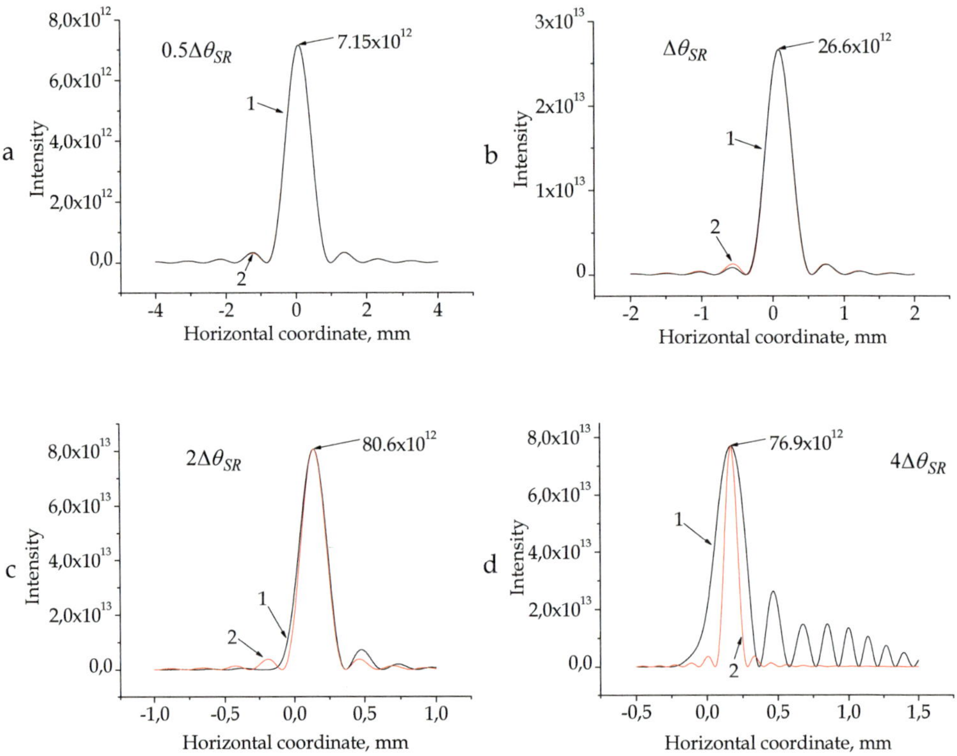

Fig. 10. Intensity distributions (photons/s/mm²/(0.1%bandwidth)) in the median plane of the focused synchrotron radiation for different horizontal angular aperture θ_{lens} of the lens: a - $\theta_{lens} = 0.5\Delta\theta_{SR}$, b - $\theta_{lens} = \Delta\theta_{SR}$, c - $\theta_{lens} = 2\Delta\theta_{SR}$ and d - $\theta_{lens} = 4\Delta\theta_{SR}$.
Curve 1 - focused synchrotron radiation, curve 2 – approximation by $(\sin(\xi)/\xi)^2$, see Eq. (5).

From Fig.10 we notice that the function $(\sin(\xi_s)/\xi_s)^2$ with $\xi_s = \frac{2\pi l_x}{\lambda L} x_s$, when normalised and horizontally shifted, adequately describes the horizontal profiles of the focused synchrotron radiation up to $\theta_{lens} = 2\Delta\theta_{SR}$. But at large horizontal aperture of the lens, the self-aberration property of synchrotron radiation changes dramatically the radiation image characteristics. For the case of point source radiation its intensity at the focused spot center varies with the square of the lens horizontal aperture, see the end of the section 2.1. This is also true for synchrotron radiation focusing by relatively small lens; at small θ_{lens} its value doubling tends to increase the maximum of intensity by a factor of four approximately. But the increasing of the lens horizontal aperture from $\theta_{lens} = 2\Delta\theta_{SR}$ to $\theta_{lens} = 4\Delta\theta_{SR}$ implies a decrease in the maximum of the focused synchrotron radiation intensity, see Fig 10.

Such kind of problems, which are caused by the self-aberration effect of synchrotron radiation (the term $\Phi_3(\vec{X})$), can be neutralized by using of the so-called "magic mirror". The idea how to bring a segment of circular orbit to an isochronous focus and profile of such mirror in the orbit plane was presented in (Lopez-Delgado & Szwarc, 1976). The expansion of formula for "magic mirror" to include the vertical direction was made in (S. Kimura et al., 2001). The "magic mirror" for synchrotron radiation is similar to the ideal lens for a point source radiation since it compensates the aberrations caused by the term $\Phi_3(\vec{X})$ in design. It is employed at some infrared beamlines (H. Kimura et al., 2001; S. Kimura et al., 2006).

6. Conclusion

The analysis of wave optical properties of synchrotron radiation given in this chapter shows their unconventionality. On the one hand, in the zero-order approximation the phases of the synchrotron radiation emitted by each of the electrons coincide with the phases of the radiation emitted by the corresponding immobile point sources. That is why standard optical equipment works well with synchrotron radiation paraxial beams. On the other hand, synchrotron radiation has the property of self-aberration. To improve beamline performance, the specialized optical elements, such as the "magic mirror", should be used in the beamline optical system. The use of the exact formula for the phase distribution of synchrotron radiation provides a way of developing new optical elements which are optimized for synchrotron radiation utilization.

7. References

Andersson, A.; Schlott, V.; Rohrer, M.; Streun, A. & Chubar, O. (2006). Electron beam profile measurements with visible and X-Ray synchrotron radiation at the Swiss Light Source, *Proceedings of the tenth European Particle Accelerator Conference (EPAC'06)*, pp. 1223-1225, ISBN 92-9083-278-9, Edinburgh, Scotland, June 26-30, 2006

Andersson, A. & Tagger, J. (1995). Beam profile measurements at MAX. *Nuclear Instruments and Methods in Physics Research A*, Vol.364, No.1, (September 1995), pp. 4-12, ISSN 0168-9002

Arp, U. (2001). Diffraction and depths-of-field effects in electron beam imaging at SURF III. *Nuclear Instruments and Methods in Physics Research A*, Vol.462, No.3, (April 2001), pp. 568-575, ISSN 0168-9002

Artemiev, N. A.; Chubar, O. V. & Valentinov, A. G. (1996). Electron beam diagnostics with visible synchrotron light on Siberia-1 ring, *Proceedings of the fifth European Particle Accelerator Conference (EPAC'96)*, pp. 340-342, ISBN 978-0-7503-0387-3, Barcelona, Spain, June 10-14, 1996

Bahrdt, J. (1997). Wave-front propagation: design code for synchrotron radiation beam lines. *Applied Optics*, Vol.36, No.19, (July 1997), pp. 4367-4381, ISSN: 1559-128X

Bahrdt, J. (2007). Wavefront tracking within the stationary phase approximation. *Physical Review Special Topics – Accelerators and Beams*, Vol.10, No.6, (June, 2007), pp. 060701-1 - 060701-15, ISSN 1098-4402

Bocci, A.; Clozza, A.; Drago, A.; Grilli, A.; Marcelli, A.; Piccinini, M.; Raco, A.; Sorchetti, R.; Gambicorti, L.; De Sio, A.; Pace, E. & Piotrowski, J. (2008). Beam diagnostics with IR light emitted by positrons at DAFNE, *Proceedings of the eleventh European Particle Accelerator Conference (EPAC'08)*, pp. 1056-1058, ISBN 978-92-9083-315-4, Genoa, Italy, June 23-27, 2008

Born, M. & Wolf, E. (1986). *Principles of optics.* (6th Edition), Pergamon, ISBN 0-08-026482-4, Oxford OX3 0BW, England

Bosch, R. A. (1999). Focusing of infrared edge and synchrotron radiation. *Nuclear Instruments and Methods in Physics Research A*, Vol.431, No.1-2, (July 1999), pp. 320-333, ISSN 0168-9002

Bowler, M.; Bahrdt, J. & Chubar, O. (2008). Wavefront propagation, In: *Modern developments in X-Ray and neutron optics,* Erko, A.; Idir, M.; Krist, T. & Michette, A. G., (Ed.), pp. 69-90, Springer, ISBN 978-3-540-74560-0, Berlin Heidelberg New York

Carr, G. L. & Dumas, P. (1999). *Accelerator-based Sources of Infrared and Spectroscopic Applications. Proceedings of SPIE,* Vol. 3775, (October 1999), ISBN 9780819432612

Chubar, O. V. (1995). Transverse electron beam size measurements using the Lloyd's mirror scheme of synchrotron light interference, *Proceedings of the sixteenth Particle Accelerator Conference (PAC'95)*, pp. 2447-2449, ISBN 0-7803-2937-6, Dallas, US, May 1-5, 1995

Chubar, O. & Elleaume, P. (1998). Accurate and efficient computation of synchrotron radiation in the near field region, *Proceedings of the sixth European Particle Accelerator Conference (EPAC'98)*, pp. 1177-1179, ISBN 0 7503 0579 7, Stockholm, Sweden, June 22-26, 1998

Chubar, O.; Elleaume, P. & Snigirev, A. (1999). Phase analysis and focusing of synchrotron radiation. *Nuclear Instruments and Methods in Physics Research A*, Vol.435, No.3, (October 1999), pp. 495-508, ISSN 0168-9002

Chubar, O.; Elleaume, P. & Snigirev, A. (2001). Phase corrections for synchrotron radiation. *Nuclear Instruments and Methods in Physics Research A*, Vol.467-468, No.1, (July 2001), pp. 932-935, ISSN 0168-9002

Clarke, J. A. (1994). A review of optical diagnostics techniques for beam profile measurements, *Proceedings of the fourth European Particle Accelerator Conference (EPAC'94)*, pp. 1643-1645, ISBN 9810219288, London, UK, June 27 - July 1, 1994

Coisson, R. & Marchesini, S. (1997). Gauss-Schell Sources as Models for Synchrotron Radiation. *Journal of Synchrotron Radiation,* Vol.4, No.5, (September 1997), pp. 263-266, ISSN 0909-0495

Duncan, W. D. & Williams, G. P. (1983). Infrared synchrotron radiation from electron storage rings. *Applied optics,* Vol.22, No.18, (September 1983), pp. 2914-2923, ISSN 1559-128X

Elleaume, P.; Fortgang, C.; Penel, C. & Tarazona, E. (1995). Measuring Beam Sizes and Ultra-Small Electron Emittances Using an X-ray Pinhole Camera. *Journal of Synchrotron Radiation,* Vol.2, No.5, (September 1995), pp. 209-214, ISSN 0909-0495

Fang, Z.; Wang, G.; Yan, X.; Wang, J.; Zhang, D.; Zhou, Y.; Zhao, F.; Xie, R.; Sun, B. & Wu, J. (1996). Monitoring the beam profile in HLS with synchrotron light. *Nuclear Instruments and Methods in Physics Research A*, Vol.370, No.2-3, (February 1996), pp. 641-643, ISSN 0168-9002

Ferrero, C.; Smilgies, D.-M.; Riekel, Ch.; Gatta, G. & Daly P. (2008). Extending the possibilities in phase space analysis of synchrotron radiation x-ray optics. *Applied Optics*, Vol.47, No.22, (August 2008), pp. E116-E124, ISSN: 1559-128X

Flanagan, J. W.; Hiramatsu, S. & Mitsuhashi, T. (1999). Optical Beamlines for the KEK B-Factory Synchrotron Radiation Monitors, *Proceedings of the eighteenth Particle Accelerator Conference (PAC'99)*, pp. 2120-2122, ISBN 0-7803-5573-3, New York, US, March 27 - April 2, 1999

Geloni, G.; Kocharyan V.; Saldin E.; Schneidmiller E. & Yurkov M. (2009a). Theory of edge radiation. Part I: Foundations and basic applications. *Nuclear Instruments and Methods in Physics Research A*, Vol.605, No.3, (July 2009), pp. 409-429, ISSN 0168-9002

Geloni, G.; Kocharyan V.; Saldin E.; Schneidmiller E. & Yurkov M. (2009b). Theory of edge radiation. Part II: Advanced applications. *Nuclear Instruments and Methods in Physics Research A*, Vol.607, No.2, (August 2009), pp. 470-487, ISSN 0168-9002

Grimm, O; Behrens, Ch.; Rossbach, J. & Schmidt, B. (2008). Longitudinal beam diagnostics application of synchrotron radiation at FLASH, *Proceedings of the eleventh European Particle Accelerator Conference (EPAC'08)*, pp. 1116-1118, ISBN 978-92-9083-315-4, Genoa, Italy, June 23-27, 2008

Guidi, M. C.; Piccinini, M.; Marcelli, A.; Nucara, A.; Calvani, P. & Burattini, E. (2005). Optical performances of SINBAD, the Synchrotron INfrared Beamline At DAFNE. *Journal of the Optical Society of America A*, Vol.22, No.12, (December 2005), pp.2810-2817, ISSN 1084-7529

Green, G.K. (1976). *Spectra and optics of synchrotron radiation*. Preprint BNL 50522, (April 15, 1976). Brookhaven National Laboratory, Associated Universities, Inc. Upton, New York, 11973

Hiramatsu, S.; Iwasaki, H.; Mitsuhashi, T.; Naitoh, T. & Yamamoto, Y. (1999). Measurement of small beam size by the use of SR interferometer, *Proceedings of the eighteenth Particle Accelerator Conference (PAC'99)*, pp. 492-494, ISBN 0-7803-5573-3, New York, US, March 27 - April 2, 1999

Hofmann, A. & Meot, F. (1982). Optical resolution of beam cross-section measurements by means of synchrotron radiation. *Nuclear Instruments and Methods in Physics Research*, Vol.203, No.1-3, (December 1982), pp. 483-493, ISSN: 0167-5087

Hs, I. C. & Huang, T. H. (1993). Design Study of Beam Profile Monitor of storage Ring by Using Synchrotron Radiation, *Proceedings of the fifteenth Particle Accelerator Conference (PAC'93)*, pp. 2465-2467, ISBN 0-7803-1203-1, Washington, D.C., US, May 17-20, 1993.

Ikemoto, Y.; Moriwaki, T.; Hirono, T.; Kimura, S.; Shinoda, K.; Matsunami, M.; Nagai, N.; Nanba, T.; Kobayashi, K. & Kimura, H. (2003). Infrared microspectroscopy station at BL43IR of Spring-8. Poster presentation at: *International Workshop on Infrared*

Microscopy and Spectroscopy with Accelerator Based Sources (WIRMS-2003), Lake Tahoe, California, US, July 8-11, 2003 Available from: http://infrared.als.lbl.gov/WIRMS/Presentations/SPring-8_poster1.pdf

Jackson, J. D. (1999). *Classical electrodynamics.* (3th Edition), John Wiley & Sons, Inc., ISBN 0-471-30932-X, New York, Chichester, Weinheim, Brisbane, Singapore, Toronto

Katoh, M. & Mitsuhashi, T. (1999). Measurement of beam size at the Photon Factory with the SR interferometer, *Proceedings of the eighteenth Particle Accelerator Conference (PAC'99)*, pp. 2307-2309, ISBN 0-7803-5573-3, New York, US, March 27–April 2, 1999

Kim, K.-J. (1986). Brightness, coherence and propagation characteristics of synchrotron radiation. *Nuclear Instruments and Methods in Physics Research A*, Vol.246, No.1-3, (May 1986), pp. 71-76, ISSN 0168-9002

Kimura, H.; Moriwaki, T.; Takahashi, S.; Aoyagi, H.; Matsushita, T.; Ishizawa, Y.; Masaki, M.; Oishi, S.; Ohkuma, H.; Namba, T.; Sakurai, M.; Kimura, S.; Okamura, H.; Nakagawa, H.; Takahashi, T.; Fukui, K.; Shinoda, K.; Kondoh, Y.; Sata, T.; Okuno, M.; Matsunami, M.; Koyanagi, R.; Yoshimatsu, Y. & Ishikawa, T. (2001). Infrared beamline BL43IR at SPring-8: design and commissioning. *Nuclear Instruments and Methods in Physics Research A*, Vol.467-468, No.1, (July 2001), pp. 441-444, ISSN 0168-9002

Kimura, S.; Kimura, H.; Takahashi, T.; Fukui, K.; Kondo, Y.; Yoshimatsu, Y.; Moriwaki, T.; Nanba, T. & Ishikawa, T. (2001). Front end and optics of infrared beamline at SPring-8. *Nuclear Instruments and Methods in Physics Research A*, Vol.467-468, No.1, (July 2001), pp. 437-440, ISSN 0168-9002

Kimura, S.; Nakamura, E.; Nishi, T.; Sakurai, Y.; Hayashi, K.; Yamazaki, J. & Katoh, M. (2006). Infrared and terahertz spectromicroscopy beam line BL6B(IR) at UVSOR-II. *Infrared Physics and Technology*, Vol.49, Nos.1-2 (September 2006), pp. 147-151, ISSN 1350-4495

Korchuganov, V.; Blokhov, M.; Kovalchuk, M.; Krylov, Yu.; Kvardakov, V.; Moseiko, L.; Moseiko, N.; Novikov, V.; Zheludeva, S.; Odintsov, D.; Rezvov, V.; Ushkov, V.; Valentinov, A.; Vernov, A.; Yudin, L. & Yupinov, Yu. (2005). The status-2004 of the Kurchatov center of SR. *Nuclear Instruments and Methods in Physics Research A*, Vol.543, No.1, (May 2005), pp.14-18, ISSN 0168-9002

Korchuganov, V. N. & Smolyakov, N. V. (2009). IR-UV edge radiation at Siberia-2 storage ring. *Nuclear Instruments and Methods in Physics Research A*, Vol.603, No.1-2, (May 2009), pp. 13-15, ISSN 0168-9002

Lopez-Delgado, R. & Szwarc, H. (1976). Focusing all the synchrotron radiation (2π radians) from an electron storage ring on a single point without time distortion. *Optics Communications*, Vol.19, No.2, (November 1976), pp. 286-291, ISSN 0030-4018

Miyahara, T. & Kagoshima, Y. (1995). Importance of wave-optical corrections to geometrical ray tracing for high brilliance beamlines. *Review of Scientific Instruments*, Vol.66, No.2, (February 1995), pp. 2164-2166, ISSN 0034-6748

Naito, T. & Mitsuhashi, T. (2010). Improvement of the resolution of SR interferometer at KEK-ATF dumping ring, *Proceedings of the first International Particle Accelerator*

Conference (IPAC'10), pp. 972-974, ISBN 978-92-9083-352-9, Kyoto, Japan, May, 23-28, 2010

Ogata, A. (1987). Focusing of synchrotron radiation. *Nuclear Instruments and Methods in Physics Research A*, Vol.259, No.3, (September 1987), pp. 566-575, ISSN 0168-9002

Ogata, A. (1991). On optical resolution of beam size measurements by means of synchrotron radiation. *Nuclear Instruments and Methods in Physics Research A*, Vol.301, No.3, (March 1991), pp. 596-598, ISSN 0168-9002

Paech, A.; Ackermann, W.; Weiland, T. & Grimm, O. (2006). Numerical simulation of synchrotron radiation for bunch diagnostics, *Proceedings of the tenth European Particle Accelerator Conference (EPAC'06)*, pp. 1031-1033, ISBN 92-9083-278-9, Edinburgh, Scotland, June 26-30, 2006

Paech, A.; Ackermann, W.; Weiland, T. & Grimm, O. (2007). Simulation of synchrotron radiation at the first bunch compressor of FLASH, *Proceedings of the twenty-second Particle Accelerator Conference (PAC'07)*, pp. 3925-3927, ISBN 1-4244-0917-9, Albuquerque, New Mexico, USA, June 25-29, 2007

Roy, P.; Rouzieres, M.; Qi, Z. & Chubar O. (2006). The AILES infrared beamline on the third generation synchrotron radiation facility SOLEIL. *Infrared Physics & Technology*, Vol.49, Nos.1-2, (September 2006), pp. 139-146, ISSN 1350-4495

Sabersky, A.P. (1973). The geometry and optics of synchrotron radiation. *Particle Accelerators*, Vol.5, (September, 1973), pp. 199-206, ISSN 0031-2460

Shintake, T. (2003). Real-time animation of synchrotron radiation. *Nuclear Instruments and Methods in Physics Research A*, Vol.507, No.1-2, (July 2003), pp. 89-92, ISSN 0168-9002

Smolyakov, N. V. (1998a) Interference diagnostics for storage ring electron beam. *Nuclear Instruments and Methods in Physics Research A*, Vol.405, No.2-3, (March 1998), pp. 229-231, ISSN 0168-9002

Smolyakov, N. V. (1998b) Wave-optical properties of synchrotron radiation. *Nuclear Instruments and Methods in Physics Research A*, Vol.405, No.2-3, (March 1998), pp. 235-238, ISSN 0168-9002

Smolyakov, N. V. (1998c) Wave-optical properties of synchrotron radiation and electron beam diagnostics, *Proceedings of the sixth European Particle Accelerator Conference (EPAC'98)*, pp. 1601-1603, ISBN 0 7503 0579 7, Stockholm, Sweden, June 22-26, 1998

Smolyakov, N. V. & Hiraya, A. (2005). Study of edge radiation at HiSOR storage ring. *Nuclear Instruments and Methods in Physics Research A*, Vol.543, No.1, (May 2005), pp. 51-54, ISSN 0168-9002

Takayama, Y.; Hatano, T.; Miyahara T. & Okamoto, W. (1998). Relationship between spatial coherence of synchrotron radiation and emittance. *Journal of Synchrotron Radiation*, Vol.5, No.4, (July 1998), pp. 1187–1194, ISSN 0909-0495

Takayama, Y.; Okugi, T.; Miyahara, T.; Kamada, S.; Urakawa, J.; Naito, T. (1999). Application limit of SR interferometer for emittance measurement, *Proceedings of the eighteenth Particle Accelerator Conference (PAC'99)*, pp. 2155-2157, ISBN 0-7803-5573-3, New York, US, March 27 - April 2, 1999

Tsien, R. Y. (1972). Pictures of dynamic electric fields. *American Journal of Physics*, Vol.40, No.1, (January 1972), pp.46-56, ISSN 0002-9505

Weitkamp, T.; Chubar, O.; Drakopoulos, M.; Souvorov, A.; Snigireva, I.; Snigirev, A.; Gunzler, F.; Schroer, C. & Lengeler, B. (2001). Refractive lenses as a beam diagnostics tool for high-energy synchrotron radiation. *Nuclear Instruments and Methods in Physics Research A*, Vol.467-468, No.1, (July 2001), pp. 248-251, ISSN 0168-9002

Williams, G. P. & Dumas, P. (1997). *Accelerator-Based Infrared Sources and Applications. Proceedings of SPIE,* Vol. 3153, (October 1997), ISBN 9780819425751

9

Passive Ranging Based on IR Radiation Characteristics

Yang De-gui and Li Xiang
Institute of Space Electronic Information Technology
College of Electric Science and Engineering, NUDT
China

1. Introduction

Using passive detecting system to range and track target is quite an important researching field in photoelectrical signal processing, and IRST (IR searching and detecting system) is such a system, thus applying which to ranging IR target is a study hotspot all the while.

In recent years, many researchers have made a lot of works and achieved remarkable results. The system effect distance is deduced from different aspects in [1, 2]. Target range is obtained based on the target movement model in [3]. [4] studies the passive ranging of ground target in mono-static and single band condition based on radiance difference between target and background. Ranging expressions are deduced and ranging error is analyzed in [5-7] based on radiance difference, ranging radiate power ratio, target contrast ratio and SNR etc. in condition of mono-static and two bands.

However, above works were not aimed at the staring IR imaging system, that there is no relationship between ranging and IR images though [8-9] deduce the effect distance of the staring IR detection system with no ranging results. Based on these works, starting with the introduction of the staring IR detection system, this paper deduces the ranging expression for point and surface targets based on signal band IR image. The real IR image data validates this method and the ranging error is analyzed.

2. Working principle of the staring IRST

With the development of IR technology, more and more IR caloric imaging systems are adopted by IRST. The caloric imaging system has developed two generations, the first is based on optical and mechanical scanning, and the second is mainly based on staring or scanning focal plane array.

Fig.1 presents the second generation of IRST, which is commonly consisted by optic system, focal surface detecting subassembly, and video signal processing and vision system, there into focal plane array [10] can greatly prompt the scanning velocity and imaging quality. Generally, target detection is not only related with the scene contrast ratio, but also the resolving and analyzing abilities of the detection system. As Fig.1 shows, that the IR radiate power of the target and its background arrive at the caloric imaging system can be called the first stage of IRST; the power arrives at the focal plane detector with attenuation, then the power signal is changed into signal voltage according to different waveband responsibility, and visual gray image at last, which course can be called the second stage of the IRST. The first stage is related with the target distance, while the second stage is decided by the responding function itself of IRST.

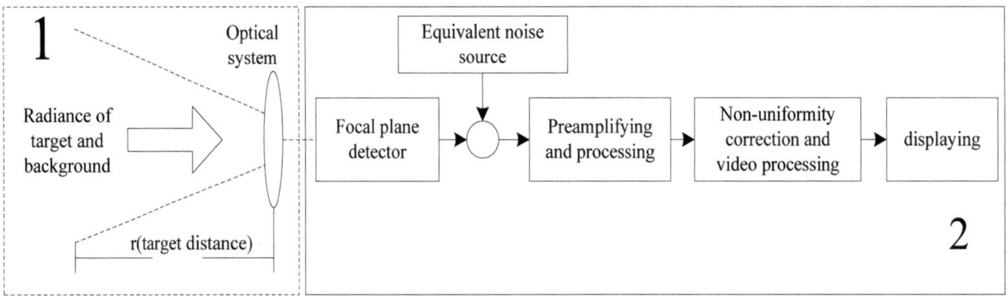

Fig. 1. The sketch map of IRST composition

At the first stage, the main influence factor is the atmosphere transmission ratio $\tau(\lambda,R)=e^{-\mu R}$, it contains an absorption coefficient $\mu=4\pi\mu_e/\lambda$. Here the μ_e is the extinction coefficient. At the second stage, the radiant power is changed into voltage, and sampled as image gray, which course can be described by the relationship between the radiance and image gray approximately as

$$L=aG+L_{off} \tag{1}$$

Where L_{off} is the bias (constant) of target radiance, a is a constant related with the radiance and the pixel gray.

3. The relationship between IR pixel gray and radiance

As we usually adopted a two-dimensional staring focal plane array in infrared imaging system, which include N × M detection cells arranged as a matrix with N rows and M columns. After correction, each detection cell in the array has same IFOV, plus and bias. As the ultimate output of imaging system is image pixel, the correction is aimed to the relationship between the radiance $L(i)$ and the pixel gray $G(i)$ (where "i" means the grade number of the pixel gray). The expression is shown as

$$L(i)=aG(i)+L_{off} \tag{2}$$

If these parameters L_{off} and a are known accurately, then we can get the radiance of the target through the image pixel gray. We can get above two parameters by calibration data of the blackbody source.

For the blackbody source, the radiance reached the sensor can be expressed as follows, the surface radiant temperature of the blackbody source is recorded as T (K)

$$L_b = \int_{\lambda 0} \{\tau_a(\lambda,r)\varepsilon M(\lambda,r) + [1-\tau_a(\lambda,r)]M(\lambda,T_a)\} R(\lambda)d\lambda \quad (3)$$

Here, $M(\lambda,T)$ is the blackbody source emittance, ε is the emissivity of the blackbody source. $R(\lambda)$ is the relative spectral response of the system, with a assumption that $R(\lambda)$ is a constant. λ is the wavelength. r is the target distance. $\tau_a(\lambda,r)$ is absorption coefficient of the atmosphere. T_a is temperature of the atmosphere. As the target is 1.13 meter away with the sensor, the atmospheric attenuation can be ignored. And it is regarded as surface target, so the radiance is shown as

$$L_b = \int_{\lambda 0} \varepsilon M(\lambda,T)d\lambda \quad (4)$$

As the IR sensors are in a special band, such as 3-5um and 8-12um, the radiance can be explained as [12]

$$L_b = \int_{\lambda 1}^{\lambda 2} \varepsilon M(\lambda,T)d\lambda = F_{\lambda 1-\lambda 2} \cdot \varepsilon\sigma T_b^4 \quad (5)$$

Here $F_{\lambda 1 \sim \lambda 2}$ denotes the ratio between the radiance of the given waveband with wavelength $\lambda 1 \sim \lambda 2$ and the one of the whole band. This can be gotten by the datasheet [13]. σ is the Stefan-Boltzmann constant, T_b is temperature of the blackbody source.

4. The blackbody source calibration test

The infrared sensors used here were cooled LWIR and MWIR sensors, they are all made in France corporation CEDIP, and the types are Jade 3 LW and Jade 3 MW F/2, and the detail technological index are shown as table 1.

sensor	Jade 3 LW	Jade 3 MW F/2
focus	50mm	50mm
NETD(25°C)	20mK	14.93mK
Noise	2.46DL	4.97DL
Sensitivity(25-26°C)	8.13mK/DL	3.01mK/DL
Image size(H×V)	320×240	320×240
Sample digit	14bit	14bit
Transmissivity of the optical system	>90%	>90%
Integral time	20, 50, 100, 200 microsecond	500, 1000, 2000, 2500 microsecond
Frame frequency	1-130Hz	1-130Hz
FOV	10.8°×8.1°	10.8°×8.1°
Spatial resolution	0.5mrad	0.5mrad
Size of image cell	25 μm ×25 μm	25 μm ×25 μm

Table 1. The technological index of LW and MW sensors

In the blackbody source calibration test, we use three blackbody source to calibrate the MW(3-5 μm)and LW(8-12 μm) staring sensors based on the GJB3756-99 criterion. And the result showed as the follow table.

	Blackbody source 1	Blackbody source 2	Blackbody source 3
Serial number	#163	#G0406011	#4845
Type	Plane	cavitary	cavitary
The range of temperature	0°C-100°C	100°C-950°C	100°C-1300°C
Uniformity of the radiant surface temperature/K	0.09	0.3	0.24(500°C)
Stability of temperature(K/h)	1.8	0.1	0.15(800°C)

Table 2. The index contrast of three blackbody sources

In the blackbody source calibration test, temperature of the blackbody source is known, so the relationship between $G(i)$ and $L(i)$ can be calculated. In this test, the two sensors are MW (3-5 um) and LM (8-12 um) staring infrared sensor, there integration time are 2500 and 200 us respectively; the blackbody source is a surface source and its temperature is adjustable from 5°C to 100 °C; the environmental temperature is 24 °C; the relative humidity is 50 %. Fig.2 shows IR images of the blackbody source in 30 °C and 40 °C respectively:

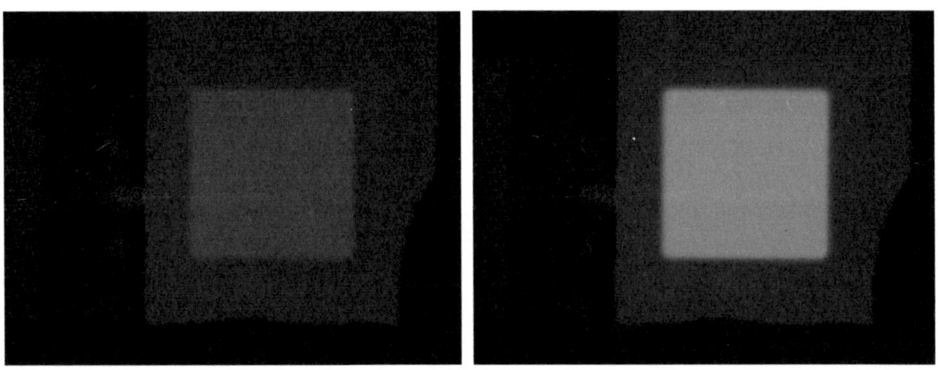

Fig. 2. MW IR image of blackbody source at 30°Cand 40°C respectively

Table3 shows the gray and temperature of the blackbody source counted by the MW and LW IR images.

	30°C	40°C
Gray of blackbody source in MW band	4333.18	5770.74
Gray variance of blackbody source in MW band	72.81	73.17
Gray of blackbody source in LW band	10257.67	11487.59
Gray variance of blackbody source in LW band	13.59	14.25

Table 3. The gray and gray variance of the target region

Therefore, based on equation (5), L_{off} and a can be counted as follows:

$$a_m = 0.0018; a_l = 0.0156; \tag{6}$$

$$L_{Moff} = -1.7908; L_{Loff} = -33.5507; \tag{7}$$

Here a_m and a_l denoted the variation ratio of MW and LM respectively, L_{Moff} and L_{Loff} denoted the radiance bias of MW and LW respectively.

After the blackbody source calibration, we can get the relationship between the target radiance, temperature and image gray, so we can use the outfield infrared images to analyze the infrared characteristic of the target.

5. Passive ranging based on single-band IRST

5.1 Radiance difference between point target and its background

When a target is quite far away from the IRST, the target image can not fill in the full detector elements, and the target can be treated as a point target[11], which is difficult to be ranged when a target is actionless as there is no shape, size etc. characteristics. Thus a target can only be ranged by using radiant and movement characteristics. At this scene, target, background and path radiance can arrive at the detector elements. The whole radiance of target and background is defined as

$$E_t = \left(I_t + L_b\left(\Omega_s - \Omega_t\right)\right)\tau\left(R_1\right)/R_1^2 + \left(1 - \tau\left(R_1\right)\right)L_a \tag{8}$$

For IR detection system, the whole radiance of target and background received by the detector is

$$L_t = \frac{E_t}{F} = \left[\left(I_t + L_b\left(\Omega_s - \Omega_t\right)\right)\tau\left(R_1\right)/R_1^2 + \left(1 - \tau\left(R_1\right)\right)L_a\right]/F \tag{9}$$

where I_t is the target whole radiant power, L_b is the background radiance, L_a denotes the atmosphere path radiance, $\tau(R_1)$ denotes the average atmosphere transmission on the transmitting path within the waveband of the detector, R_1 is the distance of a target and IRST, F is IFOV of the system, Ω_s is instantaneous scene, Ω_t is the angle between the target and the detector plane. When there is no target, the background will fill in the whole detector elements, and the received radiance is

$$L_b = \left[L_b\Omega_s\tau\left(R_1\right)/R_1^2 + \left(1 - \tau\left(R_1\right)\right)L_a\right]/F \tag{10}$$

Thus the radiance difference will be

$$\Delta L_1 = L_t - L_b = \left(I_t - L_b\Omega_t\right)\tau\left(R_1\right)/FR_1^2 \tag{11}$$

5.2 Radiance difference between surface target and background

For surface target, instantaneous scene Ω_s is smaller than angle Ω_t, thus the target image will take up several or even tens of detector elements as shown in Fig.3. Therefore the surface target image includes target edge pixels and interior pixels. These edge pixels may reflect target, background and path radiance, while interior pixels may be only related with target and path radiance.

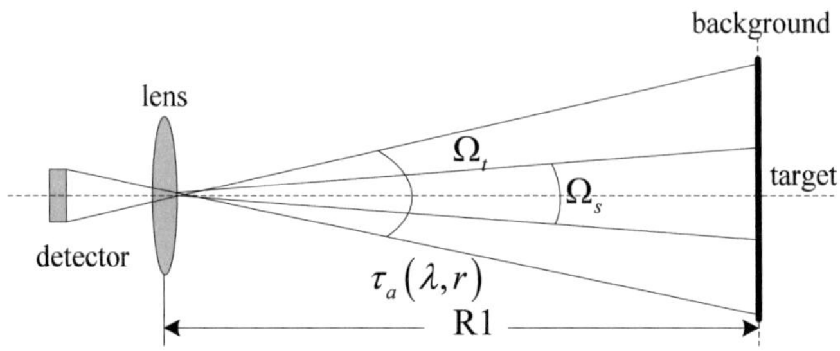

Fig. 3. The sketch map of the IR surface target's radiation

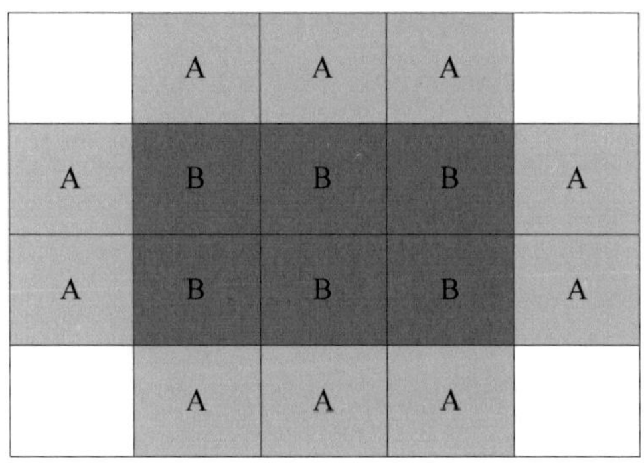

Fig. 4. The sketch map of the IR surface target

As Fig.4 depicts, surface target image includes inner pixels (denoted by "B") and edge pixels (denoted by "A"). Assuming an IR image produced by the staring IRST has N inner pixels and M edge pixels. For inner pixels, as the target is full to the detector elements, background radiance can not arrive at the detector, thus Ω_s equals to Ω_t, and the received radiance is, which is only decided by the target. For edge pixels, background radiance can arrive at the IR detector as the target is not full with the instantaneous scene. Assuming the IR radiance is L_{tB}, which reflects the sum of target and background radiance. For simplifying, we further assume that the background radiance is uniform distributed, and angles of target edge to the detector centre are all equal, we thus have

$$L_{tz} \; L_{tZ} = \sum_{i=1}^{N} \left[\left(L_{ti} \Omega_s / R_1^2 \right) \tau(R_1) + \left(1 - \tau(R_1) \right) L_a \right] \Big/ F \tag{12}$$

$$L_{tB} = \sum_{i=1}^{M} \left[\left(L_{ti} \Omega_{ti} + L_b \left(\Omega_s - \Omega_{ti} \right) \right) \tau(R_1) \big/ R_1^2 + \left(1 - \tau(R_1) \right) L_a \right] \Big/ F \tag{13}$$

Where L is the sum of target and background radiance, thus $L = L_{tZ} + L_{tB}$, and the target whole radiance power is I_t, besides

$$I_t = \sum_{i=1}^{M} L_{ti}\Omega_{ti} + \sum_{i=1}^{N} L_{ti}\Omega_s \tag{14}$$

Thus

$$L = \left[\left(I_t + \sum_{i=1}^{M} L_b\left(\Omega_s - \Omega_{ti}\right)\right)\tau\left(R_1\right) \Big/ R_1^2 + \sum_{i=1}^{N+M}\left(1-\tau\left(R_1\right)\right)L_a\right] \Big/ F \tag{15}$$

Where the radiance in the position of target corresponding pixel is L_{ti}, Ω_{ti} denotes the angle between target parts and detector element. A_t stands for the target area. When the target does not exist, the corresponding elements are filled with background radiance, defined as L where

$$L' = \sum_{i=1}^{N+M}\left[L_b\Omega_s\tau\left(R_1\right)/R_1^2 + \left(1-\tau\left(R_1\right)\right)L_a\right] \Big/ F \tag{16}$$

Therefore, the radiance difference ΔL of target and background is

$$\begin{aligned} \Delta L &= L - L' \\ &= \frac{1}{FR_1^2}\left[I_t - L_b\left(\sum_{i=1}^{N}\Omega_s + \sum_{i=1}^{M}\Omega_{ti}\right)\right]\tau\left(R_1\right) \\ &= \frac{1}{FR_1^2}\left[I_t - L_b A_t\right]\tau\left(R_1\right) \end{aligned} \tag{17}$$

Comparing with the radiance difference computation for point and surface target, the two are the same and can use a same ranging method to obtain the distance, while target area A_t reflects their difference, which is more important to surface target.

5.3 Passive ranging for the airborne target with single waveband

Let the IRST be sustained in a ranging course, that there is no necessary for movement compensation for IR image. We thus take three targets measuring for target ranging. Set the IR image be $f_i(i = 1\cdots3)$, and the interval between two measuring is same to ΔT. According to (1), the corresponding ΔL_i of the three images can be yielded from f_i as

$$\Delta L_i = a\left[G_t - G_B\right], i = 1,2,3 \tag{18}$$

Where the total gray of the target whole pixels is G_t, G_B denotes the total gray of the whole pixels when there is no target. Here the image gray of the background is set to the surroundings' for the no reiteration property of IR image. As the measuring intervals are quite short, we thus hold that the target area of each image is near equal to each other. According to (17), we can obtain that

$$\begin{cases} \xi_1 = \dfrac{\Delta L_1}{\Delta L_2} = \left(\dfrac{R_2}{R_1}\right)^2 e^{-\mu\left(R_1 - R_2\right)} \\ \xi_2 = \dfrac{\Delta L_3}{\Delta L_2} = \left(\dfrac{R_2}{R_3}\right)^2 e^{-\mu\left(R_3 - R_2\right)} \end{cases} \tag{19}$$

As the intervals of the three frame image are very short, we hold that the target move equably in this course, thus the distance difference for two near measuring is equal to ΔR , that is $R_2 - R_1 = R_3 - R_2 = \Delta R$, we can then obtain

$$\begin{cases} \xi_1 = \left(\dfrac{R_2}{R_2 + \Delta R} \right)^2 e^{-\mu \Delta R} \\ \xi_2 = \left(\dfrac{R_2}{R_2 - \Delta R} \right)^2 e^{\mu \Delta R} \end{cases} \tag{20}$$

From which the target distance R_2 and ΔR can be obtained.

5.4 Experiment results and error analysis

Experiment data is produced by long and medium wave staring IR sensors at pm in Oct.18 of year 2006. The air temperature is 15°C and it is sunshine. Civil aero-plane is used. The image format is 320×240. The atmosphere long wave attenuation coefficient is 2.17, while it is 1.45 in medium wave situation according to the Lowtran software. Fig.5 presents the IR images for both point target and surface target. The target distance is obtained by GPS and radar in the experiments.

Point target(medium wave) Point target(long wave)

Surface target(medium wave) Surface target(long wave)

Fig. 5. The two MW/LW IR outfield images

For point target, we can achieve the gray difference between target and background, making use of the gray of surroundings as the gray of background. But for surface target, compute the total pixels and the total gray of surface target when dividing surface target from image firstly, and then compute the total gray of background when replacing the gray of target with surrounding'. We achieve the gray difference between surface target and background accordingly.

As the pivotal factor, different division thresholds result in the quantitatively difference of target pixels.

Experiment data is chosen from the image data when the plane is in the climbing phase. We choose 3 frames of image data per second, which has the equal space between frames, which is 240ms. Based on the algorithm provided in the paper, the distance of one target is achieved. The total experiment time is 10 seconds. Fig.6 present the distance comparison of surface target, which is counted by IR data of actual measurement in a period of time, and by the radar data.

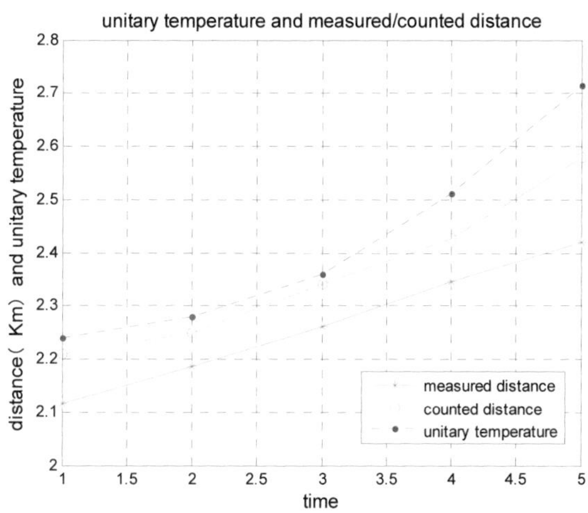

Fig. 6. Counted distance (LW/MW) and measured distance of surface target

The distance comparison of point target is shown in Fig.7, which contain the distance counted by the IR image of actual measurement in a period of time, the distance difference between frames, and the distance measured by radar. Restricted by the functionary distance of IR sensors, experiment time is only 5 seconds.

Hereinafter, the experiment results are analyzed:

1. For the actual experiment, the algorithm provided in the paper has a fixed error. The reason is that, the target is moving at a constant velocity in the hypothesis of this paper, but the phase of target climbing is an accelerated phase. Method of reduce error is reducing the gap between frames.
2. The target distance counted by long wave IR data is more accurate than which counted by medium wave data. The main reason is that, the functionary distance of long wave IR sensor is farer, and the contrast and the contour of target is clearer, which is propitious to acquire more accurate radiance difference between target and background, as to long wave IR image. The algorithm provided in the paper need to count the total pixels of target.

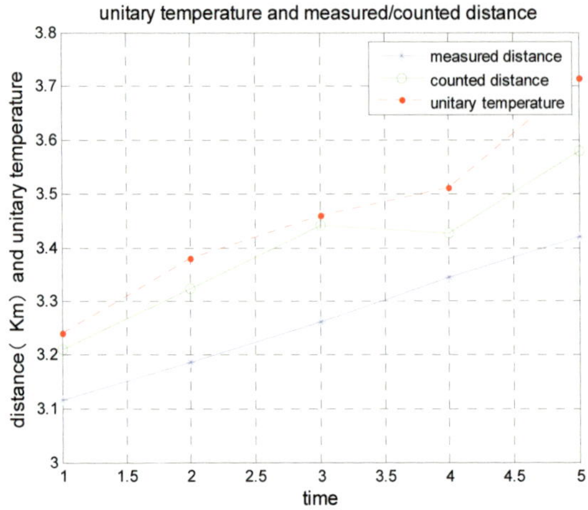

Fig. 7. Counted distance (LW/MW) and measured distance of point target

3. In the experiment, the results has a larger error counted at the time of 4 and 5, when target attitude has some change, that results in the change of target's contour. Consequently, the distance counted has some jitter. But for radar, the change of target attitude has little effect on distance in range.
4. The results of counted distance of surface target are better than point target's. And, for surface target, the farer distance of target, the larger error of results. When the target is farer, the area of target is smaller in image. For surface target, division threshold and the total pixels of target is less accurate, which results in worse result of radiance difference between target and background, further, affect distance measure. But for point target, the distance of target is counted by single pixels. If the gray of these pixels is inaccurate, there will be larger error of distance measure or ranging.
5. A great influence on the veracity of distance measure has the atmospheric attenuation coefficient. This paper uses a average coefficient of atmospheric attenuation, which introduces a fixed error, in spite of omitting the complicated integral.
6. The algorithm about passive ranging of single-band provided in the paper has the following hypotheses: one is that target is moving at a constant velocity in the short time among measures approximately; another is that the area of target is approximately constant in the period. Thus, the algorithm fits the passive ranging of air mobile target well, instead of ground mobile target.

The measure error is analyzed:

1. Influences of radiance of background

The expression of (10) is the computation of the radiance of background for surface target. Because of in the actual imaging, we achieve the average radiance of background approximately, using of the gray of background around the target, instead of computing according to (10) when the target is divided up. And then according to the total pixels of target, compute the whole radiance of background, in this way, which has a fixed error itself. When it is a surface target, the error is severe.

2. Influences of the atmospheric extinction coefficient

Another error is brought into ranging due to the atmospheric extinction coefficient, which contains the following two aspects: (1) The error is introduced when using the average atmospheric extinction coefficient to simplifying the integral. (2) The atmospheric extinction coefficient is computed by the Lowtran7 software. When the circumstance parameter is inaccurate, the atmospheric extinction coefficient will has error, results in the error of ranging.

3. Influence of target range

The farer is target range, the larger error is introduced. When the target is farer, the area of target is smaller in the image, accordingly, results in compute error in the difference of radiation power between target and background, thus, affects ranging. For point/surface target, the difference of radiance difference between target and background is presents, when R_m is the measure precision of range.

$$\Delta L = \frac{1}{FR_1^2}\left[I_t - L_b A_t\right]\tau(R_1) \tag{21}$$

When compute the derivative of R, we has

$$\frac{d\Delta L}{dR} = \frac{\left[I_t - L_b A_t\right]}{FR_1^2} e^{-\mu R}\left(-\mu - \frac{2}{R}\right) \tag{22}$$

$$dR = \frac{(d\Delta L)FR_1^2}{\left[I_t - L_b A_t\right]e^{-\mu R}\left(-\mu - 2/R\right)} \tag{23}$$

According to (17), when the farer the target range is, the worse the measure precision is; the smaller is the resolution of the pixel gray of IR image, the better is the measure precision.

6. Surface target ranging based on single-station dual-band IRST

The algorithm above is also applicable for multiband, with a presupposition of point or tiny targets. When there are surface targets, the estimation of background radiance brings severe error, which causes the algorithm invalid. Therefore, another range measurement algorithm for surface targets is proposed in the next section.

6.1 Calculation of surface target radiance

When the target is far from the infrared detector, it can be regarded as a point target [11]. There is not only the target radiation but also the background and path radiation can reach the detector cell. So the radiance flux of the detector cell is：

$$E_t = F(r)\int_{\lambda_0}\left\{\tau_a\left[\varepsilon_t M(\lambda,T_t) + \rho_l M(\lambda,T_{ae})\right] + \left[1-\tau_a\right]M(\lambda,T_a)\right\}d\lambda \tag{24}$$

In above formula, $\varepsilon_t + \rho_l = 1$, ε_t is the target emissivity, ρ_l is the target reflectivity. $F(r)$ is geometrical factor related with the target distance r . $M(\lambda, T_t)$ is the radiation flux density of the target region; $M(\lambda, T_{ae})$ is the background radiation flux density reflected by the background; $M(\lambda, T_a)$ is the radiation flux density reflected by the atmospheric path ; τ_a is the transmission ratio of the light path between the infrared detectors and the target; T_t is the temperature (K) of the target; T_a is the atmospheric temperature (K).

When the military targets (such as aircraft) near the detector, the goal may occupy several even dozens of detector cells, so it is more appropriately to regard them as surface targets. As shown in Fig.3, Ω_s (the instantaneous field of view of the detector) is smaller than Ω_t (the target stretching angle relative to the detector cell), so both target and background radiation can reach the fringe detection cell, but only the target and path radiation can reach the internal detection cell. Because the edge pixels are fewer than the internal pixels, the paper takes the internal pixels into count mainly.

According to the internal pixels, the target area fills instantaneous field of view (IFOV) of the sensor completely, $F(r)$ is IFOV of the system, it is Ω_s . The radiance of the target reached the sensor is:

$$L_t = E_t / F(r) = E_t / \Omega_s \tag{25}$$

$$L_t = \int_{\lambda_0} \{\tau_a [\varepsilon_t M(\lambda, T_t) + \rho_l M(\lambda, T_{ae})] + [1 - \tau_a] M(\lambda, T_a)\} d\lambda \tag{26}$$

In appointed band, the radiance can be described as

$$L_t = \int_{\lambda 1}^{\lambda 2} \{\tau_a [\varepsilon_t M(\lambda, T_t) + \rho_l M(\lambda, T_{ae})] + [1 - \tau_a] M(\lambda, T_a)\} d\lambda \tag{27}$$

Here the integral process can be describe as

$$L_t = F_{\lambda 1 \sim \lambda 2} \tau_a \varepsilon_t \sigma T_t^4 + F_{\lambda 1 \sim \lambda 2} \sigma T_{ae}^4 [\tau_a \rho_l + 1 - \tau_a] \tag{28}$$

In above expression, $F_{\lambda 1 \sim \lambda 2}$ denotes the ratio between the radiance of the given waveband with wavelength $\lambda 1 \sim \lambda 2$ and the one of the whole band. T_t is the target temperature, T_{ae} is the atmosphere temperature. As $\varepsilon_t + \rho_l = 1$, we can get following equation

$$L_t = F_{\lambda 1 \sim \lambda 2} \sigma T_{ae}^4 + F_{\lambda 1 \sim \lambda 2} e^{-\mu R} \varepsilon_t \sigma \left(T_t^4 - T_{ae}^4\right) \tag{29}$$

6.2 Passive ranging based on the single-station dual-band infrared images

Based on the equation (2) we can get the target radiance from the image pixel gray. In the above equation, the $F_{\lambda 1 \sim \lambda 2}$, T_{ae} and ε_t are known. The target temperature T_t is known roughly, and the target distance R is unknown totally.

As there are dual-band infrared sensors on the same platform. At the same time we can get two frames infrared image, they all aim at the same target area. According to expression (29) we can obtain the radiance of the same target area at different band.

Assume A means the same target area in the two infrared images; its temperature and distance should be equal approximately according to different band, recorded as T_A (known roughly) and r_A (unknown). We can get its radiance by LW and MW infrared image, recorded as L_M and L_L respectively. The atmospheric extinction coefficients in LW and MW bands were known as μ_M and μ_L respectively, so L_M and L_L can be expressed as follows:

$$L_M = F_{\lambda 1\sim\lambda 2}\sigma T_{ae}^4 + F_{\lambda 1\sim\lambda 2}e^{-\mu_M R}\varepsilon_t\sigma\left(T_t^4 - T_{ae}^4\right) \tag{30}$$

$$L_L = F_{\lambda 1\sim\lambda 2}\sigma T_{ae}^4 + F_{\lambda 1\sim\lambda 2}e^{-\mu_L R}\varepsilon_t\sigma\left(T_t^4 - T_{ae}^4\right) \tag{31}$$

So we use the iteration method to get the T_t and R from the above two equations.

There will be much error if we adopt only one pixel of the target in infrared image, so we must use a few pixels of the target at the same time.

Firstly after image matching, the two infrared images are aimed to the same scene; then the target is segmented and a few pixels are chose at the same position of the two images. Secondly the radiance L_i according to each target pixel can be calculated by each pixel gray G_i. Thirdly the target distance r_i can be counted by the dual-band radiance of the pixels. Lastly the distance deduced by different target pixels should be chosen in reason and their average value will be used as the ultimate distance.

The algorithm is shown as follows

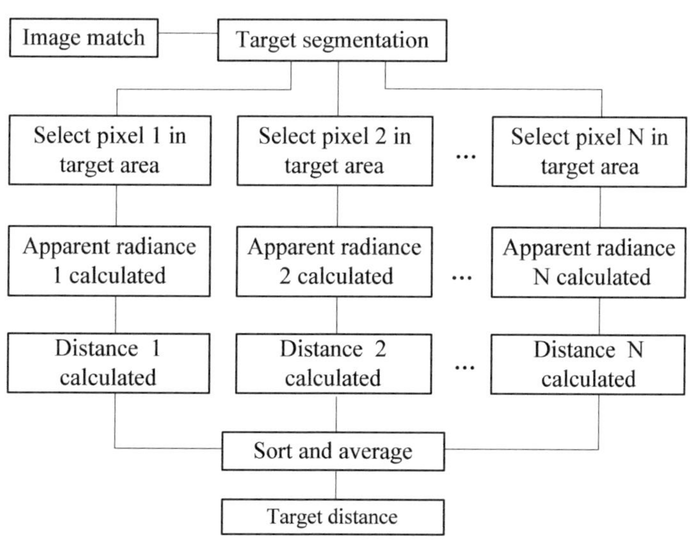

Fig. 8. The ranging process of surface target

6.3 Experiment results and error analysis

Experiments have been done to verify the ranging algorithm in October, the air temperature was 15 °C, the target was aircraft, FOV(Field Of-View) and IFOV(Instantaneous Field Of-View) of the two sensors were same as 10.8o × 8.1o and 0.5 mrad. Fig.9 show the LW and MW IR image at the same time, and the bright point was the aircraft.

Fig. 9. MW and LW infrared images of the aircraft

Table4 shows the target distance, a certain pixel gray and temperature of the same target at six different times (time interval is about one second).

Time	Pixel gray in MW band	Pixel gray in LW band	Calculated distance (Km)	Measured distance by radar(Km)	Target temperature (K)
0	16087	11230	0.9949	0.885	387.95
1	15868	11149	0.985	0.948	387.11
2	15715	11070	0.958	1.004	386.66
3	15241	10802	1.101	1.060	386.48
4	15278	10648	1.159	1.116	386.86
5	15752	10508	1.239	1.172	391.04

Table 4. Distance, temperature and metrical distance at different moments

Fig.10 is gained by above table, and we can get the movement trends of the target from the figure intuitionally. The figure shows the unitary value of measured distance, calculated distance and temperature.

Test results and analysis:

1. From time2 to time5, the average error between measurement and calculation is less. The reason is that the radar ranging is based on radar scattering centres of the target, and the infrared ranging is based on the surface of the target, there will be errors unavoidably.
2. The errors at time 0 is remarkable because that the distance is very close at each moment, the target pixels have reached a saturation level according to the integration time of the infrared sensor, so the ranging is not accurate.

3. With the distance become more and more farer, the target get smaller and smaller in the IR image. The target has become increasingly unclear according to the same sensor integration time (relative to above moment), it is very difficult to find a same pixel in two infrared images, so it will also have a greater error.
4. The target temperature is equal approximately in fig.10. In fact, as it is the same target, the temperature won't have much change when the target distance is not half far enough.

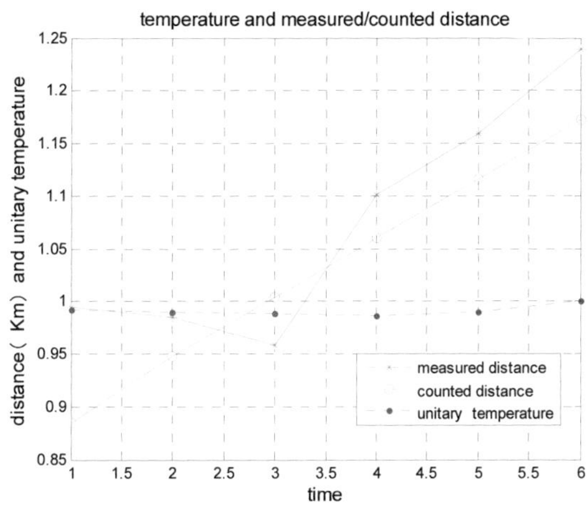

Fig. 10. Measured and calculated distance at five times

The above error is related with the sensor integration time. The paper will analyze the algorithm and find which factors will be likely to bring ranging error.

1. Will the inaccurate atmospheric extinction coefficient bring the ranging error?

From the equation 17 we can know that the error of atmospheric extinction coefficient will bring the ranging error.The error mainly includes the following two aspects: (1) the paper uses the average atmospheric extinction coefficient to simplify the integration process, it will bring ranging error. (2) The atmospheric extinction coefficient is calculated by LOWTRAN7 software in this paper, inaccurate environmental parameters will lead to inaccurate calculation, and therefore it will cause the ranging error.

2. Will the mismatch of pixels bring the ranging error?

If the chosen pixels are not in the same position of a target, their range and radiance is different, so it will bring errors in ranging process.

As the ground environment is more complex than the air environment, the algorithm has only used in airborne ground targets ranging; the author will do more tests to carry out ground targets ranging.

7. Conclusion

This paper regarded passive ranging based on IR radiation characteristics as researching background. The operating principle of staring IRST was analyzed. Then the relationship between IR pixel gray and radiance was deduced. And the parameters were got through the blackbody source calibration test. In the single-band and dual-band situations, according to point and surface target, we deduced two ranging methods respectively. Lastly the algorithm was validated for point and surface target by outfield IR image, the ranging error was analyzed also.

8. Acknowledgment

Here, I sincerely thank Dr. Vasyl Morozhenko pointed out the inadequacies in my paper.

9. References

[1] Wang Gang, Yu Bin-Xi. Approach to estimate infrared point-target detection range against sky background based on contrast. Optics and Precision Engineering, Vol.10 No.3, Jun.2002, pp276-280.

[2] Xing Qiang, Huang Hui–ming, Xiong Ren-sheng, Yu Tao. Detect ability Analyzing of IR FPA Tracking System. ACTA PHOTOICAS INICA, Vol.33 No.7, July.2004, pp894-897.

[3] Yin Shi-min, Fu Xiao-ning, Liu Shang-qian. Research for Infrared Mono-station Passive Location on Stationary Platform. ACTA PHOTOICAS INICA, Vol.33 No.2, Feb.2004, pp237-239.

[4] Lu Yuan, Ling Yong-shun, Wu Han-ping, Li Xiao-xia. Study on Passive Distance Measurement of Ground Objects by Infrared Radiation. J.Infrared Millim.Waves, Vol.23 No.1, Feb.2004, pp77-80.

[5] Lu Yuan, Ling Yong-shun, Shi Jia-ming. Measurement of Aerial Point Target Distance Using Dual-band Infrared Imaging System. Optics and Precision Engineering, Vol.12 No.2, Apr.2004, pp161-164.

[6] Feng Guo-qiang, Zou Qiang, Li Wei-ren. Algorithm of Passive Ranging by Single Station. Infrared Technology, Vol.27 No.4, July.2005, pp295-298.

[7] Anders G.M. Dahlberg, Olof Holmgren. Range performance modeling for staring focal plane array infrared detectors. Infrared Imaging Systems: Design, Analysis, Modeling, and Testing XVI, Proceedings of SPIE, 2005, Vol.5784, pp81-90.

[8] Wang Bing-xue, Zhang Qi-heng, Chen Chang-bin, Wang Jing-ru, He Xue-mei. A Mathematical Model for Operating Range of A Staring IR Search and Track System. Opto. Electronic Engineering, Vol.31, No.7, July.2004, pp8-11.

[9] Wang Wei-hua, Niu Zhao-dong, Chen Zeng-ping. Research on The Operating Range of Staring IR Imaging System in Sea-Sky Background. J.Infrared Millim.Waves, Vol.25, No.2, Apr.2006, pp150-153.

[10] Bai Yan-zhu, Jin Wei-qi. Principle and Technology of Optical-Electronic Imaging. Beijing Institute of Technology Press.

[11] Pieter A. Jacobs Write, Wu Wen-jiann, Hu Bi-ru, Man Ya-hui Translate. Thermal Infrared Characterization of Ground Targets and Backgrounds. National Defense Industry Press. 2004.

[12] Richard D. Hudson, JR. Infrared System Engineering. John Wiley & Sons, INC. 1969.

[13] Yang Chen-hua, Mei Shui-sheng, Lin Jun-ting. Laser and Infrared Technological Datasheet. National Defense Industry Press. 1990.